王迈迈英语数字化系列

四大超级功能 立体交叉效果 黄金组合模式

精彩的题型分析与技巧指导 精确的模拟训练与答案详解
图书数字化电脑互动做题 高性能软件随时核对答案
光碟特配MP3文件 专供下载录音随身听
另配4盒高质量录音磁带 满足录音机播放需求

# 四级考试核心考点
# 精确打击

主　编　王迈迈
副主编　陈维良　宁克建　王　娟　邹　燕
编　者　杨　杨　张　红　张云峰　万吉祥　张俊鹏　刘　琦　李　敏
审　订　REUBEN DAVIS(美)

CET-4

原子能出版社

**图书在版编目(CIP)数据**

四级考试精确打击 / 王迈迈主编. —北京: 原子能出版社, 2007.1
ISBN 978-7-5022-3824-7

Ⅰ. 四… Ⅱ. 王… Ⅲ. 英语—高等学校—水平考试—习题 Ⅳ. H319.6

中国版本图书馆 CIP 数据核字(2006) 第 153402 号

**四级考试精确打击**

---

**出版发行** 原子能出版社(北京市海淀区阜成路 43 号 100037)
**责任编辑** 卫广刚
**责任校对** 冯莲凤
**责任印制** 丁怀兰
**印　　刷** 枝江市新华印刷公司
**开　　本** 889mm × 1230mm 1/16
**印　　张** 17
**字　　数** 620 千字
**版　　次** 2007 年 6 月第 1 版 2007 年 6 月第 1 次印刷
**书　　号** ISBN 978-7-5022-3824-7
**经　　销** 全国新华书店
**印　　数** 1—10000 **定　　价** 23.80 元

---


 **网址**:http://www.aep.com.cn

# 前言

# PREFACE

《四级考试核心考点精确打击》是为即将参加新四级考试的考生们精心准备的。

四级考试改革之后，四级考卷的题型、难度和容量发生了巨大变化，快速阅读理解、听力长对话、选词填空、短句回答和汉译英一大批全新的题型为考生们逾越四级考试大关增添了变数。本书按照最新四级考试精神，依据最新出版的《大学英语四级考试大纲》，对考生们在四级考试过程中可能遇到的各种疑难目标，**精确定位，精确打击！**

本书对新四级考试的所有题型全面出击，重点目标是本次改革后的新增题型和传统题型中的核心考点，分析考试题型，预测命题趋势，探索命题规律；

# PREFACE

本书关注各种新四级题型的应试技巧，研究答题策略，总结考试方法，揭示高分秘诀；

本书严格按照最新《大学英语四级考试大纲》，针对四级考试的核心考点，精心设计15套新题型押题预测卷，按照各个击破的原则，分题型精确定位，按目标精确打击！

欢迎同学们就本书提出宝贵意见。

来信请寄：武汉市洪山区楚雄大道268号
武汉现代外国语言文学研究所　　邮编：430070
电话：027-88027608　　88026817　　88034727　　88026460

同学们还可以通过以下方式和我们交流：

1.语音聊天室（全天候开放，语音讲课）。

2.BBS论坛（全天候开放）。

3.通过答疑专用信箱（support@wmmenglish.com）和我们联系。

# 目录 Contents

四级考试核心考点
**精确打击**

# Part 1

## 710分新题型

# 四级考卷

# 聚焦

## 710分新题型

# 四级考卷

710 分新题型四级考试已在高校全面展开。这是进一步推进大学英语教学改革的又一项重要举措。新的四级考试和传统的四级考试相比较，从考试设计的指导思想到试卷采用的题型乃至试卷本身的题量都发生了巨大变化。按照最新《大学英语课程教学要求》的规定，"大学英语的教学目标是培养学生的英语综合应用能力，特别是听说能力"。这和 1998 年公布的《大学英语教学大纲》的内容发生了质的飞跃。原来"培养学生具有较强的阅读能力和一定的听、说、写、译能力"，明显强调了英语语言诸项能力中最容易落实也是最基础的阅读能力，其他能力轻轻带过。现在的要求显然高多了，要求培养英语的"综合应用能力"，并且特别强调了学习外语最难也是最核心的指标即听说能力。

**根据这一指导思想，新四级考卷作出了如下重大调整：**

一、大幅提升听力理解在总分中的分值，从原来的 20% 到现在的 35%，几乎翻了一番，题型和题量也增加了几乎一倍。原来的考卷听力理解部分每次只会出现 2 种题型：10 个短对话和三篇听力短文，听力短文有时和复合式听写二选一，交替出现在每次的考卷中。新四级的听力理解部分有 4 种题型：一是短对话 8 组，一问一答后接一个问题；二是长对话，这种题型在过去近 20 年的四级考试中从未出现过，2 组对话一问一答多达 15 个回合，每组对话后都有 3 ~ 4 个问题，问题既涉及对话的具体信息，也涉及整个长对话的谈话主题；三是三篇听力短文，这是四级考试的传统题型，三篇200多词的短文后，共有10道问题；四是复合式听写，一篇200多词的短文，文中留有11处空白，1~8空要求用原词填空，每空一词，9~11空要求听写的是短语或句子，既可以用原文填空，也可以用自己的语言填写大意。本部分答题时间共35分钟。

二、阅读理解的题型发生了重大变化。传统的四级阅读理解题，基本上就是一种答题形式，每次4篇300词左右的文章之后，各有5个多项选择性题目。现在，阅读理解首先在概念上被划分为两大块，一是快速阅读理解，二是仔细阅读理解。

快速阅读采用了正误辨认和完成句子两种题型，这部分答题时间为15分钟，题目为一篇1000词左右的英语文章，后面跟着10个问题，1~7题是正误辨认，8~10题为完成句子。快速阅读的概念本不陌生，但对考生而言，过去从中学到大学所有正式考试中，很少会用到这种题型，因此，此题的难度和对考生的挑战性显而易见。

仔细阅读理解，应该就是我们通常意义上的阅读理解。这部分首先保留了传统的题型，两篇300多词的文章后跟10道问题，每题附A，B，C，D四个选项。除此之外，仔细阅读理解又增加了两种新的题型：一种是选词填空，另一种是短句回答，两种题型选其一，交替出现在每次的考卷中。选词填空是一篇200多词的文章，文中留有10处空白，要求从文后所给的15个词中选出10个填空，每空一词。此题和多项选择阅读题侧重不同，多项选择考查篇章阅读理解能力，而选词填空侧重篇章词汇理解能力。短句回答也是一篇300词左右的文章，文后有5道问题，要求在阅读短文之后，根据短文内容作简短回答，回答一般要求不超过10个词。这部分的答题时间为25分钟，如果加上快速阅读的时间，阅读理解的答题时间一共是40分钟。分值也和传统四级有所不同，原占总分的40%，现为35%。

三、新四级考试大幅增加非选择性试题的比例，表明考试的标准有所提高，试题的难度有所加大。比如快速阅读的完成句子和仔细阅读的短句回答等题型，还有新四级考卷的第三大部分综合测试的汉译英，都是新增的非选择性试题。

汉译英也是新增题型，传统四级考试虽然也没有翻译题，但只是将英语译成汉语。根据2006年6月最新四级考卷和四级考试委员会公布的样卷和考试说明，汉译英测试的是句子、短语和常用表达层次上的中译英能力，本题的答题时间是5分钟，分值占总分的5%。

四、除以上变化外，新四级考试还有以下和传统四级不同的地方：

1. 试卷总分由100分变为710分。

2. 考试总时间由120分钟变为125分钟。

3. 试卷的题型排序有所不同。

传统四级为：听力理解、阅读理解、词汇与结构、完形填空（简答、英译汉）、作文；

新四级为：作文、快速阅读理解、听力理解、仔细阅读理解、完形填空、汉译英。

四级考试核心考点

**精确打击**

Part

# 快速阅读理解

## Reading Comprehension
## ( Skimming and Scanning )

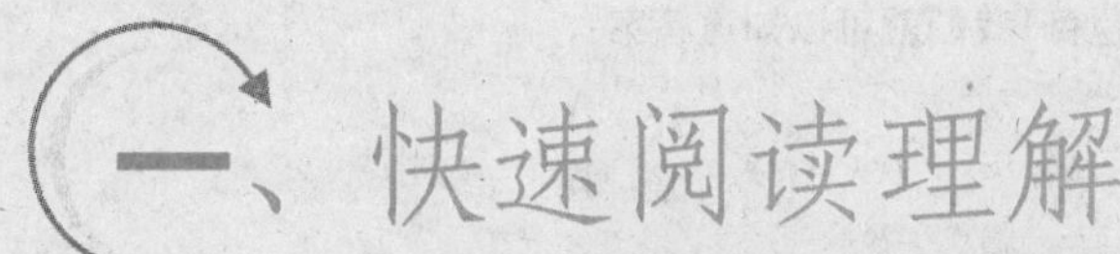

# 题型聚焦

## 1. 文章类型

快速阅读这一题型本是英语专业八级考试中的一个题型,考查快速阅读,搜索信息的能力。新四级也增加了这一题型。但不同的是新四级的快速阅读为一篇文章,长度在1000词左右。文章的体裁多样,大致可以划分为如下几种:

(1)议论文:对某一观点的赞成或反对,对于某一问题的两种看法的分别描述;

(2)记叙文:对某一事件或一系列事件的讲述。这类文章作为考题的话极有可能考的是传记,人物生平。

(3)说明文:对于某一事物或一系列事物的描述,比如样卷就是对 landfill 的具体情况的描述,是一篇典型的说明文;再就是对事物变化的描述,特别是对科技发展的描述,比如2006年6月新四级考卷里的快速阅读就是对高速公路(highway)的发展的介绍。

把握文体,有助于抓住文章脉络,比如记叙文,一般是以时间为脉络;说明文中对事物发展的描述同样以时间为线索;而对于事物的描述一般都遵循构造、原理,功用这样的结构。

## 2. 两大题型精确定位

### (1)判断题

①细节题:

快速阅读的文章后设置了10个小题,其中7个为判断正误,或判断给出信息是否在文章中出现,另3个是根据文章意思填空。我们可以发现在2006年新四级样卷的7道判断题中,除了第一题是对文章整个内容的判断。其他6题都是根据文章的某个细节作出判断。以下是样卷中的细节判断题:

【题干】样卷第7题:Hazardous wastes have to be treated before being dumped into landfills.

【原文】Along the site, there are drop - off stations for materials that are not wanted or legally banned by the landfill. A multi - material drop - off station is used for tires, motor oil, lead - acid batteries. Some of these materials can be recycled.

In addition, there is a household hazardous waste drop - off station for chemicals (paints, pesticides, other chemicals) that are banned from the landfill. These chemicals are disposed of by private companies. Some paints can be recycled and some organic chemicals can be burned in furnaces or power plants.

【解析】所谓细节题,就是指对文章中某个具体细节的考查,本题是2006年新四级样题的第7题。针对的主要问题是有害物质的处理问题,我们在文章的倒数第2和第3段可以找到答案。一段是讲轮胎,电池之类的有害垃圾,一段是讲居民生活产生的有害垃圾,如油漆,杀虫剂等。根据原文这两段,我们可以知道有害垃圾是不会进入垃圾掩埋场(landfill)的,而是由 drop - off station 来处理。这里讲到 hazardous wastes 有的能够被回收,有的能够被烧毁,但没有一定被处理的意思。因而属于 NG,而属于 NG

的题目往往是理解起来模棱两可的。

②主旨题:

样卷第 1 题 The passage gives a general description of the structure and use of a landfill.

【解析】本题和上题相反,答案无法从文章的某个部分或细节找到,而是要根据全文来得出结论。那么是否要把全文通读一遍来做这一题呢,其实大可不必,通过这篇文章的各个标题(包括文章题目)我们就可以知道答案:

Landfills

How Much Trash Is Generated?

Of the 210 million tons of trash, or...

How Is Trash Disposed of?

The trash production in the United States has almost tripled since 1960...

What Is a Landfill?

There are two ways to bury trash:...

Proposing the Landfill

For a landfill to be built, the operators have to make sure that they follow certain steps...

Building the Landfill

Once the environmental impact study is complete, the permits are granted and the funds have been raised,...

What Happens to Trash in a Landfill?

Trash put in a landfill will stay there for a very long time...

How Is a Landfill Operated?

A landfill, such as the North Wake County Landfill,...

由这些标题我们可以得知本文是讲述垃圾掩埋场(landfill)的结构和使用的,所以本题是正确的。

## (2)填空题

以下是样卷中的 3 道填空题:

样卷第 8 题:Typical customers of a landfill are ________.

【原文】A landfill, such as the North Wake County Landfill, must be open and available every day. Customers are typically municipalities and construction companies, although residents may also use the landfill.

样卷第 9 题:To dispose of a ton of trash in a landfill, customers have to pay a tipping fee of ________.

【原文】Customers are charged tipping fees for using the site. The tipping fees vary from $10 to $40 per ton. These fees are used to pay for operation costs.

样卷第 10 题:Materials that are not permitted to be buried in landfills should be dumped at ________.

【原文】Along the site, there are drop-off stations for materials that are not wanted or legally banned by the landfill. A multi-material drop-off station is used for tires, motor oil, lead-acid batteries. Some of these materials can be recycled.

【解析】三题的答案我们可以在各个段落找到(划线部分)。我们同样可以发现填空题也是对细节的考查。由此我们可以得出结论:快速阅读主要考查的是考生通过略读(Skimming)获知大意和通过查读(Scanning)了解细节的能力;这不同于我们以前见到的阅读理解题,考查的是学生的深度理解。所以,快速阅读的重点不在难度,而在速度。

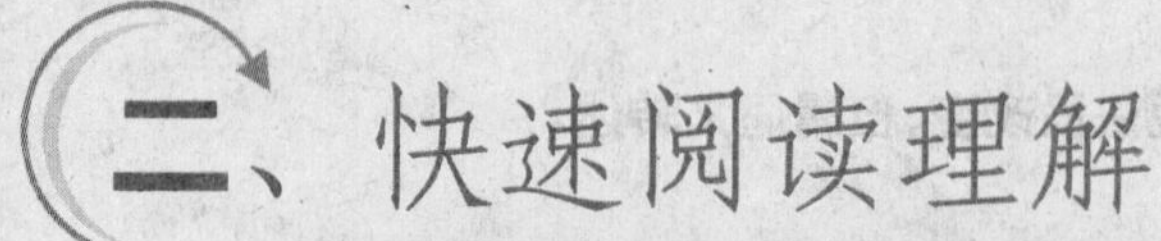

# 应试技巧精确突破

## 1. 通用技巧

### (1)略读(Skimming)

略读是指快速通览全文,了解其大意。但略读不是指将全文所有的文字都看一遍,而是应该去把握文章的标题,副标题和小标题。没有标题的时候,应把主要精力都放在文章各段的主题句上,通常快速阅读的主题句在段首,少数在段尾。比如2006年6月新四级真题中的两段:

By opening the North American continent, highways have enabled consumer goods and services to reach people in remote and rural areas of the country, spurred the growth of suburbs, and provided people with greater options in terms of jobs, access to cultural programs, health care, and other benefits. Above all, the interstate system provides individuals with what they cherish most: personal freedom of mobility.

The interstate system has been an essential element of the nation's economic growth in terms of shipping and job creation: more than 75 percent of the nation's freight deliveries arrive by truck; and most products that arrive by rail or air use interstates for the last leg of the journey by vehicle. Not only has the highway system affected the American economy by providing shipping routes, it has led to the growth of spin-off industries like service stations, motels, restaurants, and shopping centers. It has allowed the relocation of manufacturing plants and other industries from urban areas to rural.

我们可以看到第一段的主题句在段尾,第二段的主题句在段首。

除去大小标题和主题句,把握一些标志性的词也有助于我们对文章脉络的把握。比如表示时间的词。

### (2)查读(Scanning)

查读是针对细节题的一种有效解题技巧,找到题干中的关键词,然后快速在文章中搜索这一关键词,则可起到事半功倍的解题效果。充当关键词的一般有时间、人名、地名、阿拉伯数字以及缩写等。比如2006年6月新四级真题中的第2题:

General Eisenhower felt that the broad German motorways made more sense than the two-lane highways of America.

关键词是能够将题目信息和其他类似信息区分开来的词,其中人名比较适合作关键词的,因为人名醒目,易于寻找。根据本题中出现的General Eisenhower,我们很快可找到:

When General Eisenhower returned from Germany in 1919, after serving in the U.S. Army's first transcontinental motor convoy(车队), he noted: "The old convoy had started me thinking about good, two-lane highways, but Germany's Autobahn or motorway had made me see the wisdom of broader ribbons across the land."

## 2. 正误判断技巧

### (1)选 Y(YES)四大原则

**1 题目与原话相比,某些词用近义词替换,或说法虽改变,但属同义转述。**

比如样题中第5题:

【题干】In most countries the selection of a landfill site is governed by rules and regulations.

【原文】In most parts of the world, there are regulations that govern where a landfill can be placed and how it can operate.

【解析】我们可以看到 In most parts of the world 变成了 In most countries;where a landfill can be placed 变成了 selection of a landfill site;但意思仍然不变,所以正确。

**2 否定词+反义词或反义词+反义词。**

比如:

【题干】Only a few scientists are against this plan.

【原文】Most of the scientists support this plan.

【解析】most 和 few 是一对反义词,support 和 be against 是一对反义词,所以本题正确。

**3 概括关系。**

题干内容属于原文的一部分,比如:

【题干】Apple and pear are both helpful for weight loss.

【原文】Besides apple and pear, peach is also beneficial to weight loss.

【解析】原文说的是苹果,梨和桃都对减肥有益,题干说苹果,梨都对减肥有益,自然正确。

**4 逻辑推理。**

比如2006年6月新四级真题中的第1题:

【题干】National standards for paved roads were in place by 1921.

【原文】With the increase in auto production, private turnpike(收费公路) companies under local authorities began to spring up, and by 1921 there were 387,000 miles of paved roads. Many were built using specifications of 19th century Scottish engineers Thomas Telford and John MacAdam (for whom the macadam surface is named), whose specifications stressed the importance of adequate drainage. Beyond that, there were no national standards for size, weight restrictions or commercial signs.

【解析】在原文我们可以看到 Beyond that,即指 Beyond 1921,是没有国家标准的,题干说至1921年,国家标准出现,所以正确。

### (2)选 N(NO)三种技巧

**1 与原文意思相反。**

题干多了否定词,或题干中用了相反意思的表达。比如2006年6月新四级真题中的第5题:

【题干】In spite of safety considerations, the death rate on interstate highways is still higher than that on other American roads.

【原文】The death rate on highways is half that of all other U. S. roads (0.86 deaths per 100 million passenger miles compared to 1.99 deaths per 100 million on all other roads).

【解析】原文中说只占一半(is half that of)而题干的意思完全相反:is still higher than 所以判断为 NO。

又比如2006年样卷中的第3题:

【题干】Compared with other major industrialized countries, America buries a much higher percentage of its solid waste in landfills.

【原文】The United States ranks somewhere in the middle of the major countries (United Kingdom, Canada, Germany, France and Japan) in landfill disposal. The United Kingdom ranks highest, burying about 90 percent of its solid waste in landfills.

【解析】原文说的是 in the middle,而不是 much higher。

### 技巧 2 逻辑关系错误。

比如因果,条件关系颠倒。2006 年 6 月新四级真题中的第 7 题是典型的一例因果关系的颠倒:

【题干】Service stations, motels and restaurants promoted the development of the interstate highway system.

【原文】Not only has the highway system affected the American economy by providing shipping routes, it has led to the growth of spin-off industries like service stations, motels, restaurants, and shopping centers.

### 技巧 3 以偏概全。

用部分、局部代替整体,也包括由个别具体描述推广为具体情况。比如 2006 年样卷第 2 题:

【题干】Most of the trash that Americans generate ends up in landfills.

【原文】The trash production in the United States has almost tripled since 1960. This trash is handled in various ways. About 27 percent of the trash is recycled or composted, 16 percent is burned and 57 percent is buried in landfills. The amount of trash buried in landfills has doubled since 1960.

【解析】原文说是 57% 在垃圾场掩埋,显然算不上绝大多数(Most of)。

## (3)选 NG(NOT GIVEN)两种方法

判断为 NG 的题干,并没有和原文的意思发生任何的冲突,否则就应判断为 NO,这是划分两者的根本原则。

### 方法 1 与原文不相干,出现文章根本没有提及的信息。

比如 2006 年 6 月新四级真题中第 6 题:

【题干】The interstate highway system provides access between major military installations in America.

【原文】Today the interstate system links every major city in the U. S., and the U. S. with Canada and Mexico.

【解析】原文中能找到的关于州际高速公路连接的情况就这一句,并未提及能够通向主要的军事设施。所以选 NG。

### 方法 2 根据文章得出文中没有的结论。

这样的题最具迷惑性。比如下文:

A Pill to Prevent the Flu

You never got around to getting your flu vaccine, and now you're worried you'll become sick, since several of your co-workers have succumbed. Luckily, your doctor can prescribe a pill that could prevent the flu from developing. Tamiflu, an antiviral drug, is FDA-approved to prevent and treat the flu. And GlaxoSmithKline, maker of the antiviral Relenza, is seeking FDA approval for preventive use of that drug.

Already got the flu? No problem. When taken within the first 48 hours, these drugs can shorten the length of illness and reduce symptoms and complications. They may also help protect against avian flu, but more research is needed. For most of us, the drugs are not meant to replace a vaccine, says Neil Schachter, MD, author of The Good Doctor's Guide to Colds & Flu, since they're costly and have possible side effects, such as nausea(恶心), vomiting and diarrhea(腹泻). The popular herbal remedy echinacea is touted for its ability to strengthen immunity, protect against colds and shorten the duration of illness. But researchers at the University of Virginia think we should save our money, because echinacea doesn't seem to work. They divided 399 people into groups. Some took echinacea before and after being infected with a virus; some took it only after-infection; the rest got a placebo. At the end of the study, researchers concluded that echinacea did not help prevent or treat a cold.

1. If you want to defend yourself against getting flu, you can take Tamiflu, because it can protect you the same as flu vaccine.

2. Some researchers have concluded some herbal remedies did not help prevent or treat a cold.

【解析】这两题都是从文章得出的结论,但第一题与文章中 For most of us, the drugs are not meant to replace a vaccine 发生了冲突,因此应判断为 NO;而第二题看似正确,仔细一点你可以发现原文中说 echinacea 这种草药对预防治疗感冒没有效果,而题干中用的是很含糊的一些草药,文章中没有说是否是有一些,但也没说就 echinacea 这一种,所以选 NG。

## 3. 填空题三项注意

填空题为了控制答案的唯一性,一般来说都是考查细节,这样的填空细节题,往往题干在句式上和原文相比会发生变化,比如主从句位置调换,主动态被动态之间的转换,等等,所以在做填空题时还要注意语法的问题,特别是填充项包括谓语动词时。

### (1)词语位置或词形的变化

在考查细节题时,我们同样可以利用找关键词这个办法,来判断答案在哪,见 2006 年新四级样卷:

【题干】Typical customers of a landfill are ________.

【原文】A landfill, such as the North Wake County Landfill, must be open and available every day. Customers are typically municipalities and construction companies, although residents may also use the landfill.

【解析】关键词为 Typical customers,根据关键词,我们不难找到答案。同时我们可以发现原文中用的 typically,题干中是 Typical。有这么一个小小的变化。

### (2)语态的变换

【题干】People regarded William L. Shirer's The Rise and Fall of the Third Reich as ________.

【原文】William L. Shirer ranks as one of the greatest of all American correspondents. He lived and worked in Paris, Berlin, Vienna and Rome. But it was above all as correspondent in Germany for the Chicago Tribune and later the Columbia Broadcasting System that his reputation was established. He subsequently wrote The Rise and Fall of the Third Reich, which is hailed as a classic, and after the war he was awarded the Legion d'Honneur. In the post-war years he wrote in a variety of fields, and in his seventies he learned Russian, publishing a biography of Tolstoy at the age of 89. He died in 1994. His Berlin broadcasts were published posthumously by Hutchinson in 1999.

【解析】原文中使用的是被动态:which is hailed as a classic。题干使用的是主动态。需要考生细心,然后填出答案 a classic。

### (3)表达方式的变化

这样的题目稍难一点,因为答案可能并没有在原文出现,而是原文中某个词或短语表达方式的同义词。需要考生十分灵活。比如:

【题干】Those unsuccessful fathers believe that their children will ________ their dreams.

【原文】Men are ambitious animals. Each young guy having just left the campus firmly believes that he will be somebody in the future despite the fact that he is nobody at all up till now; middle aged bachelors are constantly searching for the best woman in the world and in their minds the best is always the next; a husband probably likes to make the promise that he will buy his wife a huge diamond immediately after promotion while actually what his wife really wants then is just a warm kiss; unsuccessful fathers do admit they are failures in this ruthless society, and they also insist that they are greatest trainers and their dreams will definitely come true on their telented kids; even the retired gentlemen wave their pension checks, saying perhaps there is another Harlan Sanders (founder of the Kentucky) among them.

【解析】文章中用的表达方式是 come true on kids,题干中的主语是 children,此处应灵活变换,填 realise,则意思和原文相同。

几点注意:

①答案往往不会超过 5 个词;

②题目顺序和答案在原文出现的顺序一样,这一点也适用于判断题。

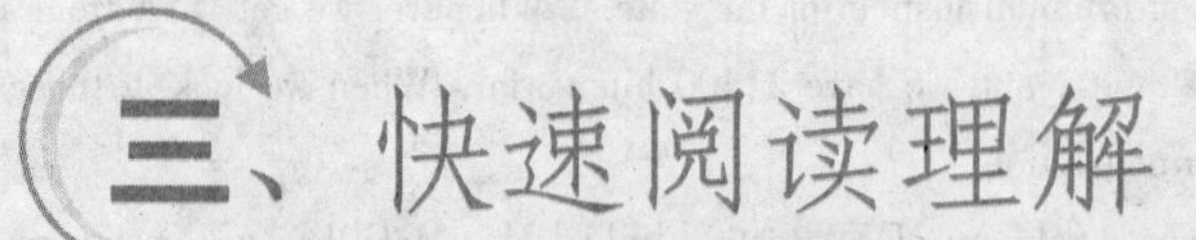

# 三、快速阅读理解

## 核心考点精确打击

**Directions**: *In this part, you will have 15 minutes to go over the passage quickly and answer the questions on* **Answer Sheet** *1.*

*For questions 1 – 7, mark*

*Y(for YES)* *if the statement agrees with the information given in the passage;*

*N(for NO)* *if the statement contradicts the information given in the passage;*

*NG(for NOT GIVEN)* *if the information is not given in the passage.*

*For questions 8 – 10, complete the sentences with the information given in the passage.*

## Unit 1

### Should DNA be collected from all criminals?

In most cities and states, vandalism, shoplifting, and loitering are *misdemeanors*(轻罪)—possibly involving community service, not jail time. But those who commit such low – level crimes in New York State may soon be required to give DNA samples to authorities—just as convicted rapists or murderers do.

If the Legislature passes the proposal, which is currently being debated, New York would be the first state in the nation to require DNA samples for all convicted offenders.

Gov. George Pataki (R), who is asking the Legislature to expand DNA collection, argues that a larger database will help solve more crimes. Supporters add it will help solve future crimes because criminals who start off committing petty crimes sometimes graduate to more serious offenses.

The Empire State's move comes as DNA work in criminal investigations is under closer scrutiny. Complaints are rising about mistakes—some inadvertent, others fraudulent—at DNA labs around the country. Groups that have championed the use of DNA to verify the guilt or innocence of convicts are now campaigning against large – scale expansion of the practice, saying it would overwhelm labs. Others cite concerns about the increasing number of innocent people whose DNA is stored in a databank without their knowledge or approval.

"People have come to appreciate the power of DNA to solve crimes. They now need to respect the care that is required to maximize its potential and avoid its abuse," says Stephen Saloom, policy director for the Innocence Project, which has used DNA testing to free 176 prisoners wrongfully convicted.

Twenty-eight states now collect samples for some misdemeanors, according to DNAResource. com, a website that tracks DNA policies. A week ago, Kansas joined California and five other states in going one step further: taking the DNA samples of some people arrested, but not necessarily charged, with a crime.

"They all tend to be violent *felony*(重罪) and burglary arrestees," says Lisa Hurst, a government - affairs consultant with Smith Alling Lane, which represents Applied Biosystems, a maker of DNA testing equipment. Smith Alling Lane also runs DNAResource. com.

New York's proposal, however, would go the furthest, requiring DNA collection for all convicted of misdemeanors. This would add about 80,000 additional DNA profiles per year to the DNA bank, say researchers at the state's Division of Criminal Justice.

It would be well worth the effort, says Chauncey Parker, director of criminal justice for the state. "Whenever we get DNA from a convicted offender, we run it against the 18,000 unsolved crimes, mostly rapes, and we have 2,400 hits so far. When we look at those hits, we find on average when an arrestee is convicted, it's his or her 12th conviction."

He cites the example of Raymon McGill arrested July 25 for attempted robbery. DNA testing linked Mr. McGill to two earlier murders and a rape, says Mr. Parker. He also notes that McGill had been arrested in 1999 for a misdemeanor.

"Had we required DNA fingerprinting back then, he would has been linked to the rape and the case solved," says Parker.

On Monday, critics of the New York proposal publicly complained that the backlog of cases is already large. Tom Duane (D), the only state senator to vote against the New York measure, worries about the cost of expansion, as well as how to ensure proper training of personnel and storage of the samples. "What good will it do if it's not done right?" he asked.

Such concerns go well beyond New York. In February, the National Association of Criminal Defense Lawyers, in their magazine, published an article detailing mistakes that have cropped up at DNA labs around the country. "Many of the mistakes arise from cross - contamination or mislabeling of DNA samples," wrote William Thompson, a professor at the University of California, Irvine.

Thompson's concerns arise just as states and cities are expanding their use of DNA testing. Los Angeles County is preparing to open a new crime lab in 2007. A proposed county budget included funding not only for the lab, but also for prosecutors and public defenders training in DNA technology. Nebraska recently passed a bill to include felony robbery and burglary convictions among those requiring DNA samples. Wisconsin is considering a bill to add some misdemeanor - related sex crimes to its DNA - collection requirements.

But with such expansions come concerns. Mr. Saloom of the Innocence Project objects to law - enforcement agencies holding the DNA of innocent people, especially those who merely cooperated with an investigation. These samples should be destroyed, he says. "The fundamental values of government accountability and personal privacy are at stake," he says.

Some of these arguments resonate with Assemblyman Joseph Lentol (D), chairman of the Assembly Codes Committee, which deals with criminal sanctions. "We are asking [DNA labs] to destroy the DNA of an innocent party," says Mr. Lentol. "This is a privacy issue."

Parker, the director of criminal justice for the state, says cooperating individuals can ask a judge to order their samples to be returned. "The burden is on the police to always demonstrate to a judge that whatever evidence they are collecting, whether it's a DNA fingerprint or a photograph or a handwriting sample, that it has been collected lawfully and that it's being used for a lawful purpose," he says.

In addition, Assemblyman Lentol also worries that private labs will use DNA information for commercial gain. "There is an almost limitless possibility of ID theft," he worries.

That concern is addressed in the *pending*(未决的)legislation, Parker says, by increasing the penalty for unlawful use of a DNA sample.

The governor and lawmakers have until the end of the legislative session, June 22, to come to terms on the proposal.

1. Presently, if someone commits shoplifting, he will not be required after being arrested to give DNA samples to authorities in New York State.
2. The proposal of requiring those misdemeanors criminals to give DNA samples to authorities has been welcomed by all the legislators.
3. According to the report, some innocent people's DNA may be collected without personally knowing the fact.
4. Stephen Saloom objects to using DNA testing in his Project.
5. Now, there are five states in US where some misdemeanor arrestees' DNA samples have been collected.
6. Chauncey Parker believes the proposal may do good to investigate some unsolved crimes.
7. New York police's expense on DNA testing is larger than that of other states.
8. Mistakes maybe occur by ________ or labelling the samples wrongly in DNA testing.
9. Mr. Saloom holds that samples of innocent people should be destroyed in order to keep their ________ and the fundamental values of government accountability.
10. ________ may face more severe punishment, according to Parker.

# Unit 2

## Why bank stocks are cash machines

**With their high yields and low P/Es, they offer the potential for solid long - term gains with little risk.**

In a world that's buzzing over the Internet's raging return and the dollar play in Asian equities, it's hard to believe that the best place to invest may well be bank stocks.

Try not to yawn. Bank stocks aren't boring, they just look that way. You might say they're so colorless they're colorful. Why? Because what they lack in glamour they more than make up in compelling numbers.

The story comes in two parts. First, the biggest banks, from Citigroup to Wachovia, are paying rich dividends and are likely to increase them steadily, practically guaranteeing double - digit returns far into the future. Second, regional banks and thrifts are prime takeover candidates as the giants rush to expand—witness Wachovia's $26 billion deal for Golden West Financial.

Says Goldman Sachs analyst Lori Appelbaum: "Bank stocks don't look sexy until you look at the potential returns."

For investors, the big diversified banks offer a rare combination: juicy yields plus significant earnings growth. And even though big - bank stocks are up 7 percent this year, they still sell at bargain prices.

It's a neat package. The five largest U. S. banks are Citi, Bank of America, J. P. Morgan Chase, Wachovia, and Wells Fargo. For the moment we'll set J. P. Morgan aside, because it's the one huge bank that's not yet highly profitable (though it's getting there). The four remaining players boast an average dividend yield of 3.6 percent. That's twice the yield of the S&P and not too far behind the three - year Treasury yield of 4.9 percent.

And while interest rates are just as likely to fall as to rise, the beauty of dividends is that they tend to keep growing. Since dividends are paid from earnings, they typically increase in step with profits. Over the past four years the big banks (again, excluding J. P. Morgan) have raised earnings at double - digit rates.

Result: BofA's dividend has jumped 14 percent a year since 2001, while Wells' has posted annual increases of 19 percent.

But wait—there's more! Let's take a cautious stance. We'll project that the big banks' earnings growth averages to 8 percent over the next seven years, far below the recent level. So dividends should also grow at 8 percent.

The banks, though, offer an added treat. While paying out around 46 percent of their annual profits in dividends, they're typically using another 20 percent or so of their earnings to buy back stock. It's a safe bet that they'll keep buying back about 1.5 percent of their shares annually. Since both earnings and dividends will be spread across fewer shares each year, the dividends per share will rise not by 8 percent a year, but by more like 9.5 percent.

By the way, these numbers tell a powerful story about banking. The banks are using two - thirds of their profits to pay dividends and buy back stock, which means they're retaining only one - third to invest in growing their business. But that's all they need, because banks generate huge returns, both on the capital they already have and on the new money flowing in as retained earnings.

The proof: On average, the four most profitable big banks boast a sumptuous return on equity of 16.5 percent over the past five years, vs. an average of 12 percent for stocks in the S&P 500.

### The payoff

Now let's get back to the payoff for investors. You're starting with a 3.6 percent dividend. It's rising at almost 10 percent a year. Those increases should far outstrip inflation. So in 2012, you'll be receiving 6.8 percent, not 3.6 percent, on your original investment, and the payouts will ratchet upward from there.

But wait—there's even more! The dividend payments are just part of your return. If earnings per share keep rising at 9.5 percent annually, the banks' stock prices will increase at the same rate (assuming the relatively low price/earnings multiples remain constant). Your returns should start in the 13 percent range—the 3.5 percent current dividend plus a 9.5 percent capital gain. But in seven years, thanks to the ever - growing dividend, that number should surpass 15 percent.

J. P. Morgan has the potential to deliver even greater returns. To buy the stock now, you have to believe that new CEO Jamie Dimon will be able to turn the bank around (and we do—see "The Contender"). If that comeback happens, the dividend—and the stock price—could soar.

Unlike its big rivals, J. P. Morgan hasn't increased its dividend in five years, chiefly because it's been saddled with weak earnings.

Dimon has been expanding in lucrative areas like credit cards and energy and mortgage - backed security trading to increase profitability. If those moves manage to raise J. P. Morgan's ROE from 8 percent to its peers' 16.5 percent average, the dividend will double along with earnings, giving investors who got in early an enormous yield of around 6 percent as well as a big capital gain.

**The risks**

Is there any risk in these equations? Of course. J. P. Morgan's comeback could stall, and the other banks' share prices could flatten or drop, saddling investors with little more than the single - digit returns provided by dividends. That's possible, but unlikely.

The reason: Bank stocks still look cheap. On average, Citi, BofA, Wells, and Wachovia have P/Es of just 12.4. That's far lower than the multiples in the other big - dividend - paying sectors—pharmaceuticals, utilities, and telecom—even though the banks' earnings are growing just as fast as profits in those three industries. In fact, it's more probable that banking multiples will rise to the 15 or so that prevails in the other big - dividend sectors, adding a big kicker for shareholders.

Investors can play the banking market a second way, by predicting which regional and smaller banks and thrifts the big boys are likely to buy. Today retail banking is highly fragmented: The big five control just 28 percent of nationwide deposits.

"We'll see tremendous consolidation in the industry in the next few years," says Meredith Whitney, an analyst with CIBC World Markets.

The key is geography. Several of the giants have gaps in their footprints. J. P. Morgan and Citi, for example, need to establish big footholds in Florida and California, while Wachovia covets more branches in the West.

Among the best candidates: SunTrust, which is a powerhouse in Florida and Georgia, where Citi and J. P. Morgan need to grow. PNC is a major player in Pennsylvania and New Jersey, which are among the nation's wealthiest markets. U. S. Bancorp boasts a solid franchise in the Pacific Northwest, California, and Colorado. Washington Mutual would hand a buyer 2,200 branches across the country, including a strong presence in California.

Watch the banks. As a famous bank robber put it, that's where the money is.

1. Golden West Financial was submerged by Wachovia.
2. J. P. Morgan Chase promises an average dividend yield of 3.6 percent.
3. Major bank's earning growth will keep on at double - digit rates in the following years.
4. The dividends per share of the big banks rise as much as the annual profit in dividends.
5. The four most profitable big banks achieved a higher return than most of the stocks in S&P 500.
6. The inflation of America is under 10%.
7. Citigroup may well also have the potential to deliver returns surpassing 15%.
8. The main reason why J. P. Morgan's dividend hasn't risen in the past five years is that it has been hobbled by ________.
9. There is unlikely any risk in these equations because ________.
10. Experts say that the key for the banks' consolidation is geography because of some of these major plays' ________.

**Blair's Gold Medal Dilemma has Two very Different Faces**

The Prime Minister is seen as a visionary hero in the US, but as a naive and *vainglorious*(虚荣的) fool in Britain

Tony Blair arrives in Washington today for talks with President Bush. They will discuss Iraq, Iran and Afghanistan, but the Prime Minister will once again leave without the Congressional Gold Medal awarded to him three years ago for his steadfast support of America after 9/11.

This is no mere oversight. Although the US Administration publicly insists that the medal is still at the design stage, senior Washington sources now acknowledge that Downing Street is deliberately dragging its feet over receiving the honour.

Aides working for Senator Elizabeth Dole, who was one of the sponsors of Mr. Blair's award, Ave told *The Times* that the problem is at the British end. It certainly does not take more than a thousand days to mint a medal: Nelson Mandela received his award in 56 days.

Mrs Dole's spokesman said that although No. 10 had approved long ago the image of Mr Blair that will appear on the front side of the medal, it was blocking agreement over the design for the obverse, which usually includes an emblem and some form of quotation.

The dispute is an appropriate symbol of the two entirely distinct images of the Prime Minister' s relationship with America since the terrorist attacks of 2001. In the US he is still seen as a clear - eyed brave heart who swiftly recognized 9/11 as an historical turning point that had profoundly affected the American psyche, and then—better than anyone, including Mr. Bush—articulated the case for war.

In Britain, however, he is more often regarded these days as a naive and vainglorious fool, the *poodle*(狮子狗) who loyally followed the President and has, consequently, tainted his own premiership with alleged falsehoods and failure.

It is this second image, the reverse side of the same coin, that lies behind his reluctance to be photographed receiving his medal from Mr. Bush and why, according to friends, he is unlikely to pick up the award until after he has left office.

The decision to award the medal was marked by Mr. Blair' s triumphant speech to both houses of Congress in July 2003. But that moment is now badly *tarnished*(失去光泽) in his memory; a few hours later he was informed that David Kelly, the government scientist named as a source for damaging stories on Iraq, had killed himself.

That visit to Washington was the seventh in only 20 months after 9/11. Since then the frequency of such trips has plummeted, along with the popularity of both leaders.

Today will be the first time that Mr. Blair has set foot in the American capital in almost a year, but British viewers will have to stay up late if they want to see the President and Prime Minister give a joint press conference scheduled to begin at 12.30 a.m. tomorrow.

One explanation for this is that the two leaders, both deeply unpopular and nearing the end of their time in office, no longer stand "shoulder - to - shoulder", but look more like two lame ducks. Downing Street is instead putting emphasis on a speech that Mr. Blair will deliver at Georgetown University tomorrow.

He will seek to set the Iraq war in the context of a "value - based foreign policy" that encompasses the well - meaning military interventions in Kosovo and Sierra Leone, as well as the broader humanitarian goals of his policy towards Africa, climate change and the peace process in the Middle East.

He will set out his belief that world problems can best be tackled by strengthening global institutions ranging from the United Nations to the World Trade Organization, the World Bank and the International Monetary Fund. Speaking only days after his return from Baghdad, Mr. Blair will reiterate forcefully that now is the time for the world community to come to the aid of Iraq' s fledgeling democratic Government, rather than to abandon it to the insurgents.

In his talks with Mr. Bush, he also plans to discuss how to secure UN sanctions against Iran, issues such as the peace process and Darfur, and the new - found aggression of a Russia led by Vladimir Putin.

But Mr. Blair knows that for all the effort he has expended in such discussions with Mr. Bush over the past 4 years, he has secured little that is tangible. The US President has backed Palestinian statehood, for instance, but Palestine is not yet a state.

The Prime Minister has said, however, that those wanting him to do a Hugh Grant impression and provide a "Love Actually moment" by taking on the American President in public will be disappointed. It is too late to start picking a fight with Mr. Bush.

Instead, his speech is intended to correct what he regards as a misleading impression of him both in the US and at home. He never was a neoconservative from the same mould as much of the Bush Administration. He arrived at the same place but from a different and much more liberal direction. Indeed, his disproportionate and sometimes disfunctional emphasis on Iraq' s supposed arsenal of WMD before the war was intended to protect multilateralism by securing UN authority for invading, rather than undermine it, as he was subsequently accused of doing.

Nor, however, was he the *craven*(怯懦的) fool taken for a ride by Mr. Bush that critics back in Britain believe him to have been. Indeed, the US Administration believes that Britain has re - emerged as a genuine global power in the past few years.

It is typical of Mr. Blair that, like his Congressional Gold Medal, he has two faces, neither of which he believes to be true images of himself.

**PEAKS AND TROUGHS**

The main meetings between Tony Blair and President Bush

Feb 23, 2001 Meeting at Camp David, discussed plans to impose sanctions on Iraq. Nicknamed "the Colgate" summit. Blair's approval rating: 48 per cent

July 18 - 20, 2001 Bush stays with Blair at Chequers before travelling to Genoa with him for G8. Rating: 55 per cent

September 20, 2001 Blair visits New York after September 11 attacks; joins Bush in Washington. Rating: 67 per cent

November 7, 2001 Meeting in White House. They insist that they will win War on Terror. Rating: 64 per cent

September 7, 2002 Meet at Camp David to discuss international coalition for Iraq invasion. Rating: 42 per cent

March 16, 2003 Azores summit, widely regarded as the end of diplomatic negotiations with Iraq and hope for a UN second resolution. Rating: 43 per cent

March 27, 2003 War summit in Washington. Rating: 43 per cent

June 9, 2004 G8 summit in Georgia. Bush and Blair discuss Iraq and say that there will be no deployment of Nato troops there. Rating: 30 per cent

November 12, 2004 Blair is first leader to meet Bush after his re – election. After Yassir Arafat's death they discuss peace process in Middle East. Rating: 32 per cent

July 7, 2005 Bush and Blair meet at G8 in Gleneagles, but the event is overshadowed by London bombings. Rating: 44 per cent

September 15, 2005 Bush and Blair meet at UN millennium event. Rating: 31 per cent

1. Tony Blair has already received the Congressional Gold Medal awarded to him for his support of America after 9/11.
2. Tony Blair is not the first person who is awarded the Congressional Gold Medal.
3. Mr. Blair is positively regarded as clear – eyed brave heart both in the US and the UK.
4. According to the report, both the President and the Prime Minister have received severe criticizing about their policy.
5. Mr. Blair may object to the suggestion that the global issues should be settled by major powers.
6. We can infer from the report that the present talk may achieve much.
7. Mr. Blair may well support the UN sanctions against Iran as it is the proposal of Mr. Bush.
8. According to the report, some critics in Britain had ever believed that Mr. Blair was a ________________ controlled by Mr. Bush.
9. The highest approval rating of Tony Blair occurred when he meet with Mr. Bush on ________________.
10. On the day when Mr. Blair meet with Bush soon after his re – election, his approval rating is ________________.

## A second baby? Russia's mothers aren't persuaded.

Cash for babies is the Kremlin's offer to women in its latest bid to reverse a population decline that threatens to leave large swaths of Russia virtually uninhabited within 50 years.

President Vladimir Putin last week defined the crisis as Russia's most acute problem, and promised to spend some of the country's oil profits on efforts to relieve it. He ordered parliament to more than double monthly child support payments to 1,500 rubles (about $55) and added that women who choose to have a second baby will receive 250,000 rubles ($9,200), a staggering sum in a country where average monthly incomes hover close to $330.

On Monday, young women at the Family Planning Youth Center, a nongovernmental clinic for northwest Moscow, said they liked the sound of more money, but suggested that Mr. Putin has no concept of their lives. "A child is not an easy project, and in this world a woman is expected to get an education, find a job, and make a career," says Svetlana Romanicheva, a student who says she won't consider babies for at least five years. She hopes to have one child, but says a second would depend on her life "working out very well." As for Putin's offer, she says "it won't change anything."

Russia's birthrate, falling for decades, has plunged in post – Soviet times, to just 1.17 in 2004 from 2.08 babies per woman in 1990—far below the 2.4 children required to maintain the population—according to the Federal State Statistics Service. The average rate from 2000 – 2005 in the US, by contrast, was 2.0, according to UN figures, while Mexico, for example, weighed in at 2.4 and Italy at 1.3.

Russia also has one of the world's highest abortion rates. In addition, the death rate has climbed to levels seldom seen in peacetime, to 16.3 in 2002 from 10.7 per thousand people in 1988. The result is a population that is shrinking by an average of 700,000 people each year—and aging. A UN report last year predicted that Russia's population, around 145 million in 2002, could fall by one – third by 2050.

Experts foretell the grim prospect of a Russia that can no longer man its factories, field a decent hockey team, or defend its borders. "I think Putin's main concern is a lack of future soldiers," says Yury Levada, head of the Levada Center, an independent polling agency. "That's a narrow perspective, but one that resonates politically."

Some women say they resent the suggestion, made explicit by many nationalist politicians, that their lack of enthusiasm for bearing children is to blame. "This problem began long ago, and even if we were to have more babies it wouldn't mean the situation... would improve," says Irina Isayeva, a medical student who volunteers at the family center. "A woman has to... ensure that her conditions are adequate to raise children. Women may want fewer children, but be able to give them better chances in life."

Young women also say that it's hard to find a good partner. Official statistics show that almost 8 of every 10 marriages end in divorce, and one - third of children are born out of wedlock. "The interests of men and women seem different, so women just depend less on men," says Olga Istomina, a student. "A lot of people live together. Partners change all the time."

Others say Putin is moving in the right direction. "Russian women typically have one child—but many of my patients would like a second if they felt they had enough support," says Galina Dedova, a gynecologist at Happy Families, a private Moscow clinic. "Most of my patients count their rubles... If they could reliably expect more money, some might [consider] more children."

Putin also doubled subsidies for foster families, to 4,500 rubles ($166) per month, a move widely welcomed by child - care experts. In recent years, Russia has cracked down on foreign adoptions, leaving 700,000 institutionalized children with few options. "I believe the situation will begin to improve after Putin's measures, and more people will see the importance of adopting," says Galina Krasnitskaya, an adviser to Russia's State Duma.

Critics point to the high male death rate, a problem Putin barely addressed. Men's ranks have been decimated by alcoholism, war in Chechnya, AIDS, and accidents. "Male life expectancy is less than 60 years," says Yevgeny Gontmakher, research head of the Center of Social Studies, an independent Moscow think tank. "Trying to stimulate the birthrate is pure populism; it's naive to think a demographic revolution can happen."

Low birthrates and high mortality could deliver an economic wallop that could dash Putin's hopes of restoring Russia as a great power. "If current trends persist, there will be four dependents for every Russian worker by 2025," Regional Development Minister Vladimir Yakovlev warned last month. "Russia needs a million new workers every year. If we don't get them, we can forget about economic growth."

Some say that Russia must open its doors to immigrants, as many Western countries have done. But Putin insists that only ethnic Russians—about 25 million remain stranded in former Soviet countries—will be eligible for easy entry. Polls show large majorities remain hostile to the idea of mass immigration of non - Slavs.

Nadezhda Kalmikova, director of the Family Planning Center, says she believes that money will solve little. "Families need to be sure there will be all the things children need," she says. "That goes beyond material requirements... We need children who will grow up well and become good citizens. You can't buy that."

1. Russia must have been experiencing a population decline these years.
2. A Russian mother will receive 1,500 rubles every month from the government.
3. Svetlana Romanicheva think that Putin's proposal turn out to be profitless.
4. If Russia wants to keep the present number of population, each woman needs to give birth to 2.08 babies as in 1990.
5. The fact that Russia has a very high abortion rate is the main reason for the population decline.
6. Some women resent the suggestion of Kremlin's offer of "Cash for babies".
7. We can infer from the report that most marriages in Russia will end in separation.
8. Putin ordered to give 4,500 rubles's subsidies for ________________, while 1,500 ruble's child support payments for families which have born babies.
9. If present trends continue, Russia will undoubtedly be caught in an ________________ which could perish their dream to restoring as a great power.
10. According to the poll, most Russians are still ________________ to the suggestion that the government permit non - Slavs immigration.

# Unit 5

## Christopher Reeve: A Hero Onscreen and Off

Christopher Reeve, a hero onscreen as Superman and in real life as an activist for stem cell research, passed away on October 10 at age

52. Just months before, Mr. Reeve spoke with our magazine for an interview that appeared in our October 2004 issue.

Dressed in a striped polo shirt, white duck pants and running shoes, Reeve spoke with Reader's Digest about the film, his advocacy and his remarkable journey. On the rebound from a recent hospital stay, he displayed his usual *tenacity*(坚韧), saying, "Your body is not who you are. The mind and spirit transcend the body."

RD: It's been more than nine years since the accident. How has it changed your perspective on life?

Reeve: I have more awareness of other people and, I hope, more sensitivity to their needs. I also find that I'm more direct and outspoken. It's important to me to say what I really mean.

RD: In your second book, you wrote about feeling angry after the accident. Have you accepted things now?

Reeve: I don't get angry, because it wouldn't do any good. I experience frustration sometimes, such as when I have a crisis, like I just did.

RD: What happened?

Reeve: I've had three bad life-threatening infections this year. This most recent was a blood infection caused by an abrasion on my left *hip*(髋部) that I probably picked up one day when I was on the exercise bike. It seemed benign but developed into strep. Then a lot of major organs shut down. We're trying to figure out what's going on. Before that one, I got a severe infection in New Orleans just a few days before shooting the movie. I was frustrated: "This is not fair; come on. Let's not fall apart. I've come too far." So sometimes I get jealous of people who take their ability to move for granted.

RD: Do you get scared?

Reeve: No, I don't.

RD: How could you not?

Reeve: It's a proven fact that you can control panic by applying rational processes. In all my days of flying and sailing and riding, every now and again I got myself into a jam. On Christmas Day in 1985 I was flying over the Green Mountains in Vermont. Thick clouds, snowing. And the warning light went on. I looked out and saw oil all over the wing. I knew I had to shut down that engine and fly to Boston on the other. You're hoping it doesn't develop a problem too. But the chance of a multi-engine failure is very, very remote. Literally, you use your brain to stop panic. I've had a lot of training in that area from my life before the injury.

RD: It's almost as if everything in your life up to the accident was preparation for this phase.

Reeve: That's probably true. I'm glad I didn't know it at the time.

RD: So what's the latest product of your determination, in terms of regaining movement or sensation?

Reeve: There hasn't been any new recovery since what was published in 2002. But I've been able to maintain most of what I achieved.

RD: Are you still optimistic you will walk again?

Reeve: I am optimistic. But I also know that, with time, I'm beginning to fight issues of aging as well as long-term paralysis. So it seems more difficult to project than it was five years ago. But I haven't given up.

RD: Has there been a change in your optimism?

Reeve: Hope, to me, must be based—now knowing as much as I do—on a projection derived from solid data. But, yes, there's been a change in my state of mind, because in May of next year it will be ten years since the accident, and I doubt if by that time there's going to be a procedure suitable for me. At 52, knowing that a safe trial for me may still be years away has changed my perspective. I didn't think it would take this long.

RD: What's been the hardest part?

Reeve: Watching the slow progress of research in this country. I don't know if it would have made me walk sooner, but I would have had the satisfaction of knowing we're all on the same page. Groups of people who have differences about all kinds of issues are united to fight against AIDS. Wouldn't it be great if we were as united about biomedical research for diseases that affect 128 million Americans?

RD: Tell me about the Christopher Reeve Paralysis Act.

Reeve: It's broken into three parts. One is for biomedical research. The other is rehabilitation research. The third is for quality-of-life programs. It would create five centers across the country, to make sure that there is support for people living with paralysis. Patients do better the sooner you get them up and moving. Put them in pools, on treadmills, on exercise bikes—anything to keep the systems of the body from breaking down. No magic pill will cure spinal cord injury. It'll be a combination of a drug therapy, or procedure, plus rehabilitation.

RD: Does the Act have broad support?

Reeve: I'm quite optimistic that it will pass, because there's nothing controversial about it. It doesn't even mention embryonic stem cells.

RD: What's your position on embryonic stem cell research?

Reeve: I advocate it because I think scientists should be free to pursue every possible avenue. It appears though, at the moment, that embryonic stem cells are effective in treating acute injuries and are not able to do much about chronic injuries.

RD: How have political decisions slowed stem cell research?

Reeve: The religious right has had quite an influence on the debate. I don't think that's appropriate. When we're setting public policy, no one segment of society deserves the only seat at the table. That's the way it's set in the Constitution. So debate all we want, hear from everybody. And then allow our representatives to weigh the factors and make laws that are going to be ethically sound, moral, responsible, but not the result of undue pressure from any particular entity.

RD: Is it hard to be patient?

Reeve: I've lasted more than nine years, so I can wait a little longer. I also realize that a lot of people are watching me, to see what I'm going to do. I want to make sure I'm making a smart choice. I'm not at a point of desperation where I'd say, "Just somebody fix me, anywhere."

RD: Did you ever feel that way?

Reeve: I was much more impatient five years ago. I started out saying, "What do you mean you can't fix the spinal cord?" I remember telling a neurosurgeon, "Don't give me too much information, because at the moment my ignorance is my best asset." Then, over time, as you learn more about the complexities of the central nervous system, and you learn to balance your life—even to get a life back—your perspective changes.

RD: You've talked people out of suicide who've just suffered the kind of injury you had. How do you do it?

Reeve: I tell them about a lot of things that are available, what's happening with research, particularly for the acute phase of injury, and what opportunities there are for rehabilitation. And I tell them that even though you can't imagine building a new life now, you need to wait until everything stops spinning and you can look at it more clearly. In the meantime, don't do anything rash. I haven't lost anybody yet. And it probably happens, gosh, sometimes as much as once a week.

1. Christopher Reeve had never get any frustration.
2. Christopher Reeve had experience many fearful things before his injury.
3. Christopher Reeve's health have been turning better since 2002.
4. According to the interview, Christopher Reeve's was injured at about 42 years old.
5. Christopher Reeve said he was confident about the fact that he would walk again soon.
6. Presently, Reeve is participating in the campaign of Christopher Reeve Paralysis Act.
7. Christopher Reeve Paralysis Act will surely pass in US.
8. According to Christopher Reeve, embryonic stem cells are effective in ________________ while not much in chronic injuries.
9. According to Christopher Reeve, the reason why stem cell research are advancing slowly is the fact that ________________.
10. According to the interview, Reeve had once persuaded some people from ________________.

## Unit 6

### Courtroom Chaos Halts Saddam's Trial

Saddam Hussein's trial is on hold once again, this time until early next month.

The Iraqi chief judge in Baghdad *adjourned*(延期) the stormy trial until April 5 after the obstinate former dictator outright refused to answer prosecutors' questions. Wednesday marked the first time Saddam testified at his trial, and he did so with fiery political speeches that prompted the chief judge to close the courtroom, CBS News chief foreign correspondent Lara Logan reports.

Saddam called on Iraqis to stop a bloody wave of *sectarian*(宗派主义的) violence and instead fight American troops. He also encouraged Iraqis to "unite in a *jihad*(讨伐异教徒) against the occupiers," Logan reports.

Even as the judge repeatedly yelled at him to stop, Saddam read from a prepared text, insisting he was still Iraq's president.

"Let the (Iraqi) people unite and resist the invaders and their backers. Don't fight among yourselves," he said, praising the *insurgency*(叛乱). "In my eyes, you are the resistance to the American invasion."

Finally chief judge Raouf Abdel-Rahman ordered the session to continue in secret, telling journalists to leave the chamber. The video and audio broadcast of the trial was cut off. Logan reports the blackout lasted almost two hours and raised questions about the transparency of the court.

After nearly two hours, reporters were called back into the court. Saddam was sitting alone in the defendants' pen in front of the judge. The former Iraqi leader then refused to answer questions from the chief prosecutor, demanding to see a copy of his testimony given to investigators before the trial began. The prosecution agreed to the demand and said they would question Saddam in the next session.

Saddam was the last of the case's eight defendants to be called to testify. Though he has spoken frequently since the trial began in October, Wednesday's session was to be the first chance for the judge and prosecutors to directly question him on charges of killing 148 Shiites and imprisoning and torturing others during a 1982 crackdown against the Shiite town of Dujail.

Instead, Saddam, dressed in a black suit, read from his statement, insisting he was Iraq's elected president and calling the trial a "comedy".

He addressed the "great Iraqi people", a phrase he often used in his speeches as president, and urged them to stop the wave of Shiite-Sunni violence that has rocked the country since the bombing of a major Shiite shrine last month.

"What pains me most is what I heard recently about something that aims to harm our people," Saddam said. "My conscience tells me that the great people of Iraq have nothing to do with these acts," he said referring to the bombing of the shrine in the city of Samarra.

Abdel-Rahman interrupted saying he was not allowed to give political speeches in the court.

"I am the head of state," Saddam replied.

"You used to be a head of state. You are a defendant now," Abdel-Rahman said.

The judge repeatedly closed his microphone to prevent his words from being heard and told him to address the case against him. But Saddam ignored him, continuing to read from his text.

"What happened in the last days is bad," he said, referring to the recent violence. "You will live in darkness and rivers of blood for no reason."

"The bloodshed that they (the Americans) have caused to the Iraqi people only made them more intent and strong to *evict*(驱逐) the foreigners from their land and liberate their country," Saddam said.

At one point, Abdel-Rahman screamed at him, "Respect yourself." Saddam shouted back: "You respect yourself."

"You are being tried in a criminal case for killing innocent people, not because of your conflict with America," Abdel-Rahman told him. "What about the innocent people who are dying in Baghdad? I am talking to the Iraqi people," Saddam replied.

Finally, Abdel-Rahman ordered the session closed to the public. "The court has decided to turn this into a secret and closed session," he said.

One of Saddam's lawyers told Logan he has little doubt the court has already decided Saddam is guilty. "I fully expect him to be dead by the end of the year, executed by this court in this show trial," he said, adding that Saddam himself was convinced the court was determined to see him dead and would find him guilty. "They can kill his body, but not his spirit," Saddam's lawyer said.

The stormy session was a stark contrast to the past three hearings, when each of Saddam's seven co-defendants has appeared, one by one, and was questioned by Abdel-Rahman and the chief prosecutor.

Saddam and the seven former members of his regime face possible execution by hanging if they are convicted in connection to the crackdown launched in Dujail following a July 8, 1982, shooting attack on Saddam's motorcade in the town.

Last month, Saddam stood up in court and boldly acknowledged that he ordered the 148 Shiites put on trial before his Revolutionary Court, which eventually sentenced them all to death. But Saddam insisted it was his right to do so since they were suspected in the attempt to kill him.

Before Saddam's testimony, his half-brother Barzan Ibrahim, who headed the feared Mukhabarat intelligence agency at the time of the Dujail attack, was questioned for more than three hours by the chief judge and prosecutor, who presented him with half a dozen Mukhabarat documents and memos about the crackdown.

Barzan, a secular Sunni, interspersed his commentary with passages from the Koran, Logan reports. At times, this drew laughter from the Iraqi journalists listening in the gallery, Logan says.

One after another, Ibrahim insisted that the documents were fake and that his signatures on them were forged. "It's not true. It's forged. We all know that forgery happens," he said.

In previous sessions, Dujail residents have testified that Ibrahim personally participating in torturing them during their imprisonment at the Baghdad headquarters of the Mukhabarat intelligence agency, which Ibrahim headed. One woman claimed Ibrahim kicked her in the chest while she was hung upside down and naked by her interrogators.

But Ibrahim insisted the Mukhabarat agency was not involved in the investigation into the attack on Saddam and denied any personal role in the crackdown.

"I didn't order any detentions. I didn't interrogate anyone," he said, adding that he resigned from the Mukhabarat in August 1983. "There is not a single document showing that I was involved in the investigation."

Chief prosecutor Jaafar al – Moussawi showed the court a series of Mukhabarat documents on the Dujail case from 1982 and 1983, some of which bore signatures he said were Ibrahim's. One of them was a memo from Ibrahim's office asking Saddam for rewards for six Mukhabarat officers involved in the Dujail crackdown.

"This is not my signature. My signature is easy to forge, and this is forged," Ibrahim said.

He said the same of another document that listed Dujail families whose farmlands were *razed*(夷为平地) in retaliation for the shooting. Another document, signed by an assistant to Ibrahim, talked of hundreds of Dujail detainees being held by the Mukhabarat at its headquarters and at the notorious Abu Ghraib prison. Ibrahim said that memo as well was false.

At the end of Wednesday's session, Abdel – Rahman ordered forensic tests on the signatures to determine their veracity.

Ibrahim insisted that the General Security agency, not the Mukhabarat, carried out the Dujail crackdown. He said his sole role came on the day of the shooting, when he went to the village and ordered security officials to release Dujail residents who had been arrested.

The defense has argued that Saddam's government acted within its rights to respond after the assassination attempt on the former Iraqi leader. The prosecutor has sought to show that the crackdown went well beyond the authors of the attack to punish Dujail's civilian population, saying entire families were arrested and tortured and that the 148 who were killed were sentenced to death without a proper trial.

1. The chief judge put off the trial because Saddam Hussein kept on giving political speeches in the courtroom.
2. Saddam Hussein continued his speech during the two hours when video and audio broadcast of the trial was cut off.
3. The booming of a major Shiite shrine last month had caused Shiite – Sunni violence which was disastrous to the country.
4. According to Saddam's speech, he might believe that the bombing of the shrine in the city of Samarra was not done by Iraqis.
5. One of Saddam's lawyers may hope that Saddam die soon.
6. Saddam did not accept the charge of slaughtering 148 Shiites in 1982.
7. The woman who claimed Ibrahim kicked her in the chest had been imprisoned in Dujail.
8. According to Ibrahim's words in the court, we learn that we may use ________________ to refer to fake things.
9. One of Ibrahim's assistants said that Ibrahim had imprisoned many civilians in ______________, besides headquarters of Mukhabarat.
10. According to the prosecutor, after being arrested, the 148 civilians must have been ______________ and then sentenced to death.

## IAEA head to probe Iran's nukes claims

The head of the U. N. nuclear watchdog, Mohamed ElBaradei, expressed optimism about his visit to Iran on arriving there for talks aimed at defusing tension over Tehran's nuclear program.

"The time is right for a political solution and the way is negotiations," the head of the International Atomic Energy Agency told journalists at Mehrabad International Airport in Tehran just after midnight Wednesday local time.

ElBaradei's visit began not long after Tehran's announcement Tuesday that the country had successfully enriched uranium, a key step to producing peaceful nuclear energy or nuclear weapons.

"I would like to see Iran come to terms with the requests of the international community," he said, explaining the purpose of his trip as being "to clarify remaining outstanding issues on the nature of the Iranian program."

Earlier Wednesday, the country's deputy nuclear chief said Iran intends to move toward large – scale uranium enrichment involving 54,000 centrifuges, signaling its resolve to expand a program the international community has insisted it halt.

That will be hundreds of times more than what the country has now, reports CBS News correspondent Jim Axelrod. If true, Axelrod adds, that would be enough to produce hundreds of nuclear warheads.

Iran's president had announced Tuesday that the country had succeeded in enriching uranium on a small scale for the first time, using 164 centrifuges. The U. N. Security Council has demanded that Iran stop all enrichment activity because of suspicions the program's aim is to make nuclear weapons.

"We will expand uranium enrichment to industrial scale at Natanz," Deputy Nuclear Chief Mohammad Saeedi told state – run television Wednesday.

He said Iran has informed the International Atomic Energy Agency that it plans to install 3,000 centrifuges at Natanz by late 2006, then expand to 54,000 centrifuges, though he did not say when.

He said using 54,000 centrifuges will be able to produce enough enriched uranium to provide fuel for a 1,000 – megawat nuclear power plant like the one Russia is currently putting the finishing touches on in southern Iran.

Iran's claims brought it fresh international condemnation as allies Russia and China joined several European countries and the United States in expressing their disapproval over the nuclear activities.

Already the U. N. Security Council had given it until April 28 to clear up suspicions that it wants to become a nuclear power. It has asked Tehran to suspend enrichment and allow unannounced IAEA inspections.

The White House is pressing for U. N. sanctions against Iran.

ElBaradei said he hoped the visit "would bring Iran in line with the requests of the international community to take confidence—building measures regarding its activities including suspension of enrichment and related activities until outstanding issues are clarified."

A team of five IAEA inspectors arrived in Iran late last week.

On Thursday, ElBaradei is expected to meet Iran's nuclear chief Vice President Gholamreza Aghazadeh.

"I don't know anyone on god's green earth that actually thinks that's what the Iranians are interested in," Richard Haass, president of the Council on Foreign Relations told CBS News' *The Early Show* co – anchor Hannah Storm. "They don't need nuclear power for electricity because they've got all this oil and gas. So this is a serious moment. But, again, the world still has time for diplomacy to try to put a ceiling on what they're doing," Haass said.

Iranian President Mahmoud Ahmadinejad announced the enrichment success Tuesday in a nationally televised ceremony, saying the country's nuclear ambitions are peaceful and warning the West that trying to force Iran to abandon enrichment would "cause an everlasting hatred in the hearts of Iranians."

The United States is not taking Iran's claim at face value, CBS News senior White House correspondent Bill Plante reports, and top officials tell CBS News that they just can't be sure. But the announcement quickly raised condemnations from the U. S., who said the claims "show that Iran is moving in the wrong direction."

Denouncing Iran's successful enrichment of uranium as unacceptable to the international community, Secretary of State Condoleezza Rice said Wednesday the U. N. Security Council must consider "strong steps" to induce Tehran to change course. She said "this latest announcement... will further isolate Iran."

Rice also telephoned Mohamed ElBaradei, the head of the International Atomic Energy Agency, to ask him to reinforce demands that Iran comply with its nonproliferation requirements when he holds talks in Tehran on Friday.

The U. S. remains convinced Iran wants to build a nuclear weapon.

"This is not a question of Iran's right to civil nuclear power," Rice said while greeting President Teodoro Obiang Nguema Moasogo of Equatorial Guinea. "This is a question of,... the world does not believe that Iran should have the capability and the technology that could lead to a nuclear weapon."

Russia also criticized the announcement Wednesday, with Foreign Ministry spokesman Mikhail Kamynin saying, "We believe that this step is wrong. It runs counter to decisions of the IAEA and resolutions of the U. N. Security Council."

Former Iranian president Hashemi Rafsanjani, a powerful figure in the country's clerical regime, warned that pressuring Iran over enrichment "might not have good consequences for the area and the world."

If the West wants "to solve issues in good faith, that could be easily possible, and if they want to... pressure us on our nuclear activities, things will become difficult and thorny for all," Rafsanjani said in an interview with the Kuwaiti newspaper Al – Rai Al – Aam, published on Wednesday.

Rafsanjani, who heads Iran's Expediency Council, a powerful body that arbitrates between the parliament and the clerical hierarchy, said planned talks between Iran and the United States on stabilizing Iraq could lead to discussions on the nuclear dispute.

"We don't have a mandate to discuss the nuclear issue with the Americans... but if the talks on Iraq go in the right direction, there might be a possibility for that issue," Rafsanjani said in an interview with the Al - Hayat daily. "There have been many cases where big and wide - ranging decisions had small beginnings."

Iranian and U. S. officials have insisted the talks will deal only with Iraq. So far, no date for the talks has been set.

Enrichment is a key process that can produce either fuel for a reactor or the material needed for a nuclear reactor. But thousands of centrifuges, arranged in a network called a "cascade," are needed for either purpose, and getting any number of centrifuges to work together is a very delicate and difficult task.

1. The head of the International Atomic Energy Agency went Iran for talks because Tehran announced that they could make nuclear weapons.
2. Mohamed ElBaradei didn't think that Iran would make nuclear bombs.
3. If Iran did not halt its program, the country may soon get enough material for large amount of nuclear weapons.
4. Iranian leaders did not comment on the intention of their expanding uranium enrichment to 54,000 centrifuges.
5. ElBaradei expected that Iran would cooperate with the international community.
6. Richard Haass did not believe that Iran would develop the program only for electricity.
7. US are not sure about Iran's intention in the future.
8. According to Rice, if Iran did not halt its program, the U. N. Security Council should carry out ____________ to stop it.
9. Besides America, ____________ also criticized the announcement.
10. According to Rafsanjani, the precondition is that ____________ go in the right direction.

# Unit 8

## World's churches seek best ways to counter the "Code"

Beyond that almost self - evident truth, however, church leaders worldwide are divided over just how they should respond to "The Da Vinci Code," as the blockbuster film opens around the planet this week.

Some are demanding that censors ban the film, or cut scenes that they say undermine Christian beliefs.

Others are angrily advocating boycotts to protest what they see as an attack on the Roman Catholic church.

And then again, priests from both the Protestant and Catholic traditions are seizing the occasion as a "teachable moment," using everything from scratch - cards in Britain to an animated version of Leonardo da Vinci's "Last Supper" in Australia to make their points.

"We see it as an opportunity to put God and Jesus on the agenda," explains Margaret Rodgers, a spokeswoman for the Anglican church in Australia, which has launched a cinema advertising campaign around the film. "Over the tea trolley at work, it's going to be a hot topic of conversation for a while."

"It's a match we could win or lose," says Philippe Joret, a French evangelical preacher who has been challenging The Da Vinci Code's theories at public meetings since September. "But it's a match worth playing."

The ideas explored in the fictional thriller, which has sold more than 40 million copies in more than 40 languages, are expected to find an even wider audience with the film starring Hollywood actor Tom Hanks and French actress Audrey Tautou.

The book has upset many Christians by suggesting that Jesus fathered a child by Mary Magdalene, the first in a line of Christly descendants still extant today, and that the Catholic church has covered this up for 2,000 years. Mixing fact and fiction in a way that religious leaders say might confuse readers, the book also suggests that Christ's divinity was an idea that the Emperor Constantine imposed on the Council of Nicea in 325 AD for political reasons.

Branding the novel "obstinately anti - Christian," top Vatican official Archbishop Angelo Amato—a close confidante of Pope Benedict XVI—called three weeks ago for a boycott of the film.

Cardinal Francis Arinze, the Nigerian who heads the Vatican's office on liturgy, went even further in a church - backed documentary released Tuesday titled "A Masterful Deception." Christians should not just "forgive and forget" insults to the founder of their religion, he said, but should react, possibly by taking legal action against the film.

Vatican officials have been "unusually outspoken and aggressive" in their attacks on "The Da Vinci Code," says John Allen, Vatican analyst for the National Catholic Reporter.

They have not gone as far as church leaders in Jordan, however, where Archbishop Hanna Nour has called on the government to ban the film, or in Thailand, where a group of Protestant leaders has asked government censors to cut the last 15 minutes of the movie, which concludes that Jesus has heirs alive today.

In India, one Catholic activist has gone on what he says is a "hunger strike until death" unless the film is banned.

Indian Christians opposed to the film have won support from an umbrella organization of Islamic clerics in Bombay who labeled the film "blasphemous" because it spreads "lies" about Jesus Christ. "Muslims in India will help their Christian brothers protest this attack on our common religious belief," Maulana Mansoor Ali Khan, general secretary of the All – India Sunni Jamiyat – ul – Ulema, told Reuters news agency.

**A different approach**

In the Philippines, however, the Roman Catholic Archbishop of Manila has taken a different approach, even though he calls the film "blasphemous."

"Like in anything negative, let us take this occasion to convert the cinema industry's money – motive production into a pastoral challenge, an evangelization and catechistical moment of grace," Cardinal Gaudencio Rosales said in a statement to priests last week.

That is a tack that Catholic priests elsewhere have taken. Just up the coast from Cannes, the French resort where "The Da Vinci Code" premiered at the film festival there Wednesday, a priest in Nice was planning to lead a public debate after a screening of the film in a local cinema.

"The film asks ridiculous questions, mixing the historical with the non – historical," says the Rev. Vincent – Paul Toccoli. "But the church has left itself open to this sort of thing: People are disoriented, but the church doesn't help them much. She just repeats the same old things. She needs a kick in the behind like this."

**An opening for dialogue**

Mr. Joret, the evangelical pastor, also sees the film's release as a useful opportunity for Christians to examine the roots of their faith. Even more important, he says, "it offers a real chance for believers to have a dialog with nonbelievers, and to explain what the Scriptures say. We will have an opportunity to address people who wouldn't normally read books about Jesus or about the church."

The important thing, he says, is that "we should not panic" about the damage the film might do to Christianity's image. "If you start off with a bad attitude based on fear, being aggressive or defensive, you'll get it all wrong," he argues.

On the other side of the world in Australia, the Anglican Bishop of South Sydney, Robert Forsyth, has adopted a similar approach.

"We decided to be tongue – in – cheek rather than hysterical or anxious," says the bishop, who heads a media group that has set up challenging davinci.com, a website answering the questions that the film raises about early Christian history.

The church is advertising the site with a 15 – second video clip airing in cinemas that plays on Leonardo da Vinci's "Last Supper," a painting crucial to The Da Vinci Code's plot. In the animated—and updated—painting, Jesus is reading the novel at the center of the storm, and rolls his eyes incredulously.

In Britain, Protestant churches are using equally innovative tools to get their message out. One group has distributed 270,000 scratch – cards to cinemas where the film will be shown, asking moviegoers to say whether 10 claims made in the film are true or false.

Beside the statement "the marriage of Jesus and Mary Magdalene is a matter of historical record," for example, are two boxes: scratch the one marked "Fact" and you will discover an "X." Scratch "Fiction" and you uncover a check mark.

The cards are the work of the Rev. Mark Stibbe, author of a booklet called "Cracking the Da Vinci Code," who will be visiting Paris this weekend to examine the film's claims in more detail at events organized by a local Anglican church, St. Michael's.

"Being in Paris, where so much of the film is set, we felt it was quite crucial to do something" says Dan Ritchie, a pastoral assistant at the church. "We see it as a good *evangelistic*(福音传道者的) tool."

Taking a tack diametrically opposed to the boycott advocated by Archbishop Amato, St. Michael's is organizing a night out at the movies for its *congregation*(集会) on Friday, encouraging worshippers to go to see the controversial film.

"It's a good social event, an opportunity to bring family and friends along and show people what church is," says Mr. Ritchie. "All churches have to fight misconceptions that they are cults hidden behind church doors: This is a good chance to show who we are."

1. Church leaders are all angry about the newly-cast film "The Da Vinci Code".

2. "We see it as an opportunity to put God and Jesus on the agenda" indicates that priests want to use the occasion as a good time to preach.

3. According to the report, "The Da Vinci Code" may attract more audiences because it is starred by some famous actors and actress.
4. Producers of the film will face strict legal charge.
5. We can infer from the report that Christian belief even has some followers in Muslim world.
6. A priest was thinking of sponsoring a parade to forbid the screening of the film in Cannes.
7. Mr. Joret, the evangelical pastor holds that it is not reasonable to be afraid of the negative impact caused by the film.
8. Robert Forsyth proposed to be ______________ instead of being anxious.
9. Protestant in Britain using scratch - card to address the movie viewer that "the marriage of Jesus and Mary Magdalene is a ______________.
10. Archbishop Amato, St. Michael's will arrange ______________ as the main content for the congregation on Friday.

# Unit 9

### It's all about me: Why E - mails Are So Easily Misunderstood

Michael Morris and Jeff Lowenstein wouldn't have recognized each other if they'd met on the street, but that didn't stop them from getting into a shouting match. The professors had been working together on a research study when a technical *glitch*(小故障) inconvenienced Mr. Lowenstein. He complained in an e - mail, raising Mr. Morris's *ire*(愤怒). Tempers flared.

"It became very embarrassing later," says Morris, when it turned out there had been a miscommunication, "but we realized that we couldn't blame each other for yelling about it because that's what we were studying."

Morris and Lowenstein are among the scholars studying the benefits and dangers of e - mail and other computer - based interactions. In a world where businesses and friends often depend upon e - mail to communicate, scholars want to know if electronic communications convey ideas clearly.

The answer, the professors conclude, is sometimes "no." Though e - mail is a powerful and convenient medium, researchers have identified three major problems. First and foremost, e - mail lacks cues like facial expression and tone of voice. That makes it difficult for recipients to decode meaning well. Second, the prospect of instantaneous communication creates an urgency that pressures e - mailers to think and write quickly, which can lead to carelessness. Finally, the inability to develop personal *rapport*(和谐,亲善) over e - mail makes relationships fragile in the face of conflict.

In effect, e - mail cannot adequately convey emotion. A recent study by Prof. Justin Kruger of New York University and Nicholas Epley of the University of Chicago focused on how well sarcasm is detected in electronic messages. Their conclusion: Not only do e - mail senders overestimate their ability to communicate feelings, but e - mail recipients also overestimate their ability to correctly decode those feelings.

One reason for this, the business - school professors say, is that people are egocentric. They assume others experience stimuli the same way they do. Also, e - mail lacks body language, tone of voice, and other cues—making it difficult to interpret emotion.

"A typical e - mail has this feature of seeming like face - to - face communication," Professor Epley says. "It's informal and it's rapid, so you assume you're getting the same paralinguistic cues you get from spoken communication."

To avoid miscommunication, e - mailers need to look at what they write from the recipient's perspective, Epley says. One strategy: Read it aloud in the opposite way you intend, whether serious or sarcastic. If it makes sense either way, revise. Or, don't rely so heavily on e - mail. Because e - mails can be ambiguous, "criticism, subtle intentions, emotions are better carried over the phone," he says.

E - mail's ambiguity has special implications for minorities and women, because it tends to feed the preconceptions of a recipient. "You sign your e - mail with a name that people can use to make inferences about your ethnicity," says Epley. A misspelling in a black colleague's e - mail may be seen as ignorance, whereas a similar error by a white colleague might be excused as a typo.

If you're vulnerable to this kind of unintentional prejudice, pick up the phone: People are much less likely to prejudge after communicating by phone than they are after receiving an e - mail. Kruger and Epley demonstrated this when they asked 40 women at Cornell to administer a brief interview, 20 by phone and 20 by e - mail. They then asked a third group of 20, the "targets," to answer the phone interviewers' questions. They sent a transcription of the targets' answers to the e - mail interviewers.

The professors then handed each interviewer what they said was a photo of her subject. In reality, each got a picture of either an Asian or an African – American woman (in reality, all were white).

E – mail interviewers who thought the sender was Asian considered her social skills to be poor, while those who believed the sender was black considered her social skills to be excellent. In stark contrast, the difference in perceived sociability almost completely disappeared when interviewer and target had talked on the phone.

E – mail tends to be short and to the point. This may arise from the time pressures we feel when writing them: We know e – mail arrives as soon as we send it, so we feel we should write it quickly, too. On the other hand, letters depend on postal timetables. A letter writer feels he has a bigger window of time to think and write.

Psychologists Massimo Bertacco and Antonella Deponte call this characteristic "speed facilitation," and they believe it influences our episodic memory—our ability to recall events. They found that e – mailers wrote shorter messages and were less likely to "ground their communications" in memories of shared experience than letters writers were.

The brevity of e – mail and the absence of audiovisual cues can endanger business and personal relationships unless e – mail is supplemented with the rapport that comes from more personal communication.

"Rapport creates a buffer of positive regard," says Professor Morris, "and when it's not there negotiation becomes brittle, vulnerable to falling apart."

Morris, who studies negotiation at Columbia, led a study that found that negotiators exchange more than three times the information in face – to – face interactions as they do via e – mail. Though Morris and his colleagues concluded that e – mail lets negotiators make "more complex, multiple – issue offers," they ultimately built less rapport, thereby increasing tensions and lowering the average economic value of the agreements.

Rapport "is an interpersonal resonance of emotional expression," Morris says, "involving synchronous gesture, laughing, and smiling together. Once this rapport exists, it's a buffer against a moment in the negotiation when there's some friction." This buffer is hard to develop without speaking over the phone or in person. Those who negotiated by e – mail in Morris's study trusted each other less and weren't as interested in working together again.

But the pitfalls of e – mail interaction were easily overcome by a single phone call. Morris ran a second round of negotiations, all conducted via e – mail, but made half of the corresponding pairs chat on the phone before negotiating—"just for five or 10 minutes," Morris explains, "and the key thing is we told them, Don't get into the issues. It's just an icebreaker." The result was dramatically improved agreements.

So if you want to buy something on Craig's List, Morris says, "make a brief phone call, even if it's not practical to do the whole negotiation by phone. You can establish a favorable bias with someone and then proceed in a less rich medium, but it's very hard to just get right into the negotiation on a medium that isn't rich."

| How well do we communicate? | | |
|---|---|---|
| FREQUENCY THAT... | E-MAIL | PHONE |
| Communicator believes he is clearly communicating | 78% | 78% |
| Receiver believes he is correctly interpreting | 89% | 91% |
| Receiver correctly interprets message | 56% | 73% |

1. Mr. Morris was angry about Mr. Lowenstein's sending him a e – mail because he didn't know the latter.
2. Morris and Lowenstein believed that electronic communications cannot say things explicitly.
3. E – mails are often written in a hurry, which some mistakes may appear.
4. Emotion cannot be well conveyed by e – mail because both the senders and the recipients may overestimate their respectively ability to communicate and decode feelings.
5. If you do not want be unintentionally prejudiced, you'd better phone the recipients after sending your e – mail.
6. Most people prefer to sending e – mail instead of writing letters because it tends to be short and to the point.
7. The content of a e – mail tends to be less memorable than that of a letter.
8. In the sentence "and when it's not there negotiation becomes brittle, vulnerable to falling apart.", here there refer to ________________.
9. When negotiate by e – mail, those negotiators tend to be less interested in ________________.
10. In Morris' words, he call e – mail ________________.

# Unit 10

## For Environmentalists, a Growing Split over Immigration

To environmentalists worried about population growth, people are people.

Even if they do their best to live lightly on the land, their rising numbers are a growing burden on Earth's resources. And whether they sing the "The Star - Spangled Banner" in English or in Spanish really doesn't matter.

As politicians and the public heatedly debate immigration, so, too, are environmental activists.

The flow of people into the United States is troubling some environmentalists for two reasons. First, more Americans means more people living in one of the world's most resource - consuming cultures. Second, there's new evidence that Hispanic women who move to the US have more children than if they stayed put.

"We've got to talk about these issues—population, birth rates, immigration," says Paul Watson, founder of the Sea Shepherd Conservation Society, which confronts whalers, seal hunters, and those who poach wildlife in the Galapagos Islands. "Immigration is one of the leading contributors to population growth. All we're saying is, those numbers should be reduced to achieve population stabilization."

Mr. Watson also was a Sierra Club board member. Last month, he resigned in protest just before his three - year term ended because he thinks the organization ignores immigration as a major factor in population growth.

Beneath the dispute is a political subtext. Environmentalists generally see themselves as political progressives; they don't want to be bedfellows with anti - immigrant activists sometimes labeled as *xenophobic*(恐惧或憎恨外国人的,恐外的) or racist. Very few greens raise a supportive fist when they see "Stop the Invasion" billboards sprouting from California to Florida. For the most part, they skirt the issue.

"The leadership and the membership have said we want to be neutral on this," says Eric Antebi, national press secretary for the Sierra Club in San Francisco, one of the largest and oldest grassroots environmental groups in the country. It's a global issue, says Mr. Antebi, caused by environmental degradation and poverty that need to be solved so people won't have to look elsewhere for a better life. Other large environmental groups take the same position.

Yet the US population is far from stabilized, and immigrants (legal and illegal) are one of the main reasons. There are about 11 million illegal immigrants in the US today, 57 percent from Mexico, and another 24 percent from other Latin American countries, according to the Pew Hispanic Center. Of the US foreign - born population, nearly 30 percent is illegal, according to Pew.

The US Census Bureau this week reported that Hispanics—the largest minority at 42.7 million—are the nation's fastest - growing group. They are 14.3 percent of the overall population, but between July 2004 and July 2005, they accounted for 49 percent of US population growth. Of the increase of 1.3 million Hispanics, the Census Bureau reported, 800,000 was because of natural increase (births minus deaths), and 500,000 was due to immigration.

"The Hispanic population in 2005 was much younger, with a median age of 27.2 years compared to the population as a whole at 36.2 years. About a third of the Hispanic population was under 18, compared with one - fourth of the total population," according to the Census Bureau report. That means such younger people are just entering (or will remain longer in) the years in which they have children of their own.

Steven Camarota, director of research at the Center for Immigration Studies in Washington, finds that once women emigrate to the US, most tend to have more children than they would have in their home countries. "Among Mexican immigrants in the United States fertility averages 3.5 children per woman compared to 2.4 children per woman in Mexico," he wrote in a study last October. And the same is true among Chinese immigrants. Fertility is 2.3 in the US compared with 1.7 in China. However, typically these high fertility rates decline in the successive generations as immigrants assimilate into America.

"New immigrants (legal and illegal) plus births to immigrants add some 2.3 million people to the United States each year," Camarota writes, "accounting for most of the nation's population increase."

Over the past 60 to 70 years, US population doubled to nearly 300 million. If current birth and immigration rates were to remain unchanged for another 60 to 70 years, US population again would double to some 600 million people—the equivalent of adding another state the size of California every decade.

"You just can't deal with that issue without dealing with immigration," says Bill Elder of Issaquah, Wash., a former Sierra Club activist now organizing prominent conservation leaders to focus on population.

Though China and India have much larger populations, the US has the highest population growth rate of all developed countries. Also, experts say, Americans on average have greater environmental impact. The equation for this is I = PAT (Impact = Population × Affluence × Technology), with such impact being the main thing determining whether an area's "carrying capacity" has been exceeded.

Harvard University ecologist Edward Wilson figures that the "ecological footprint"—which he defined in a Scientific American article in 2002 as "the average amount of productive land and shallow sea appropriated by each person in bits and pieces from around the world for food, water, housing, energy, transportation, commerce, and waste absorption"—is about 5 acres per person worldwide. In the US, each individual's ecological footprint is about 24 acres, according to Dr. Wilson.

"Our responsibility for pollution and resource use is all out of proportion to our numbers," says Alan Kuper, a retired physicist in Cleveland and founder of Comprehensive US Sustainable Population. The group publishes a "Congressional Environmental Scorecard" on lawmakers' votes about conservation, consumption, and population, including immigration. "It's not a matter of where or how people come, it's the growth that we have to be concerned with," says Dr. Kuper. "If you're going to be an environmentalist, you have to be concerned about the numbers as well as the usual issues—public lands, energy, pollution, and so forth—because the numbers will just wipe you out."

1. Among those who discuss about the topic of the flow of people into the United States, there are environmentalists.
2. Spanish women bear more children than women in the US do.
3. Paul Watson may believe that too many immigration may be one of the factors which causes population increasing in the US.
4. The Sierra Club believes that immigration may cause environmental degradation and poverty in the US.
5. Hispanic population has increased 49 percent between 2004 and 2005.
6. Hispanic population's high ratio of young people means rapidly Hispanic population growth in the coming years.
7. The number of the children each women immigrant has may be more and more.
8. Keeping the present rate of birth and immigration, US population may be doubled in ________________.
9. The formula I = PAT means the ________________ is determined by Population, Affluence and Technology.
10. The group gives the information about conservation, consumption, population and immigration to the lawmakers by publishing a ________________.

# Unit 11

## Lessons on retirement from the experts: retirees

Trudy and Paul Schuett are still a decade away from retirement. But that defining event has been on their minds, at least indirectly, ever since they moved from Detroit to Yuma, Ariz., in their mid-30s. Surrounded by older neighbors, friends, and relatives, they absorbed, as if by *osmosis*(渗透性), the positive and negative comments people made about retirement choices.

"It's totally changed our outlook on retirement," says Mrs. Schuett, a library aide. "We're already planning right now what we want to do. We're talking it over. Do we want to stay here? Do we want to move up North to be with our son? A lot of people have regretted leaping in without really seeing where they're going."

As the first baby boomers turn 60 and approach a new stage beyond full-time careers, retirement advisers are trying to spread an important message: Leaping in without planning could lead to uninformed or impulsive decisions and the possibility of mistakes and regrets. Although some baby boomers are already considering future finances, housing, and activities, others are taking a mañana approach, largely ignoring these issues.

"Many people don't give any thought to retirement until they get there," says Bruce Juell, who leads retirement-planning seminars at the University of Southern California. "They think, 'I'll love this. I don't have to get up in the morning. I don't have to shave. I can play golf, go on trips. But after a few months, they're bored. If they had a choice, they'd go back to work."

In an age of shrinking pensions, rising healthcare costs, and increasing longevity, many current and future retirees rank money as their biggest concern. In a Met Life study released last month, nearly two-thirds of 55- to 59-year-olds and almost half of 60- to 70-year-olds wish they had done better financial planning.

"The people who are most regretful are those who have retired without benefits," says David DeLong, who conducted the study. Although he notes that retirement planning is more complicated today than in the past, he adds, "It isn't always the people who are richest who are happiest. Some people living on fairly limited incomes are actually having the time of their lives. Income is not the only predictor of satisfaction in retirement."

Anne Hartman of Truro, Mass., a retired career counselor who now advises retirees, hears other kinds of dissatisfaction. "The most regrets seem to come from people who enter retirement abruptly, without daydreams, thoughts, or plans," she says. "They haven't talked with their life partner to learn what is on his or her mind. They have neglected to explore what might satisfy them. They retreat. Or they go into the community with unrealistic expectations about work, about volunteering, about community opportunities, only to be disappointed."

Mr. Juell, author of "The Retirement Activities Guide," urges people to "find a need and fill it." A passionate advocate for productivity in the later years, he sees "a lot of wasted talent." He even proposes that the government appoint an unpaid "retirement productivity czar" to encourage involvement in meaningful volunteer work.

Some people learn from their parents' mistakes. "In a typical case, their parents had put off retirement, and then it was too late," says Robert Weiss, author of "The Experience of Retirement." "They had plans to travel or maybe just take it easy. But by the time they retired, they didn't really have good years. People who could tell that story about their parents used it as an *admonition*(警告) to themselves: 'Don't you repeat that.'"

David Rourke, who heads a retirement-planning firm in Needham, Mass., offers similar "don't wait too long" advice to clients worrying needlessly that they can't afford to retire. "They save and save, and don't want to start drawing from their savings," he says. "A lot of single women think they can't touch it."

In his meetings with clients, other subjects inevitably come up. "They want to know, should they stop working, what do they do after they stop working, and should they downsize their house. That's a big one."

Housing is also a big subject among Schuett's acquaintances in Yuma, especially those who have had second thoughts about living in a mobile home or recreational vehicle. "We've encountered people who really missed their house," she says. "It's a lot smaller and a lot less convenient than they were led to believe by brochures and salesmen. It's hard, especially if you're used to having a basement and an attic."

Others have told Schuett they moved to the Sun Belt too impulsively. "They didn't really know what they were getting into. It's so hot in the summer." She urges prospective retirees considering a move to a different area to take a short-term *rental*(租住) for a month or two to see what the community and climate are like.

Yet even planning, however important, has its limits. "There's a pretty significant gap between what people expect to do and what they end up doing," says Mr. DeLong, author of "Lost Knowledge: Confronting the Threat of the Aging Workforce." At 55 they might say, "Oh, I'm going to work until I'm 75. That may not happen."

For Albert Sutkus of Bedford, Mass., there was no time to plan when his employer, a wire and cable company, announced a downsizing. "They told us to take the severance package and run." In the 20 years since then, he has engaged in varied volunteer activities. He now runs the fix-it shop at the Bedford Council on Aging.

"Everyone has great expectations," Mr. Sutkus says. "They project their life and future life, and it doesn't always work out the way they planned. I'm alive, I've got a good wife, good children. Minus some medical problems, what else can I ask for?"

While some people forget to plan, Hartman finds others making another mistake. "They ignore the magic of serendipity—that opportunity that appears when you are out looking for and doing something else. There is that sense of possibility, adventure, exploration." She and her husband have both found serendipitous activities in retirement, he as a town selectman and she as a member of a chorale. For the Schuetts, who expect to retire at 65, the next 10 years will involve "working on having savings for our backup money." She adds, "Both of us are making sure we have things to do. We'll probably end up doing volunteer work. That's what my parents did. They were a lot happier having something to do." Already the couple has spent years volunteering for programs that help seniors.

"For some people, retirement is the best thing they ever did," Schuett says. "Some people just blossom. They're not doing their same old job. They find a volunteer job they really enjoy. They're having a whale of a time."

1. After retirement, Trudy and Paul Schuett will move from Detroit to Yuma.
2. According to retirement advisers, it is better for people plan in advance before retirement.
3. Mañana approach may refer to the attitude that people don't care about the future life after retirement.
4. Most aged people in the year from 60 to 70 may find some financial shortage in life.
5. David Rourke thought that people need not a financial plan for their retirement.
6. Robert Weiss may object to the idea of putting off retirement in order to earn more money.

7. Many prospective retirees would be interested in the idea of renting a short - term rental for a month or two before a move.

8. According to Mr. Juell, the government need to carry out a plan as ________________ to advocate those retirees to involve in some volunteer work without pay.

9. Albert Sutkus has been involved in all kinds of ________________ for the past 20 years.

10. According to this passage, seniors may feel much happier when ________________ than having considerable savings.

# Unit 12

### Mummy of Tattooed Woman Discovered in Peru Pyramid

An exquisitely preserved and elaborately *tattooed*(文身的) mummy of a young woman has been discovered deep inside a mud - brick pyramid in northern Peru, archaeologists from Peru and the U. S. announced today.

The 1,500 - year - old mummy may shed new light on the mysterious Moche culture, which occupied Peru's northern coastal valleys from about A. D. 100 to 800.

In addition to the heavily tattooed body, the tomb yielded a rich array of funeral objects, from gold sewing needles and weaving tools to masterfully worked metal jewelry.

Such a complete array has never been seen before in a Moche tomb.

Surprisingly, the grave also contained numerous weapons, including two massive war clubs and 23 spear throwers.

The unusual mix of ornamental and military artifacts has experts speculating about the woman's identity and her role in Moche society.

"The war clubs are clear symbols not only of combat but of power," said John Verano, an anthropologist at Tulane University in New Orleans, Louisiana, who is part of the research team.

Peruvian archaeologists, under the direction of lead scientist Régulo Franco, made the discovery last year at an ancient ceremonial site known as El Brujo.

The tomb lay near the top of a crumbling pyramid called Huaca Cao Viejo, a ruin near the town of Trujillo that has been well known since colonial times.

Verano said the finding is the first of its kind in Peru, and he likens it to the discovery of King Tut's tomb in Egypt.

"We have an entire repertoire of a very high status tomb, preserved perfectly," Verano said.

"It's as if she was wrapped up yesterday - no information has been lost."

The Peruvian team is funded by the Augusto N. Wiese Foundation and Peru's National Institute of Culture.

**Mummy an "Astonishing" Find**

Verano, who has been working with the El Brujo project since 1995, said the area is "one gigantic cemetery" that has been scoured by grave - robbers for centuries.

But the newly found funerary chamber had been sealed from both looters and the elements since around A. D. 450.

The Peruvian team found the complete burial array intact and perfectly preserved, down to the white cotton wrappings of the mummy bundle.

"It's astonishing," said Moche authority Christopher Donnan, an anthropologist at the University of California, Los Angeles, who was not part of the excavation.

"This is far and away the best preserved Moche mummy that has ever been found."

Verano arrived on the scene shortly after the discovery.

"The tomb was a rectangular chamber, sealed under many meters of adobe brick," he said.

"There was a very large cotton bundle with a large embroidered face, a cane mat on top, and a pillow underneath, and a skeleton lying beside it that was a sacrifice made to accompany her to the afterlife."

"I took the sacrifice out of the tomb. The excavation team built a very large frame and lifted the mummy out by hand and carried it down the hill to the laboratory."

The size of the mummy alone told investigators it was a member of the Moche elite.

But the full richness of the tomb's contents did not become apparent until the bundle was unwrapped, a process that took months.

"Every layer, every twist of cloth was recorded," Verano said.

"It was hundreds of yards of cotton in thin strips, and there were hundreds of objects inside the bundle."

Donnan praised the meticulous care taken by the archaeologists.

"The team that unwrapped this was absolutely first rate," he said.

**Jewelry and Weapons**

Moche culture is known for its sophisticated art and *metallurgy*(冶金术).

The El Brujo mummy was accompanied by numerous necklaces, nose ornaments, and earrings finely wrought in gold, gilded copper, and silver.

The wooden weapons were sheathed in gold, with finely carved designs and bird or human heads.

"These are among the largest and most beautiful war clubs and most elaborate spear throwers we've ever seen," Donnan said.

When the investigators pulled back the last layers of wrapping, a final surprise awaited.

The woman, thought to have been in her late 20s when she died, had long braided hair and a series of intricate tattoos covering much of her arms, legs, and feet.

Verano said the tattoos were probably done using charcoal pigment inserted beneath the skin with a needle or cactus spine. The tattoos included both geometric designs and images of spiders and mythical animals.

"Who would have thought the Moche were tattooing this extensively?" Donnan said.

"I'm looking forward to sitting down with the evidence and comparing the tattoo patterns with what we see in Moche art."

**Mummy a Mystery**

But the big question remains: Who was she?

Answering that question will help experts better understand the political and religious structure of Moche society, particularly the role of women.

Donnan said recent discoveries of two tombs, apparently of high-level priestesses, have shown that "some women were extremely high status and major figures in the Moche state religion."

But the El Brujo tomb is very different, richer in gold ornaments and symbols of power, suggesting a different type of authority.

"In my opinion this woman from El Brujo is even higher status," Donnan said.

"But we don't know what role she played or why she was buried with war clubs."

Verano added, "She's an unknown character at the moment, with no clear parallel in Moche art."

He finds the large number of spear throwers found in the tomb particularly intriguing.

"They are all very similar, as if they might have come from a single workshop," he said. "Maybe they were carried by her entourage at the time of her death."

1. The mummy may help to learn more about the ancient Moche culture in Peru.
2. There also a mix of ornamental and military artifacts in the tomb as in other Moche tombs.
3. The pyramids in Peru are preserved more perfectly than those in Egypt.
4. The newly discovered tomb must have been scoured hundreds years ago.
5. The excavation of the tomb is carried out with the joint efforts from different countries.
6. The archaeologists learned that the mummy was a capable woman when she was alive by referring to the historical records.
7. The ancient people with Moch culture had created many metal weapons.
8. The final surprise on the mummy may be the extensive ________________.
9. By seeking the identity of the mummy will throw light on ________________ of Moche society.
10. Some experts guess the spear throwers were carried by her ________________.

# Unit 13

## 20 Fishy Facts

Little - known tidbits to help you impress your buddies and catch more fish.

Maybe you will never be able to go out fishing with your best fishing buddies, but you might be able to one - up them in the fish trivia department with the following 20 *tidbits*(珍闻). No doubt your friends will be amazed by your vast archive of wisdom. Otherwise, these friends might simply provide a way to kill time when the fish aren't biting. Who knows? Some might even help you put a few more fish in your boat. Have fun!

**1. On the Cool Side**

White and striped *bass*(鲈鱼) are members of the temperate bass family, as opposed to black bass, which belong to the sunfish family. The term temperate bass refers to the moderate water temperature preference of members of this family. As a rule, they gravitate toward temperatures a little lower than those favored by largemouth or smallmouth bass.

**2. See In the Dark**

Walleyes are known for their marble - like eyes, which let them see well in dim light. Their retinas have a layer of reflective pigment, called the tapetum lucidum, that intensifies any light the eye receives. It's the same membrane that causes a cat's eyes to glow yellow. But the sauger, the walleye's close relative, has even better night vision because the tapetum covers a much larger portion of its retinas.

**3. Black, But Just Barely**

Ever wonder where the term black bass came from? The fry of smallmouth bass turn coal black within a few days after they hatch. Even though the fry of largemouths and other bass species do not turn black, all members of the group (genus Micropterus) are referred to as black bass.

**4. Orange Delight**

Researchers studying walleye vision found that orange is the color most visible to walleyes, followed by yellow and yellow - green. Small wonder so many chartreuse and orange lures fill the tackle boxes of savvy walleye fishermen.

**5. An Ear That Can't Hear**

The ear of a sunfish is really not an ear at all, but merely an extension of the gill cover that varies in color from species to species. The redear sunfish, for example, gets its name from the distinct red margin on its ear.

**6. Vive la Difference**

Everyone knows that smallmouth bass love rocks, but in waters where the bottom is almost all rock, they could be anywhere. In such lakes or rivers, smallmouths will often gravitate to dissimilar structure such as a sandy bottom with weeds or wood cover.

**7. Mud Cats in Name Only**

Flatheads are often called mud cats, giving anglers the impression that they scavenge dead food items off the bottom. But flatheads are more apt to eat live fish than any other catfish species. Channel cats are most likely to consume dead, stinky food (and bait) and blue cats are intermediate in their food preference.

**8. Lights Out? Let's Eat**

Research has shown that a sudden decrease in light level triggers walleyes to bite. That explains why the fish usually turn on just as the sun is disappearing below the horizon and the light intensity is rapidly decreasing. It also accounts for the hot bite that starts when the dark clouds roll in before a thunderstorm.

**9. Not So Special**

Trout are the only kind of fish with an adipose fin, right? Wrong. Several other fish species, including catfish, bullheads, madtoms, smelt, ciscoes and whitefish, also have an adipose fin (the small fin on the back just in front of the tail).

**10. In - Between**

Whisker heads Catfish are generally considered to be bottom feeders, but that's not necessarily true for blue cats. Blues tend to roam open water more than other catfish species, and commercial fishermen often catch more blues on trotlines fished near the surface than on those fished tight to the bottom.

**11. Hungry Mothers**

The best time to catch a trophy walleye is five to seven weeks after the fish have completed spawning. That's when the big females, famished after not having eaten for nearly two months, go on the prowl for food. And with the natural supply of baitfish at its annual low, they're likely to hit almost anything you throw at them.

**12. On the Edge**

Ever wonder why the teeth of a pike or muskie easily shear your line, while those of an equally toothy walleye rarely do? It turns out that walleye teeth are round, while those of a pike or muskie have razor - sharp edges. So don't forget to tie on a wire leader when you're chasing after those esocide.

**13. Bladder Control**

You might have noticed that some fish pulled from deep water can be released and will swim right back down, but others have trouble descending. That't because some fish (physostomous species such as trout) can't up air as they're being pulled up, relieving the gas buildup in their swim bladders, while others (physostomous species like walleyes) have no connection between the swim bladder and the gut. Consequently, their swim bladders expand, greatly increasing their buoyancy. Eventually, gases that build up in the bladder dissipate and the fish can once again return to its deepwater haunt.

**14. Choice Trout Bait**

One of the least known but most effective baits for trout is the water worm, which is the larval form of the crane fly (those big, slow - moving bugs that look like giant mosquitoes). You can often find water worms by digging through the mud and sticks of a beaver dam on your favorite trout stream.

**15. Water dogs as Puppies**

Water dogs, which are the larval form of tiger salamanders, make good bait for bass, walleyes and pike. But once the young salamanders lose their gills and turn into adults, they don't work nearly as well. To prevent water dogs intended for bait from developing into adults, keep them refrigerated in water at a temperature of no more than 50 degrees.

**16. Yummy! A Worm Inside Out**

Channel cats love catalpa worms, which can be found on the leaves of catalpa trees from late spring through summer. To make a catalpa worm even more irresistible, cut off its head and then poke a matchstick through the body to turn the worm inside out. The extra scent exuded by the juicy morsel helps catfish find the bait more quickly.

**17. Madtom Mealtime**

One of the best baits for walleyes and smallmouth bass in the northern parts of their range is the tadpole madtom, also known as the willow cat. These tiny catfish look a lot like bullheads, but the dorsal, caudal (tail) and anal fins are all connected. Be very careful when hooking a willow cat to use as bait. Like those of its bigger cousins, a willow cat's needle - sharp pectoral fins can prick you worse than a bee sting.

**18. Catfish in a Can**

You might have trouble finding a bait shop that carries willow cats, but you might be able to catch your own supply of bait. String together some empty pop cans and then sink them in a lake or river backwater inhabited by the tiny catfish. Leave the cans overnight and pick up the madtoms the next morning; like other catfish, willow cats like to swim into holes, so the pop cans make an inexpensive trap.

**19. Never Trust the Ice**

You're probably heard that it's safe to walk on ice when it's at least 3 inches thick or drive a car on ice that's at last 10 inches thick. But is the ice ever really safe? No. Schools of carp can gather under the ice and wear it away, occasionally even opening a hole. And groundwater might well up from the depths and melt the ice, as famously happened on several Minnesota lakes during the winter of 2002.

**20. Plankton Paddles**

Paddlefish get their name from their long, flattened, paddle - like snouts, which many anglers assume they use to dislodge food from the bottom. But the paddle is really not used for digging; it's equipped with super - sensitive nerve endings that enable it to feel for suspended plankton, the fish's main food.

1. The 20 tidbits will surely help you to catch more fish.
2. Largemouth or smallmouth bass both belong to sunfish family.
3. Fisherman usually uses orange lures to fish walleyes.
4. We can always fish more smallmouth basses in the area where is many rocks.
5. We may fish more walleyes at night.
6. Blue cats are not bottom feeders as other whisker heads.
7. Walleye rarely shears your line because it has no teeth.
8. If we see some water worms in a fisher's baits, we may conclude that he wants to fish ________.

9. Willow cat has another name, that is ________________.

10. Paddlefish uses paddle - like snout to find food because ________________.

# Unit 14

## Festival of Fun For More Than 27,000 Pupils in Berlin

On 10 and 11 May 2006, Berlin will play host to a massive festival to mark the end of the "Talent 2006: The FIFA World Cup at school" campaign, which has involved more than 11,000 schools and exceeded all the expectations of the organisers. More than 27,000 pupils have already registered to spread the FIFA World Cup spirit around Berlin's Olympic Stadium two months before the Final is due to be played at this historic venue. "The work submitted on the theme of 'A time to make friends' can only be described as outstanding," said patron Rudi Völler.

The former German coach and current sporting director of Bayer Leverkusen will be on hand to personally congratulate the winners, who have been invited to Berlin. A total of 20 schools have won prizes in the competition categories of the performing arts, media arts, visual arts, music and creative writing. They will unveil their projects in Berlin and will also play an integral role in a highly attractive interactive programme of events with no charge for entry. The winning schools come from a variety of countries including Malaysia, Paraguay, Russia and the USA.

Federal Minister of the Interior Dr. Wolfgang Schäuble, FIFA General Secretary Dr. Urs Linsi, Berlin's Senator for Sport Klaus Böger and OC Vice - President Wolfgang Niersbach will appear alongside Rudi Völler to officially open the festival on Wednesday 10 May 2006 at 11.30 a.m.

This would be an ideal opportunity for a class trip, and spontaneous visitors are also more than welcome.

**Details of official opening ceremony**

Official welcome to festival visitors at 11.30 a.m. on 10 May 2006:

- Dr. Wolfgang Schöuble—Federal Minister of the Interior
- Dr. Urs Linsi—FIFA General Secretary
- Wolfgang Niersbach—Vice - President OC 2006 FIFA World Cup
- Klaus Böger—Senator for Education, Youth and Sport
- Rudi Völler—Sporting director at Bayer 04 Leverkusen and patron of "Talent 2006"
- Karin Wolff—Culture minister of Hesse and patron of "Talent 2006"
- Michael Preetz—FIFA World Cup Ambassador for Berlin
- GOLEO VI and Pille—Official Mascots of the 2006 FIFA World Cup

All representatives of the media are cordially invited to this event. In order to obtain accreditation, please send an e - mail to the following address by 05.05.2006: miriam.herzberg@ok2006.de. You can pick up your accreditation at the festival site from "Info Point 2" (Osttor [East Gate] entrance).

**Free entry to both days of festival**

The programme for Wednesday 10 May 2006 (9 a.m. - 6 p.m.) will focus on children in classes up to Year 6, while the spotlight on Thursday 11 May 2006 (9 a.m. - 6 p.m.) will move to Years 7 to 13. The festival is open to everyone, not just school classes. Registration is not required.

In addition to the appearances of the school classes, festival visitors can also look forward to the following attractions on 10 and 11 May 2006:

- "FIFA World Cup Tour" with the original FIFA World Cup trophy and the official mascots GOLEO VI and Pille
- Adidas Football Park with a wide range of football activities
- Coca - Cola fan truck with football quiz and karaoke show
- Massive climbing wall
- "Chili TV - Show" from KI. KA, the children's channel of ARD and ZDF

- German final of football talent competition in the hockey stadium
- Federal Centre for Health Education's "smoke – free" beach lounge
- "Toshiba Air Dome"
- Balloon – based activity centre sponsored by German Centre for Tourism
- and much, much more besides

**Workshops and visitor participation**

As well as offering a wide range of sporting activities, the festival is also calling on the creativity of the pupils. Young artists will be able to put their skills to the test in workshops. These will include D!'s Dance Club Workshop, where visitors will be able to learn Hip Hop and Street – Style moves with Detlef D! Soost, and a session with the cheerleaders from Berlin Thunder. Alternatively festival visitors can try their hand at composing in the song – writing workshop or join the circus for the day thanks to a workshop with the "Cabuwazi" troop.

There are various other ways to get involved above and beyond the workshops, with no charge for any of the events. Visitors might like to create their own fan outfit from balloons or try out more than 40 musical instruments in the "ringing mobile phone". Should participants have any creative energy left, they can also have their faces painted in the colours of their team as a reminder of the "Talent 2006 Festival" or even design their own T – shirt.

Further highlights of the festival programme include:

**Talks/workshops with patrons**

10.05.2006:

12:30 p.m: Workshop with Otmar Alt (duration approx. 2 hours)

1:30 p.m: Talk with Anke Engelke (duration approx. 1 hour)

11.05.2006:

10:30 a.m: Workshop with Otmar Alt (duration approx. 2 hours)

11:30 a.m: Talk with Anke Engelke (duration approx. 1 hour)

12:30 p.m: Talk with Sönke Wortmann (duration approx. 1 hour)

1:30 p.m: Reading with Benjamin Lebert (duration approx. 1 hour)

2:30 p.m: Talk with Sasha (duration approx. 1 hour)

**"RAPSOUL"—Top music act to appear on stage at 4 p.m. on 11 May**

The band "RAPSOUL" will take to the stage at 4 p.m. on 11 May to mark a fitting end to the stage programme. The Frankfurt trio of Jan, Steve and CJ Taylor will unveil their latest single "Gott schenk ihr Flügel" (God give her wings), which is currently number 1 in two major music charts in Germany.

**German final of football talent competition**

The prize – giving ceremony for the winners of the German final of the "Football talent competition" (football competition organised under the auspices of the German Football Association involving a 4 – a – side tournament) will be held on 11.05.2006 at 5 p.m. in the hockey stadium.

An overview of the festival (programme, map of site, information on workshops, etc.) can be found on the Internet at www.FIFAworldcup.com/talente2006 in the "Festival" section.

**Should you have any questions, please do not hesitate to contact Miriam Herzberg (Telephone: +49 341—149 33 56, E – mail: miriam.herzberg@ok2006.de).**

1. The campaign of "Talent 2006: The FIFA World Cup at school" will last for two days in Berlin.
2. There at least one winning school comes from Asia.
3. Visitors of the festival will be welcomed by many officials including German President.
4. You can enter the stadium free of charge but need to register.
5. Some super football stars will give shows during the two days.
6. Some pupils will show their skills of dancing.
7. All the events held in the two days will be free of charge.
8. Talk with Sasha will be held at ________________.
9. "Football talent competition" will be held in ____________ on 11.05.2006 at 5 p.m.
10. If you don't know how to get to the site where the events will be carried out, you may get the information by sending an e – mail to ________________.

# Unit 15

## The Historic Center of Macao

### Appraisal

"The Historic Center of Macao" includes the oldest Western architectural heritage on Chinese soil today. Together with Macao's traditional Chinese architecture, it stands witness to successful East – West cultural pluralism and architectural traditions.

"The Historic Center of Macao" is solid testimony of the city's missionary role in the Far East while also reflecting the dissemination of Chinese folk beliefs to the Western world.

"The Historic Center of Macao" is the product of East – West cultural exchanges, constituting the most unique blend of cultural heritage existing in China's historic cities.

"The Historic Center of Macao" presents a complete social infrastructure that has encompassed and sustained the living traditions of different cultures.

### History

Macao, a lucrative port of strategic importance in the development of international trade, was under Portuguese administration from the mid – 16th century until 1999 when it came under Chinese sovereignty.

The emergence of Macao with its dual function as a gateway into China, and as Ming China's window onto the world, reflected a relaxation of certain restrictions combined with a degree of open – mindedness that offered a creative way to supplement China's vassal – state trading system and marked a turning point in the history of both China and Europe.

The settlement of Macao by Portuguese navigators in the mid – 16th century laid the basis for nearly five centuries of uninterrupted contact between East and West. The origins of Macao's development into an international trading port make it the single most consistent example of cultural interchange between Europe and Asia.

For almost three centuries, until the colonization of Hong Kong in 1842, Macao's strategic location at the mouth of the Pearl River meant that it retained a unique position in the South China Sea, serving as the hub in a complex network of maritime trade that brought tremendous wealth and a constant flow of people into the enclave.

### Cultural Pluralism

Macao, as the West's first established gateway into China, was remarkable in setting off a succession of connections and contacts that progressively enriched both civilizations across a huge range of human endeavor, both tangible and intangible.

People of different nationalities came, bringing their own cultural traditions and professions, permeating the life of the city as can been seen in both intangible and tangible influences. This is evident in the introduction of foreign building typologies such as western – style fortresses and architecture.

Macao also inherited various cultural experiences and regional influences, further developing these in conjunction with the local Chinese culture and blending them to produce the rich texture seen in the city's exceptional heritage.

Meanwhile, "The Historic Center of Macao" coincides with the heart of the Western settlement area, also known as the "Christian City" in history. Exposure to diverse cultures in this lasting encounter between the Eastern and Western worlds has therefore benefited Macao in assimilating a rich array of cultural heritage.

### "Firsts" for China in Macao

During the late Ming (1368 – 1644) and early Qing (1644 – 1911) dynasties, missionaries from different European religious orders, such as the Jesuits, the Dominicans, the Augustinians, and the Franciscans, entered China through Macao. They made efforts to engage in missionary work and brought with them a certain cultural influence.

The missionaries introduced Western concepts of social welfare and founded the first Western – style hospitals, dispensaries, orphanages, and charitable organizations. Besides, they brought in the first movable – type printing press to be used on Chinese soil, and published the first paper in a foreign language.

As Macao was the base for the Jesuit mission in China and other parts of East Asia, Jesuit priests entering into China service would always come first to Macao where, at St. Paul's College, they would be trained in the Chinese language together with other areas of Chinese knowledge, including philosophy and comparative religion. Macao was thus the training ground for the Jesuit's mission to China and other parts of Asia.

St. Paul's College was the largest seminary in the Far East at the time, acclaimed as the first Western – style university in the region.

Other achievements of Christian missionaries in Macao include the production of the first English – Chinese Dictionary and the first Chinese translation of the Bible by Robert Morrison.

The worship of A – Ma in Macao originated with the folk beliefs of fishermen living along the coast of South China. Due to Macao's special position in channeling cultural exchanges between East and West, the A – Ma Temple has played a prominent role as the earliest reference to A – Ma worship abroad.

**Strolling through "The Historic Center of Macao"**

The A – Ma Temple is located on the southwestern tip of the Macao Peninsula overlooking Barra Square and the seashore. Around the corner of the A – Ma Temple is the Moorish Barracks situated on Barra Street. Further up the road, the narrow street suddenly opens onto Lilau Square, the first residential district of the Portuguese settlers in history where the Mandarin's House is just tucked behind the pastel facade across the street.

Further up the road, Barra Street runs into Padre António Street and Loureno Street where St. Lawrence's Church stands. Behind the church, Prata Street leads to the junction of St. José Street where the grand entrance to St. Joseph's Seminary and Church is located.

Walking alongside the granite wall on Prata Street and the adjoining Seminário Street, one arrives at the junction of Gamboa Lane. Climbing up the hill from there, the path leads to St. Augustine's Square enclosed by a cluster of monuments—St. Augustine's Church, the Sir Robert Ho Tung Library, and the Dom Pedro V Theatre.

Moving down Tronco Velho Lane to Almeida Ribeiro Avenue, the narrow streetscape opens onto the main city square—Senado Square. Situated at one end, the "Leal Senado" Building has a commanding view overlooking the entire square, flanked on both sides by South European – style buildings with the glimmering white facade of the Holy House of Mercy standing in its midst.

Tucked behind the commercial shop fronts to the left of the Leal Senado Building is the Sam Kai Vui Kun Temple. Climbing up the slope alongside the Holy House of Mercy, one eventually lands at Cathedral Square on the hilltop where the Cathedral is located. Turning back down to St. Dominic's Square, one passes a typical Chinese courtyard house compound—the Lou Kau Mansion.

St. Dominic's Church is located at the junction of Senado Square and Dominic's Square. Ascending from the base of Mount Hill from this urban piazza along Palha Street, the bluestone cobbled road leads to the grand facade of the Ruins of St. Paul's, with Mount Fortress to the side of it.

Behind the majestic church front is the miniature Na Tcha Temple and Section of the Old City Walls. Further down the hill, the linear route ends at St. Anthony's Church, the Casa Garden and the Protestant Cemetery. Standing on the highest hill of Macao Guia Hill, Guia Fortress, Chapel and Lighthouse are visible along the skyline of the peninsula.

1. Western architecture appeared in Macao earlier than Chinese architecture.
2. Maocao's cultural heritage is very unique to other Chinese historic cities by its cultural exchanges between the East and the West.
3. Portuguese navigators might just want to use Macao as a military basis by settling there.
4. Chinese culture had been enriched much by Western culture through Macao as well as the Western culture had by Chinese culture.
5. We call Macao as the "Christian City" because it has become one of the "holy cities" for Christians.
6. Missionaries from Europe had contributed to introducing Western culture into China.
7. A missionary might also learn some knowledge about other Asian countries in Macao.
8. The name of the missionary who must have contributed much to translating the Bible into Chinese is ________________.
9. If you want to see all the scenery of the Senado Square with a single glance, you may look from ________________.
10. There is a church standing in the midst of Senado Square and ________________.

# 四、快速阅读理解

# 答案详解

## Unit 1

1.【答案】Y

【详解】文中 those who commit such low - level crimes in New York State may soon be required to give DNA samples to authorities 一句说那些犯轻罪的人即将要求被提取 DNA 样本,说明目前这个做法还没实行。

2.【答案】N

【详解】文章第二段说了 the proposal is currently being debated。由 debate 可以发现本论断太绝对了。

3.【答案】Y

【详解】可由 Others cite concerns about the increasing number of innocent people whose DNA is stored in a databank without their knowledge or approval. 一句做佐证。

4.【答案】N

【详解】他只是觉得应该尽量避免夸大 DNA testing 的作用,以防 its abuse,并非他就反对在自己的 Innocence Project 采用这个方法,何况,他 has used DNA testing to free 176 prisoners wrongfully convicted。

5.【答案】N

【详解】根据 A week ago, Kansas joined California and five other states in going one step further 做一下计算,原先就有 5 个,加上 Kansas,California,一共是 7 个。

6.【答案】Y

【详解】根据他提供的数据:we run it against the 18,000 unsolved crimes, mostly rapes, and we have 2,400 hits so far. 他认为这会 be well worth the effort。

7.【答案】NG

【详解】文中并没明确指出是否纽约的 DNA testing 的花费就比其他州高。

8.【答案】cross - contamination

【详解】文中说 Many of the mistakes arise from cross - contamination or mislabeling of DNA samples,显然此处应填入 cross - contamination。

9.【答案】personal privacy

【详解】文中说 The fundamental values of government accountability and personal privacy are at stake,可见此处应填入 personal privacy。

10.【答案】Unlawful use of a DNA sample

【详解】根据 Parker says, by increasing the penalty for unlawful use of a DNA sample. 内容填充。

## Unit 2

1.【答案】Y

【详解】本题关键是理解 Wachovia's $26 billion deal for Golden West Financial 的含义,根据前文 the giants rush to expand 的意思,显然,承接这个内容,这句话意思是说 Wachovia 以 $26 billion 收购了后者。

2.【答案】NG

【详解】文中提到的 3.6 percent 是除 J. P. Morgan 以外的 four remaining players 鼓吹的回报率。

3.【答案】N

【详解】与 We'll project that the big banks' earnings growth averages to 8 percent over the next seven years, far below the recent level. 一句意思相反。

4.【答案】N

【详解】由 Since both earnings and dividends will be spread across fewer shares each year, the dividends per share will rise not by 8 percent a year, but by more like 9.5 percent. 来判断,share 的增幅应该高些。

5.【答案】Y

【详解】通过各自回报率的对比,16.5vs12,答案十分明显。

6.【答案】Y

【详解】文中说 It's(dividend)rising at almost 10 percent a year. 接下来说 Those increases should far outstrip inflation,显然本论断正确。

7.【答案】NG

【详解】文中只交代了 J.P. Morgan has the potential to deliver even greater returns,至于 Citigroup 是否也会,找不到确切的佐证。

8.【答案】weak earnings

【详解】根据 chiefly because it's been saddled with weak earnings. 内容作答。

9.【答案】Bank stocks still look cheap

【详解】报道在谈到 The risks 时,说了,Is there any risk in these equations?... That's possible, but unlikely,接下来讲了原因:Bank stocks still look cheap.

10.【答案】gaps in their footprints

【详解】根据 The key is geography. Several of the giants have gaps in their footprints. 内容作答。

## Unit 3

1.【答案】N

【详解】关键是理解 the Prime Minister will once again leave without the Congressional Gold Medal 一句的意思,言外之意是他还未正式得到。

2.【答案】Y

【详解】Nelson Mandela received his award in 56 days. 我们根据本文可以判断至少还有另一个人。

3.【答案】N

【详解】根据报道内容,In the US he is still seen as a clear-eyed brave heart... 和 In Britain, however, he is more often regarded these days as a naive and vainglorious fool...,我们可以判断两国对他的评价不尽相同。

4.【答案】Y

【详解】这与 the two leaders, both deeply unpopular 传达的信息一致。

5.【答案】Y

【详解】根据 He will set out his belief that world problems can best be tackled by strengthening global institutions ranging from... 的内容,我们可判断他认为国际事务由国际组织来处理会效果更佳。

6.【答案】N

【详解】由 But Mr. Blair knows that for all the effort he has expended in such discussions with Mr. Bush over the past 4 years, he has secured little that is tangible. 一句传达的信息,我们完全可以判断这次会谈会照样无法达成有意义的结果,这正好与本论断结果相反。

7.【答案】NG

【详解】这在文中没有明确说明,只说他们会进行这方面的讨论。

8.【答案】craven fool

【详解】根据 was he the craven fool taken for a ride by Mr. Bush that critics back in Britain believe him to have been. 一句内容填充。

9.【答案】September 20, 2001

【详解】可到报道后的附件 The main meetings between Tony Blair and President Bush 中查找。

10.【答案】32 per cent

【详解】通过附件内容对比查找,应是32%。

## Unit 4

1.【答案】Y

【详解】文章第一段交代了"Cash for babies is the Kremlin's offer to women in its latest bid to reverse a population decline",意思是说他们正在通过现金奖励来鼓励妇女生育,以改变人口负增长的局面。所以本论断所说俄罗斯近年来人口减少正确。

2.【答案】N

【详解】本文是一个细节性的题目,关键是搞清楚文中所说的"1,500 rubles"是"double monthly",是每两个月得到"1,500 rubles",平均到一个月只有750卢布了。

3.【答案】Y

【详解】根据她的评论,"As for Putin's offer, she says 'it won't change anything.'"可以为本论断找到依据。

4.【答案】N

【详解】文中"far below the 2.4 children required to maintain the population — according to the Federal State Statistics Service."一句说明,俄罗斯要维持现在的人口数量,每个妇女必须生2.4个小孩。

5.【答案】NG

【详解】文中谈了几点俄国人口减少的原因,包括高堕胎率,高死亡率等,但并没指出哪个是主要原因。

6.【答案】N

【详解】本论断主要是迷惑考生与"Some women say they resent the suggestion, made explicit by many nationalist politicians, that their lack of enthusiasm for bearing children is to blame."一句混淆,这句话是说妇女对某些政治家的关于她们对抚养孩子没

激情的言论不满,可见她们并非对普京总统的建议不满。

7.【答案】Y

【详解】本论断可以以"Official statistics show that almost 8 of every 10 marriages end in divorce"一句作为依据。

8.【答案】foster families

【详解】本句主要是搞清楚4500卢布的补贴给予的对象,可通过"Putin also doubled subsidies for foster families, to 4,500 rubles ( $166) per month, a move widely welcomed by child - care experts."了解。

9.【答案】economic wallop

【详解】本句实际上是对"Low birthrates and high mortality could deliver an economic wallop that could dash Putin's hopes of restoring Russia as a great power."的释义。

10.【答案】hostile

【详解】根据"Polls show large majorities remain hostile to the idea of mass immigration of non - Slavs."内容填充。

1.【答案】N

【详解】采访中,他说I experience frustration sometimes, such as when I have a crisis, like I just did,显然本论断是错的。

2.【答案】Y

【详解】Christopher Reeve 谈到I've had a lot of training in that area from my life before the injury,言外之意当然就是本论断的意思。

3.【答案】N

【详解】本论断正好与情况相反,根据Christopher Reeve的说法,There hasn't been any new recovery since what was published in 2002. 从2002年后,情况就再也没有新的进展。

4.【答案】Y

【详解】这一点完全可以从采访中找到佐证,文中交代了明年5月就是他病倒10年了,而我们可以了解到他现在将近52岁。

5.【答案】N

【详解】我们可以通过他的I don't know if it would have made me walk sooner一句来了解本论断有点言过其实了。

6.【答案】Y

【详解】我们可以通过他对该法案的细节了解程度看出来,且不说法案是以他的名字命名的。

7.【答案】NG

【详解】文中只是Reeve表示I'm quite optimistic that it will pass,至于是否一定获通过我们不知道。

8.【答案】treating acute injuries

【详解】根据that embryonic stem cells are effective in treating acute injuries and are not able to do much about chronic injuries. 内容作答。

9.【答案】the religious right has had quite an influence on the debate

【详解】可以以他回答记者的提问How have political decisions slowed stem cell research? 的内容填充。

10.【答案】suicide

【详解】根据记者最后一个访谈内容确定答案。

1.【答案】N

【详解】根据报道可知,这是对Saddam Hussein的第二次审判,第一次因为他拒绝回答问题,法官推迟了庭审。而这次法官并未推迟审判,只是关闭了法庭,进行了秘密审判。

2.【答案】NG

【详解】因为记者都被请出了法庭,两个小时的秘密审判后才重新被法官叫回来,因此这两个小时内发生的事我们不得而知。

3.【答案】Y

【详解】根据文中stop the wave of Shiite - Sunni violence that has rocked the country since the bombing of a major Shiite shrine last month. 所说,可以肯定冲突是由清真寺的爆炸引发的。

4.【答案】Y

【详解】Saddam说the great people of Iraq have nothing to do with these acts,言外之意当然是指这次爆炸与伊拉克人无关。

5.【答案】N

【详解】虽然这个律师说I fully expect him to be dead by the end of the year, executed by this court in this show trial,看上去似乎他的想法和本题论断一样,但看完他接下来的话They can kill his body, but not his spirit我们不难明白他的真实想法。他之所以那样说是因为他清楚the court has already decided Saddam is guilty,这个审判只不过是个show trial。

6.【答案】Y

【详解】根据But Saddam insisted it was his right to do so since they were suspected in the attempt to kill him. 可判断。

7.【答案】N

【详解】根据Dujail residents have testified that Ibrahim personally participating in torturing them during their imprisonment at the Baghdad headquarters of the Mukhabarat intelligence agency一句我们知道,当时Dujail residents是在巴格达被关押的,不是在他们自己的家乡。

8.【答案】forgery

【详解】根据It's not true. It's forged. We all know that forgery happens一句的上下文理解。

9.【答案】(notorious) Abu Ghraib prison

【详解】可根据 hundreds of Dujail detainees being held by the Mukhabarat at its headquarters and at the notorious Abu Ghraib prison. Ibrahim said that memo as well was false. 内容来填充。

10.【答案】tortured

【详解】文章最后一句话交代了 prosecutor 说的这些平民被抓后的遭遇:saying entire families were arrested and tortured and that the 148 who were killed were sentenced to death without a proper trial。

## Unit 7

1.【答案】N

【详解】根据文章内容,伊朗只是宣布 the country had successfully enriched uranium,而这只是 a key step to producing peaceful nuclear energy or nuclear weapons。

2.【答案】NG

【详解】Mohamed ElBaradei 说他此行的目的是 to clarify remaining outstanding issues on the nature of the Iranian program,这是一句典型的外交辞令,我们无法辨别他的真实用意。

3.【答案】Y

【详解】根据 CBS News correspondent Jim Axelrod 的报道:If true, Axelrod adds, that would be enough to produce hundreds of nuclear warheads 本论断应该正确。

4.【答案】N

【详解】根据 He said using 54,000 centrifuges will be able to produce enough enriched uranium to provide fuel for a 1,000 – megawat nuclear power plant like the one Russia is currently putting the finishing touches on in southern Iran 的内容,很明显,他们表达了自己的目的是发电。

5.【答案】Y

【详解】报道中说 ElBaradei 希望自己的访问能 bring Iran in line with the requests of the international community to take confidence... until outstanding issues are clarified, 和本论断意思一致。

6.【答案】Y

【详解】Richard Haass 的理由是:They don't need nuclear power for electricity because they've got all this oil and gas。

7.【答案】N

【详解】由 The United States is not taking Iran's claim at face value 可推断,美国显然认为伊朗致力于发展核武器,并不是不确定。

8.【答案】"strong steps"

【详解】本句与 Secretary of State Condoleezza Rice said Wednesday the U. N. Security Council must consider "strong steps" to induce Tehran to change course. 意思一致。

9.【答案】Russia

【详解】报道中有 Russia also criticized the announcement Wednesday 作为依据。

10.【答案】the talks on Iraq

【详解】Rafsanjani 说 We don't have a mandate to discuss the nuclear issue with the Americans... but if the talks on Iraq go in the right direction,意思就是说我们不想和美国谈核问题,但是如果伊拉克的和谈进展顺利的话可以考虑,也就是说他把这作为前提条件。

## Unit 8

1.【答案】N

【详解】文章第一段交代了"church leaders worldwide are divided",接下来又具体阐释了原因:"Some are demanding that censors ban the film, or cut scenes that they say undermine Christian beliefs 和 Others are angrily advocating boycotts to protest what they see as an attack on the Roman Catholic church"。

2.【答案】Y

【详解】从上文"priests from both the Protestant and Catholic traditions are seizing the occasion as a 'teachable moment,'"一句可以为本论断找到依据。

3.【答案】Y

【详解】"... are expected to find an even wider audience with the film starring Hollywood actor Tom Hanks and French actress Audrey Tautou"说的就是这个意思。

4.【答案】NG

【详解】文中只是引用了"Cardinal Francis Arinze"的话,"should react, possibly by taking legal action against the film",但是否制片商真会受到法律指控不得而知。

5.【答案】Y

【详解】文中多次提到了亚洲的几个穆斯林国家(菲律宾等)的信徒们对电影的反应,本论断正确。

6.【答案】N

【详解】根据"a priest in Nice was planning to lead a public debate after a screening of the film in a local cinema"的意思,他正计划的是进行辩论,并非游行。

7.【答案】Y

【详解】可根据他的谈话"we should not panic"的内容判断。

8.【答案】tongue – in – cheek

【详解】可从原文"We decided to be tongue – in – cheek rather than hysterical or anxious,"中找到答案。

9.【答案】Fiction

【详解】报道中描述了 Protestant in Britain 的做法,根据他们的做法,在 scratch - cards 上的 statement "the marriage of Jesus and Mary Magdalene is a matter of historical record," 旁,刮开 "Fact" and you will discover an "X.",但是 Scratch "Fiction" and you uncover a check mark,用意显然是指后者是对的。

10.【答案】a night out at the movies

【详解】文中说"Archbishop Amato, St. Michael's is organizing a night out at the movies for its congregation on Friday",所以本题填入 a night out at the movies 与之相符。

# Unit 9

1.【答案】N

【详解】"Mr. Morris"对"Mr. Lowenstein"生气的原因不是因为他不认识而后者给他发了个邮件,而是邮件中抱怨的话。

2.【答案】N

【详解】根据"The answer, the professors conclude, is sometimes 'no.'"一句,他们只是认为有时候表达不清。本论断太绝对了。

3.【答案】Y

【详解】根据"the prospect of instantaneous communication creates an urgency that pressures e - mailers to think and write quickly, which can lead to carelessness"的意思判断。

4.【答案】Y

【详解】本论断可以作为第五段的大意,显然与第五段内容相符。

5.【答案】N

【详解】本论断"People are much less likely to prejudge after communicating by phone than they are after receiving an e - mail"内容相反,正确的做法是"sending e - mail"前先"make a phone".

6.【答案】NG

【详解】"E - mail tends to be short and to the point"没错,但是文中并没讲"Most people prefer to sending e - mail instead of writing"。

7.【答案】Y

【详解】文中指出:"Psychologists Massimo Bertacco and Antonella Deponte" 发现"were less likely to 'ground their communications' in memories of shared experience than letters writers were"。

8.【答案】e - mail

【详解】这可以根据上文交代的 e - mail 的特点来判断。上文"the inability to develop personal rapport makes relationships fragile in the face of conflict over e - mail"一句说明只有在"e - mail"里,不能制造出"rapport"的气氛。

9.【答案】working together again

【详解】本句内容与"Those who negotiated by e - mail in Morris's study trusted each other less and weren't as interested in working together again"意思一致。

10.【答案】less rich medium

【详解】在"make a brief phone call, even if it's not practical to do the whole negotiation by phone. You can establish a favorable bias with someone and then proceed in a less rich medium"一句中,显然"less rich medium"是相对于"phone call"来说的另一种交流方式,根据上下文,显然就是指"e - mail"。因此本题答案为"less rich medium"。

# Unit 10

1.【答案】Y

【详解】这句话实际上是 As politicians and the public heatedly debate immigration, so, too, are environmental activists. 的同义表达。

2.【答案】NG

【详解】本题主要考查学生是否弄清楚了从西班牙来的移民和西班牙人的区别,文章只说 Hispanic women who move to the US have more children than if they stayed put,至于西班牙妇女是否比美国妇女生的小孩多,本文未讨论。

3.【答案】Y

【详解】根据 Immigration is one of the leading contributors to population growth. 可判断。

4.【答案】N

【详解】文章 It's a global issue, says Mr. Antebi, caused by environmental degradation and poverty that need to be solved so people won't have to look elsewhere for a better life. 一句分析了出现移民的原因是环境恶化和贫穷,如果把这句话理解成移民在美国造成了环境恶化和贫穷就错了。

5.【答案】N

【详解】文中说的 49 percent 是指美国增加的人口中,Hispanics 占了 49%,西班牙人口虽然增长得快,但还不足以达到 49%。

6.【答案】Y

【详解】文章在分析了西班牙裔移民的年轻化比率很高后,阐释了其背后的意义:That means such younger people are just entering (or will remain longer in) the years in which they have children of their own,当然他们人口增长率会高。

7.【答案】N

【详解】本论断正好与 However, typically these high fertility

rates decline in the successive generations as immigrants assimilate into America. 内容相悖。

8.【答案】60 to 70 years

【详解】根据 If current birth and immigration rates were to remain unchanged for another 60 to 70 years, US population again would double to some 600 million people 内容,保持现有移民和出生率,美国人口会在 60 至 70 年后翻番。

9.【答案】environmental impact

【详解】本句就是把 Impact = Population × Affluence × Technology 用文字描述出来而已。

10.【答案】Congressional Environmental Scorecard

【详解】根据 The group publishes a "Congressional Environmental Scorecard" on lawmakers' votes about conservation, consumption, and population, including immigration. 内容作答。

7.【答案】NG

【详解】这只是 Schuett 的建议,至于是否大家都感兴趣文章未交代。

8.【答案】retirement productivity czar

【详解】本句与"He even proposes that the government appoint an unpaid 'retirement productivity czar' to encourage involvement in meaningful volunteer work."意思一致,显然这个计划被他称作"retirement productivity czar"。

9.【答案】volunteer activities

【详解】本句是"In the 20 years since then, he has engaged in varied volunteer activities."的同义表达。

10.【答案】having something to do

【详解】根据结尾部分的"They were a lot happier having something to do."一句填充。

## Unit 11

1.【答案】N

【详解】文章第二句中说自从他们从 Detroit 搬到 Yuma 后,他们开始考虑退休的问题,并不是本论断所说的退休后才搬家。

2.【答案】Y

【详解】根据文章所说,"Leaping in without planning could lead to uninformed or impulsive decisions and the possibility of mistakes and regrets",言外之意当日是指提前计划可以避免这些问题。

3.【答案】Y

【详解】根据该词所出现的句子的句式:some..., others...,我们知道 mañana approach 应该正好与前文说的"considering future finances, housing, and activities"相对应,当然是"ignoring these issues"。

4.【答案】N

【详解】文中"half of 60 – to 70 – year – olds wish they had done better financial planning"一句的确告诉我们有人感觉到财政短缺,否则他们怎么会希望当初有更好的计划呢? 但是本句错误主要是比率,most 显然不确切,almost half 只能说有"将近一半",还不能说"大多数"。

5.【答案】N

【详解】本论断说他认为人们不需要做好退休后的财政计划,原文说有些人"worrying needlessly that they can't afford to retire"意思不一致,文中是说他们过于担心退休后负担不起开支,两者不一样。

6.【答案】Y

【详解】根据他的观点"In a typical case, their parents had put off retirement, and then it was too late,"可以很确定。

## Unit 12

1.【答案】Y

【详解】根据 The 1,500 – year – old mummy may shed new light on the mysterious Moche culture 内容判断。shed light on sth. 使某事清楚明白地显示出来。

2.【答案】N

【详解】文中交代了 mix of ornamental and military artifacts 是 unusual 的,显然本论断错误。

3.【答案】NG

【详解】文中 Verano 说了一句话,We have an entire repertoire of a very high status tomb, preserved perfectly,但这只是说明了 Peru 的 tombs 保存得很完好,但文中并未提及两个地方的金字塔比较谁更完好。

4.【答案】N

【详解】该论断与 But the newly found funerary chamber had been sealed from both looters and the elements since around A. D. 450. 意思不符。

5.【答案】Y

【详解】我们至少可以通过文中提到的几个专家判断出有美国和秘鲁的科学家在合作。

6.【答案】N

【详解】根据 The size of the mummy alone told investigators it was a member of the Moche elite. 的内容,考古学家们根本不需要查历史记录。

7.【答案】Y

【详解】Moche culture is known for its sophisticated art and metallurgy. 一句是最好的佐证。

8.【答案】tattoos(covering much of her arms, legs, and feet)

【详解】这一点可以通过 a final surprise awaited 后面的内容填充。

9.【答案】political and religious structure

【详解】通过 Answering that question will help experts better understand the political and religious structure of Moche society 一句内容填充。

10.【答案】entourage

【详解】文中 Maybe they were carried by her entourage at the time of her death. 交代得很清楚。

9.【答案】tadpole madtom

【详解】根据 One of the best baits for walleyes and smallmouth bass in the northern parts of their range is the tadpole madtom, also known as the willow cat. 一句告诉我们 tadpole madtom 又叫做 willow cat，所以本句答案当然是颠倒一下，填 tadpole madtom。

10.【答案】it's equipped with super - sensitive nerve endings

【详解】可根据全文最后一句话 But the paddle is really not used for digging; it's equipped with super - sensitive nerve endings that enable it to feel for suspended plankton, the fish's main food. 推断。

## Unit 13

1.【答案】N

【详解】通过开头一段的最后一句 Otherwise, these factoids might simply provide a way to kill time when the fish aren't biting. Who knows? Some might even help you put a few more fish in your boat. Have fun! 来判断，作者说这也许只是在鱼儿不咬钩时提供一点谈资。显然作者并不肯定是否可以钓到更多鱼。

2.【答案】Y

【详解】文章在第一条里介绍了 black bass 是 sunfish family，在第三条里提到了这两种都属于 black bass，black bass 是 sunfish family，largemouth or smallmouth bass 也属于此类。

3.【答案】Y

【详解】这一点在第 4 条里做了介绍，因为 orange is the color most visible to walleyes。

4.【答案】N

【详解】虽然 smallmouths 喜欢有石头的地方，但在 in waters where the bottom is almost all rock 的水域，它们常 gravitate to dissimilar structure such as a sandy bottom with weeds or wood cover。

5.【答案】NG

【详解】第八条告诉我们只是在光线突然变暗的情况下会促使 walleyes 咬钩，但虽然晚上光线很暗，但是否能钓更多并未说。

6.【答案】N

【详解】这个论断对了一半，blue cats 的确喜欢靠近水面，但说它像其他 catfish 一样就有失偏颇了，因为 Catfish are generally considered to be bottom feeders。

7.【答案】N

【详解】walleye 很少咬断线是因为它们的 teeth are round，并非因为没牙。

8.【答案】trout

【详解】第 14 个小知识介绍了 One of the least known but most effective baits for trout is the water worm。

## Unit 14

1.【答案】N

【详解】根据第一段内容，10 and 11 May 2006 只是 mark the end of the "Talent 2006: The FIFA World Cup at school" campaign 的两天，并非这个活动只进行了两天。

2.【答案】Y

【详解】文中指出：The winning schools come from a variety of countries including Malaysia, Paraguay, Russia and the USA，既然有马来西亚的学校，当然本论断正确。

3.【答案】N

【详解】我们从 Official welcome to festival visitors 的名单中找不到。

4.【答案】N

【详解】文中明确指出了 Registration is not required。

5.【答案】NG

【详解】文章列举了 attractions on 10 and 11 May 2006，虽然找不到本句所说，但 and much, much more besides 提示我们还是有可能性的，所以选择 Not Given 更合适。

6.【答案】Y

【详解】根据 Young artists will be able to put their skills to the test in workshops. These will include D!'s Dance Club Workshop 可判断。

7.【答案】Y

【详解】根据 There are various other ways to get involved above and beyond the workshops, with no charge for any of the events. 可判断。

8.【答案】2. 30 p. m

【详解】根据 Talks/workshops with patrons 的安排确定答案。

9.【答案】the hockey stadium

【详解】根据 German final of football talent competition 介绍的安排确定答案。

10.【答案】miriam. herzberg@ ok2006. de

【详解】文章最后一句给出了组织者的联系方式。

# Unit 15

1.【答案】NG

【详解】文中只是交代了 Macao 的历史中心包括中国现在最古老的西方建筑,至于澳门的西方建筑是否就比中国建筑出现的早并不清楚。

2.【答案】Y

【详解】本论断可以通过"The Historic Center of Macao" is the product of East - West cultural exchanges, constituting the most unique blend of cultural heritage existing in China's historic cities. 一句内容判断。

3.【答案】N

【详解】根据对澳门 history 的介绍,澳门的作用讲的更多的还是它作为通商口岸的作用:Macao's strategic location at the mouth of the Pearl River meant that it retained a unique position in the South China Sea, serving as the hub in a complex network of maritime trade that brought tremendous wealth and a constant flow of people into the enclave。

4.【答案】Y

【详解】根据 Macao, as the West's first established gateway into China, was remarkable in setting off a succession of connections and contacts that progressively enriched both civilizations 一句内容可以判断本句内容与之一致。

5.【答案】N

【详解】这一论断相对是主观臆断了,根据文章内容,澳门被称做"Christian City",是因为它 coincides with the heart of the Western settlement area。

6.【答案】Y

【详解】文章介绍了 missionaries 所做的一系列有关文化传播的事,包括 founded the first Western - style hospitals, dispensaries, orphanages, and charitable organizations, brought in the first movable - type printing press to be used on Chinese soil, and published the first paper in a foreign language 等等,由此可见本论断是正确的。

7.【答案】Y

【详解】文章交代了因为 Macao was the base for the Jesuit mission in China and other parts of East Asia,所以 Macao was thus the training ground for the Jesuit's mission to China and other parts of Asia。

8.【答案】Robert Morrison

【详解】文中说 Other achievements of Christian missionaries in Macao include... the first Chinese translation of the Bible by Robert Morrison, 显然 Robert Morrison 肯定为翻译圣经作出了贡献。

9.【答案】the "Leal Senado" Building

【详解】根据 Situated at one end, the "Leal Senado" Building has a commanding view overlooking the entire square 内容可填充。

10.【答案】Dominic's Square

【详解】根据 St. Dominic's Church is located at the junction of Senado Square and Dominic's Square. 一句内容填充。

# Part 3

# 听力理解

## Listening Comprehension

# 命题规律与应试技巧

## 1. 七大考点精确定位

### (1) 地点

根据对话内容判断对话发生的地点或对话中所提事件发生的地点是四级英语听力测试中常见的也是比较重要的一个题型。

### (2) 职业、身份

根据说话内容判断说话者的身份和职业是四级听力测试中又一常见题型。

### (3) 计算题

计算题在四六级英语听力中属于比较难的题型,要求同学们不仅能分辨不同的时间、金钱等数量概念,还应能将听到的各个数量联系起来进行加减运算。

### (4) 言外之意、弦外之音

推测说话者话中之话是听力中必考的题型。从试题的设计特点来看,大体有以下几种:

**①对虚拟语气的考查**

包含虚拟语气的听力考题中一般都有"是非"相对的选项,这就要求同学们了解虚拟语气的表意功能,根据虚拟语气判断正确选项。

**②对建议的考查**

建议题要求大家掌握建议的各种表达方式,根据建议选择正确选项。

建议的常用表达方式有:

Why not. . . ? /What do you think of. . . ?

If I were you/If I were in your shoes, I would. . .

Shall we. . . ? /I suggest. . .

You'd better/You ought to. . .

**③对话题的考查**

话题指会话双方所谈论的话题,要求同学们具有概括能力。

**④同义表达方式的考查**

同义表达方式,也就是同学们对具体句式、短语、词组甚至单词的理解。在听力选择中,相当一部分情况下,答案就是对话信息的同义表达方式。

**⑤上下文概念的考查**

有些问题问的是含义之类,但其实考查的是对上下文关系的掌握。

### (5)肯定与否定类题型往往涉及考生对以下词语和句型的理解。

①含否定语义的副词和形容词,如:hardly, barely, scarcely, rarely, little, seldom, few 等。

②含否定语义的代词和连词,如:nobody, nothing, neither, nor 等。

③含否定意义的词缀,如:im-,un-,mis-,dis-,-less 等。

④含否定语义的动词、动词词组及介词词组,如:fail, miss, avoid, deny, hate, stop, refuse, doubt, far from, anything but, instead of, rather than 等。

⑤含 too...to 的结构。

⑥强调否定句,句首的否定词多数是 never, little, rarely,并且句子必须倒装,如:Never before have I..., Little did they..., Rarely do we...等。

⑦双重否定句,如 not uncommon, no one can deny..., not careless 等。

⑧注意缩写形式的否定读音,如:aren't, don't, hasn't, weren't, wasn't 等。

### (6)倍数

表达倍数的题目常常使用以下词汇:

times 乘,quarter 四分之一,twice 两倍,one-third/fourth 三/四分之一,couple 双,三两个,discount 折扣,half(of) ……的一半,half as much/ many as 加半倍,一倍半,double 使加倍,增加一倍,twice as much/ many as 是……的两倍,percentage 百分比,off 减、降、少,pair(a pair of) 一对,一双

### (7)比较与选择

最常见的词汇句型和搭配有:

as...as 和……一样,not as/not so... as...和……不一样,twice (half) as... as...是……两倍/一半 三组表示同级比较。more/less...than,not more...than,not/no+比较级(=最高级)三组表示不等比较:"……比……更"。类似词还有 never better, nothing better, than ever before, than anyone else, than anything else 等等。

其他:the more...,the more...表示越……越……,not so much... as (=less... than...)与其说……不如说……

其他暗含的比较形式还有:动词 prefer... rather than/to..., reduce...(to),形容词 top, favorite,句型 would rather... than...(宁可……也不……),the last (=the least likely),not the less (=none the less)仍然,依然

## 2.十二类疑难句式各个击破

### (1)六类肯定句式表否定

句式①I'd like to/I'd love to,but...表否定,重点放在 but 后面。如:

-The students' English club is having a party on Saturday night. Can you come?

-I would like to,but I work at a restaurant on weekends.

句式②使用虚拟语气的句子。在很多四级听力考试试题当中,经常采用虚拟语气,这些句子形式上是肯定的,但往往表达一种"本应该,本可以(却没有),希望"的情绪,放在一定的语言环境中表否定。而这一点也是考生常忽略的,要特别注意。如:

-If the traffic wasn't so bad,I could have been home by 6:00.

-What a pity! John was here to see you.(意思是如果早到家的话,就可以见到约翰了,但是事实是因为交通堵塞,没能及时回家)

句式③I'm sorry,...这种句型是委婉的否定,其后一般接原因。如:

-Hello,may I speak to John Smith, please?

-I'm sorry,nobody by that name works here.(意思是说没有叫约翰·史密斯的人)

句式④由形容词 last 构成的特殊句型。这种结构的字面意思是"……是最后一个",但真正的意思是"……是最不可能的"。如:

-Would you like to go mountain climbing with us?

-That's the last thing in the world I want to do.(字面意思是爬山是这世界上我最后想做的事,反过来讲就是根本不想去爬山)

句式⑤anything but 句型。此句型在四级考试中语法与结构部分考过,如果出现在听力理解部分,应该来说比较难,但是如果我们掌握了它的基本意思"除……以外的任何事物"或"根本不",并在听的过程中多加小心的话,还是可以做对的。如:

-Everyone is helping out with dinner. Could you make the soup?

- Anything but that.（意思是我不可能做汤）

句式⑥由一些特殊短语构成的句型。这种短语很多，因此就需要学生平时不断地积累，这里举两个短语为例：

- You're not much of a rock and roll fan, are you?

- It's far from being my favorite kind of music, that's for sure.（这里的 be far from 短语意思是“远非”，可理解为“这根本不是我所喜欢的音乐”。）

## (2)六类否定句式表肯定

句式①Why don't you/Why not...？这种句型相对来说比较简单，意思是“为什么不呢?”，但我们还是应该注意它真正表达的是一个肯定概念，即“建议做某事”。如：

- John, I don't know what to get for your father. He has just about everything, doesn't he?

- Do you have any suggestions?

- Why don't you get him a pocket calculator?

句式②Do you mind...？问句的回答用 No, of course not. 或者 Not at all.。虽然字面上是否定的，而且也理解为“不介意”，但考虑到具体的语境，通常都应理解为肯定的，意思是对方可以做其想做的。

如：

- Do you mind if I borrow your note?

- No, of course not. They are on my desk.（从侧面讲可以使用）

句式③not...until...句型。此句型一般都应理解为“直到……才”，因此是肯定的。如：

- When can the doctor see me?

- He won't be free until tomorrow.（意思是直到明天才能见你）

句式④not...more/better 构成的特殊句型。

此句型意思是不可能有比这样更好的情况发生了，反过来讲就是指这样很好，说话者对此表示赞许，所以此句型仍然表示的是肯定意思。如：

- I think it's high time we turned our attention to the danger of drunk driving now.

- I can't agree with you more. You see countless innocent people are killed by drunk drivers each year.（意思是说后者非常同意第一人的观点）

句式⑤Without a doubt; Don't mention it; No problem 等作为回答的否定句型。这些我们常用的作为回答的句型，其实在语境中通常是用作肯定的回答，意思是“没问题”，肯定是这样的。如：

- Do you think we have to review the chapter of Industrial Revolution?

- Without a doubt, it will be on the exam.

句式⑥由一些除 not 和 never 之外的否定词如 hardly, seldom, scarcely, rarely 等构成的句型。这种句型其实本身就应归结为否定句，但在此将其放到这里是想强调这些句子如果出现在听力理解当中，学生经常容易疏忽这些词的否定意思，因此需特别留意。如：

- What a surprise! Tim has improved his English so much after a holiday abroad.

- I can hardly hear an accent.（意思是几乎听不出有任何地方口音。）

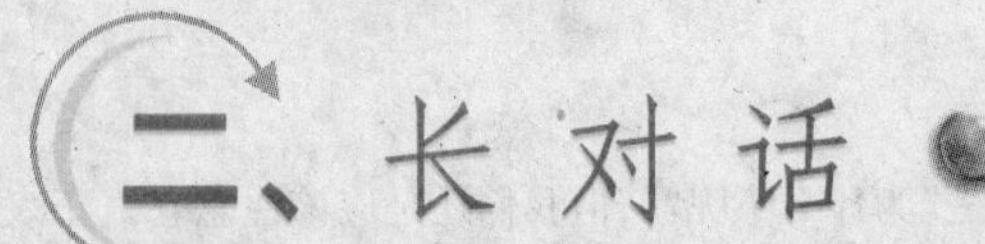

# 二、长对话

# 命题规律与应试技巧

## 1. 三大题型精确定位

长对话这一题型源于托福。新四级中的长对话这一题型包含两篇对话，对话长度250词左右，一般由两人轮流式交谈，对话回合为6～10次，说话长度为一至两句。内容为日常生活中发生的事，因此一般不会出现考生没有接触过的词。问题共设置7个，通常一篇3个，一篇4个。问题的顺序对应文章的开头，中间部分和结尾。对2006年样题长对话中的问题加以分析，我们可以看到大致有三类问题：

### (1)主旨题

【问题】What are the two speakers talking about?

【选项】A. The benefits of strong business competition.
B. A proposal to lower the cost of production.
C. Complaints about the expense of modernization.
D. Suggestions concerning new business strategies.

【解析】这类题目要在听过全篇文章后才能作出判断，选项的内容不会以原话的形式出现在对话中，且选项的意思高度浓缩。这类题目的提问方式往往如下：

What are the two speakers talking/discussing about?

What's the main idea/topic of the conversation?

### (2)细节题

这类题通常是针对对话中的某一句话，或某一个片断的。问题通常是问时间(when)、地点(where)、人物(who)、内容(what)、原因(why)、方式(how)等等，答案在对话中可以直接找到。比如：

①问内容：

【问题】What does the woman suggest about human resources?

【选项】A. The personnel manager should be fired for inefficiency.
B. A few engineers should be employed to modernize the factory.
C. The entire staff should be retrained.
D. Better-educated employees should be promoted.

【录音】W: We should also consider human resources. I've been talking to personnel as well as our staff at the factory.
M: And what's the picture?
W: We'll probably have to hire a couple of engineers to help us modernize the factory.

【解析】问内容的题总是以what开头，通常的问法有：

What is... doing?

What is his/her/their opinion/idea/suggestion?

②问为什么：

【问题】Why does the woman suggest advertising on TV?

【选项】A. Their competitors have long been advertising on TV. B. TV commercials are less expensive.

C. Advertising in newspapers alone is not sufficient. D. TV commercials attract more investments.

【录音】M: What about advertising?

W: Marketing has some interesting ideas for television commercials.

M: TV? Isn't that a bit too expensive for us? What's wrong with advertising in the papers, as usual?

W: Quite frankly, it's just not enough anymore. We need to be more aggressive in order to keep ahead of our competitors.

【解析】显而易见，问为什么大多数情况下以 why 开头，有时也会问 for what reason。但值得注意的是这类题的答案在录音中并不总是由 because 引出，比如上题。

③问地方：

【问题】Where can the man find the relevant magazine articles?

【选项】A. At the end of the online catalogue. B. At the Reference Desk.

C. In The New York Times. D. In the Reader's Guide to Periodical Literature.

【录音】W: Oh... another thing you might consider... have you tried looking for any magazine or newspaper articles?

M: No, I've only been searching for books.

W: Well, you can look up magazine articles in the Reader's Guide to Periodical Literature. And we do have the Los Angeles Times available over there. You might go through their indexes to see if there's anything you want.

M: Okay. I think I'll get started with these books and then I'll go over the magazines.

W: If you need any help, I'll be over at the Reference Desk.

M: Great, thanks a lot.

【解析】这类题相对容易解答，因为答案总是由表示方位的介词如 at、in、on 等等引导，十分明显。

### (3)推理题

这类题目同样也是针对文章中的某句话或某个片断，但和细节题不同的是，这类题需要在理解之后才能作出回答，也就是说答案不会直接在对话中出现。

【问题】What is the man doing?

【选项】A. Searching for reference material. B. Watching a film of the 1930s'.

C. Writing a course book. D. Looking for a job in a movie studio.

【录音】W: Sir, you've been using the online catalogue for quite a while. Is there anything I can do to help you?

M: Well, I've got to write a paper about Hollywood in the 30s and 40s, and I'm really struggling. There are hundreds of books, and I just don't know where to begin.

【解析】我们可以看到，在录音中根本没有 Searching for reference material 这个短语。要正确判断出这个答案，必须要听懂这个对话片断。首先我们从女子口中得知男子是在网上查资料(Sir, you've been using the online catalogue for quite a while)，接着根据男子所说 I've got to write a paper 可知他查资料是为了写文章。第一步可以初步断定答案是 A，第二步排除其他选项。

## 2. 听力技巧精确突破

### (1)听音技巧

#### 技巧1 把握对特殊句式的理解。

如反问句，设问句，感叹句等等。比如：

【问题】Why do the speakers want to change their schedule?

【选项】A. Because tomorrow it won't rain.

B. Because tomorrow it will rain.

C. Because they don't know how tomorrow's whether is about.

D. Because they are sure about tomorrow's whether.

【录音】W: It must be going to rain tomorrow, won't it?

M: Well, let's make a change of our schedule.

【解析】本题的关键是把握对反意疑问句的理解。本题的反意疑问句中有一个表示推测的 must,所以这个反意疑问句的意思是:明天肯定要下雨,是吗? 由此可以判断选 B。

### 技巧 2 关键词的捕捉。

要特别注意录音中的时间、地点、人物、数字、主题等等。比如:

【问题】How often will the woman's son have piano lessons from next week on?

【选项】A. Once a week.　　B. Twice a week.

C. Three times a week.　　D. Four times a week.

【录音】M: Your son seems to have made much progress in playing the piano. Does he attend any piano classes?

W: Yes, he takes lessons twice a week, but from next week on, he will go to the class on Saturday evenings, too.

【解析】本题要特别注意录音中的数字,因为选项中都是数字。而数字相对于考生来说是容易听出来的。夸张的讲,本题只要听到 twice 这个表示倍数的词,其余的都没听出来也不要紧。

### 技巧 3 语气的把握。

通过语气的把握,可以帮助了解说话者的态度,观点。

【录音】A: What's your name?

B: Monica Simonson.

A: Where are you from?

B: Sweden.

A: What are you doing here?

B: I'm on holiday.

A: How long are you staying?

B: Six weeks.

【解析】如果以一种生硬的语气去读上述的提问,那么大家很容易猜测这可能是使馆签证人员或边境检查人员对游客的检查。但大家如果用一种轻松、欢快的口气去读,而回答用生硬的口气去读,你可能会认为这是一个小伙子在向一位姑娘搭讪,而姑娘对他不理不睬。

### 4 对俗语、短语和俚语的理解。

俗语、短语和俚语往往是学生不很熟悉的一部分,一旦听不懂,往往会卡了壳,比如下面:

【录音】W: We should also consider human resources. I've been talking to personnel as well as our staff at the factory.

M: And what's the picture?

W: We'll probably have to hire a couple of engineers to help us modernize the factory.

这里的 what's the picture? 表示的是情况如何的意思,如果不懂这个俚语,考生不要慌张,通过上下文可以把它的意思推测出来。因为一旦慌张,影响后面所有的听音,则损失惨重。

## (2)读题技巧

### 1 听音前,熟悉所有选项。

这样能够让学生了解听力材料的背景,有助于理解。另外,提前熟悉所有选项,有助于在做题时立即作出反应。比如 2006 年 6 月的真题:

12. A. She's worried about the seminar.　　B. The man keeps interrupting her.

C. She finds it too hard. D. She lacks interest in it.

13. A. The lecturers are boring. B. The course is poorly designed.

C. She prefers Philosophy to English. D. She enjoys literature more.

14. A. Karen's friend. B. Karen's parents.

C. Karen's lecturers. D. Karen herself.

15. A. Changing her major. B. Spending less of her parents' money.

C. Getting transferred to the English Department. D. Leaving the university.

【解析】第12题告诉我们说话女子可能在某一方面有问题,这个方面有可能是与男子的关系,也可能是在 seminar 方面;13题谈的都是学科的问题(lecturers, course, Philosophy, literature ),是女子对学科的好恶。第15题A,C,D三项都表明她对现在的学科或学习不满。由此,我们可以大胆推测:一、这篇文章跟大学学习相关;二、跟学科相关;三、这名女子可能要换专业,要么是因为没兴趣,要么是因为太难。结果证明,这些推测都是正确的。

## 技巧2 把握选项中反复出现的词,同时分析选项中的相反项。

【问题】What does the woman say about the equipment of their factory?

【选项】A. It cost much more than its worth. B. It should be brought up - to - date.

C. It calls for immediate repairs. D. It can still be used for a long time.

【录音】W: Sure. I've been trying to come up with some new production and advertising strategies. First of all, if we want to stay competitive, we need to modernize our factory. New equipment should've been installed long ago.

【解析】该题四个选项中A和D都是对it的正面的评价;而B和C则是负面的。四个选项由此可以化成两组选项,这样就可以降低题目的难度。

## 技巧3 通过选项对问题进行推测。

【问题】Where can the man find the relevant magazine articles?

【选项】A. At the end of the online catalogue. B. At the Reference Desk.

C. In The New York Times. D. In the Reader's Guide to Periodical Literature.

【录音】W: Well, you can look up magazine articles in the Reader's Guide to Periodical Literature. And we do have the Los Angeles Times available over there. You might go through their indexes to see if there's anything you want.

M: Okay. I think I'll get started with these books and then I'll go over the magazines.

W: If you need any help, I'll be over at the Reference Desk.

M: Great, thanks a lot.

【解析】四个选项都是表示地点的介宾短语。我们完全可以肯定问题是问地点。又因为这一题是最后一题,因此在听录音时我们在最后阶段应把注意力主要放在地点上,这样就不难找到答案 in the Reader's Guide to Periodical Literature。

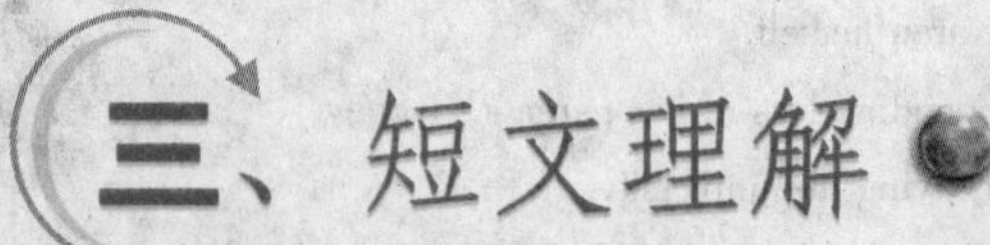

# 三、短文理解

# 命题规律与应试技巧

## 1. 两大命题规律精确定位

**短文理解共有短文三篇，设 10 个小题，主要考查两方面的内容。**

### (1) 对主题的判断

①提问的方式

主旨题通常以下列方式提问：

What is the main idea/topic of this passage?

What does the passage mainly discuss?

What can we learn from the passage?

What is the passage mainly about?

**②选项特点**

主题的四个选项一般都以短语的方式出现。

**③解题技巧**

对于这类题，在听的时候应特别注意文章的首句和尾句，首句一般开篇点明主题，而尾句则总结全文，根据这两句进行推断一般可以确定文章的主题。

### (2) 对具体事实的判断

**1) 提问的方式**

对细节的提问通常以 wh - question 的方式出现，主要针对文章的有关人物、事件、地点、时间、原因、目的、数据等。

**2) 选项特点**

这种细节判断题以辨认题居多。有可能几个选项在文章中都有所提及，但只要仔细听，注意分别，就能从原文中找到出处。如有一次的考题，第 1 题可以在文章中找到 The chief duty of every government is to protect persons and property，这同样也是第二段的主题句，是整段要说明的内容。第 2 题也能从文章中找到 Years ago the government made money from the sale of public lands. 一句为佐证。

从考查的范围来看，短文听力中原因的考查最多，其次是对具体所发生的事情的判断，也包括依据事实所做的推理判断。一般原因考查题多出现在故事短文中，只要能够理清故事，这种原因题就不难找到正确的答案。

**3) 解题技巧**

一篇文章，一个主题，内容上完整统一，所有的细节都是围绕主题展开，为主题服务。听力理解所要求的是对文章的整体把握，细节也应该是用于说明主题的主要细节。也就是说，每篇短文的几个题是相互关联、相互说明的。选择细节理解题时应注意：

①注意抓主要细节；

②注意信息的直接辨认，从听力文章中找到信息句；

③注意各题之间的关联，保证相互说明，不能相互矛盾。

### 2. 六种解题方法精确突破

短文部分主要在于其整体难度较大。如果说对话部分侧重语句水平，那么短文部分则侧重于语篇水平，更强调理解的整体性和逻辑性，强调隐含信息的推理、综合信息的归纳。做短文部分听力测试题时，应注意应用以下方法。

(1)先浏览问题，再根据问题预测内容；

(2)许多问题的出现顺序与文章内容的顺序基本一致；

(3)短文中的问题多为对细节内容的考查，因此选项加上题干的内容大致与文章的内容相符；

(4)记住事件发生的时间和地点；

(5)综合记忆短文中的事实和理由；

(6)通过所给信息判断人物的身份及相互间的关系。

## 四、复合式听写

# 命题规律与应试技巧

### 1. 题型聚焦

**复合式听写共有短文一篇，设 11 个空。1 ~8 空要求用原词填写，9 ~11 要求听写短语或句子，既可用原文，也可以填写大意。**

#### (1)1 ~8 题命题规律

①所填单词以实词为主。

② 8 个单词以评价性词汇为主，也就是说可以从上下文找到说明的信息。

③表示信息复现的词汇为题眼，有些词语即使听不清楚同样可以填出。

④表示信息同现的词汇为题眼，复合式听写所填词汇一部分是同现词汇。

⑤对文章叙述逻辑的考查，叙述逻辑即上下文的因果、转折、递进、解释等关系。如果能看出这些关系，则不用听就可以将所缺单词填上。

## (2)9～11 题命题规律

**①用于说明主题的细节**

这部分听写一般是段落的主题已经给出，要求是补全细节。

**②概括性的结论或主题**

例如有一次四级考卷的第 10 句为结论句(Michael is smart, but he is like every other kid.)是对全文的一个概括。主题句还没有考过，但主题与结论作为文章的重点之笔应该是听力理解的重点，也应该是复合式听力所应包含的东西。

# 2. 解题技巧精确突破

## (1)单词听写

这部分侧重检测考生对单词的音、形、义、用的综合掌握能力，一般说来，做听写填空题，可按下列步骤和技巧进行：

**①听抄**

要从音、形、义、用四个方面入手，并结合上下文听懂该单词，如有一次考试复合式听写空格中，No working day is identical to any other, so there is no“typical” day for a police officer. 许多考生都听懂了“typical” 一词，但却将其拼写为“topical”，显然不了解 typical 派生于 type (类型)一词，还有些考生将其拼为 difficult，也许是由于未能扎实掌握形容词后缀 - al 的缘故。

**②检查**

A. 时态、语态是否正确?

B. 语意是否通顺? 英语中有很多读音相似或同音异形、异义词，选择哪一个词应由整句话的句意及上下文的连贯性来决定。语句结构是否完整? 听的时候，如冠词、介词可能听得不是很清楚，检查时要根据上下文决定是否需要。

## (2)句子听写

**①培养和锻炼逻辑思维能力**

一是要培养直接用英语进行逻辑思维的能力，二是要锻炼根据上下文用逻辑推理来预测事物的发展情况的能力。

**②使用速记方法**

学生在听写时往往会出现记下了听写的第一个单词，而后面的几句匆匆而过，来不及填写后面内容。针对这个问题，在考试中应采用速记方法，迅速记下每个听到的单词。所谓的速记就是用一些简单的符号缩写、字母记下所听到的内容，不让每个单词漏网。

例如：∵ → because ；∴ → so ； = → equal；sth. → something；sts→ students；ads→ advertisement；fridge - refrigerator；demo→ demonstration ；esp. → especially；somebody→ sb. 等等。先速记，然后再展开这些单词，这样所听的内容就不易漏掉了。

## (3)几点建议

根据上面所谈的复合式听写的特点，建议在做复合式听写时最好做到以下几点。

①注意话题知识的运用。

文章的话题规定了用词范围，根据话题判断单词有助于明确词汇。

②利用词汇的同现和复现关系。

③根据上下文推测词汇的运用。

④在做 9～11 题时在不能将原句完全记下的情况下，将关键词记下来，然后根据关键词，结合文章叙述的逻辑，重新编写句子。

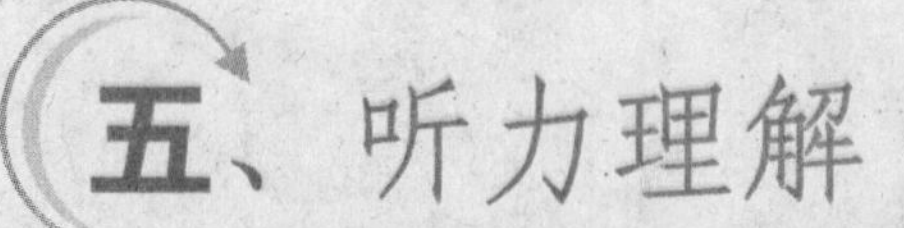

# 五、听力理解

# 核心考点精确打击

## Unit 1

### Section A

**Directions:** *In this section, you will hear 8 short conversations and 2 long conversations. At the end of each conversation, one or more questions will be asked about what was said. Both the conversation and the questions will be spoken only once. After each question there will be a pause. During the pause, you must read the four choices marked A, B, C and D, and decide which is the best answer. Then mark the corresponding letter on* ***Answer Sheet*** *2 with a single line through the centre.*

1. A. She is often late for meals. B. She is expecting a letter from abroad.
   C. She wrote to her mother last month. D. She is anxious to go back home.
2. A. He is modest. B. He is satisfied. C. He is proud. D. He is upset.
3. A. Here. B. Europe. C. Australia. D. Austria.
4. A. The train is crowded. B. The train is on time.
   C. The train is out of order. D. The train is late.
5. A. 3 pills. B. 4 pills. C. 9 pills. D. 12 pills.
6. A. He enjoys writing home every week. B. He doesn't write home once a week now.
   C. He never fails to write a weekly letter home. D. He has been asked to write home every week.
7. A. The teacher postponed the meeting.
   B. There won't be a test this weekend.
   C. The students will be attending the meeting.
   D. The students will take a foreign language test this weekend.
8. A. Last year. B. Two years ago. C. Three years ago. D. Early last year.

**Questions 9 to 12 are based on the conversation you have just heard.**

9. A. The definition of eccentricity. B. Essentiality.
   C. How to keep pets. D. How to enjoy special food.
10. A. Being unusual and strange. B. Charming and special.
    C. Aggressive and hardworking. D. Common and usual.
11. A. A poor British man. B. A rich American.
    C. A rich British man. D. A poor American.

12. A. The Victorian surgeon lived at Buckland.
    B. Howard was always a hermit.
    C. A hermit is a person who enjoys communicating with others.
    D. Howard Hughes became a recluse because he was tired of high living.

**Questions 13 to 15 are based on the conversation you have just heard.**

13. A. The items are far beyond his financial means.
    B. He feels his daughter really doesn't need them.
    C. The family already owns some of these supplies.
    D. The supplies will be out of date quickly.
14. A. They are on sale until the end of the week.
    B. Her teachers require them as part of the curriculum.
    C. She volunteers to use some of her own money.
    D. She would pay the money back.
15. A. He discovers he had more money than he thought.
    B. He concludes that she will provide for him when he is older.
    C. The girl promises to help her mother in exchange for the supplies.
    D. He loves her daughter very much and doesn't want to make her unhappy.

**Directions:** *In this section, you will hear 3 short passages. At the end of each passage, you will hear some questions. Both the passage and the questions will be spoken only once. After you hear a question, you must choose the best answer from the four choices marked A, B, C and D. Then mark the corresponding letter on* ***Answer Sheet*** *2 with a single line through the centre.*

**Passage One**

**Questions 16 to 18 are based on the passage you have just heard.**

16. A. How to collect and utilize body heat.
    B. How to use the conventional fuel.
    C. How to design the modern building in the university.
    D. How to absorb heat given by lights.
17. A. None. B. Two. C. Four. D. Six.
18. A. A fat female who studies hard.
    B. A thin female who studies hard.
    C. A fat male who does not study.
    D. A thin male who studies hard.

**Passage Two**

**Questions 19 to 21 are based on the passage you have just heard.**

19. A. To explain the meaning of the word "tip".
    B. To illustrate why difficult customers give bad tips.
    C. To put forward reasons for a salary increase.
    D. To indicate how one can get better service.
20. A. The person who gets least service.
    B. The person who gets most service.
    C. The person who orders a lot of food and drinks.
    D. The person who gets unsatisfactory food.
21. A. To make sure that waiters do things right away.
    B. To make sure that waiters do things perfectly.
    C. To insure that waiters get proper amount of tips.
    D. To insure that waiters have enough money to live.

**Passage Three**

**Questions 22 to 25 are based on the passage you have just heard.**

22. A. They feel guilty and waste even more time worrying about it.
    B. They wish they had good intentions to work more efficiently.
    C. They wish that they had someone else to organize their time.
    D. They hope those who distracted them will feel guilty.
23. A. Creative. B. Rare. C. Extreme. D. Well-organized.

24. A. Because he does research in many different subjects.
    B. Because he cannot bear to throw them away.
    C. Because he gets ideas from them for his writing.
    D. Because he is a very untidy, disorganized person.
25. A. Too tight. B. Irregular. C. Inefficient. D. Too loose.

## Section C

**Directions:** *In this section, you will hear a passage three times. When the passage is read for the first time, you should listen carefully for its general idea. When the passage is read for the second time, you are required to fill in the blanks numbered from 26 to 33 with the exact words you have just heard. For blanks numbered from 34 to 36 you are required to fill in the missing information. For these blanks, you can either use the exact words you have just heard or write down the main points in your own words. Finally, when the passage is read for the third time, you should check what you have written.*

An old lady who lived in a (26) ________ went into town one Saturday. After she had (27) ________ fruit and (28) ________ in the market for herself and for a friend who was ill, she went into a shop which (29) ________ glasses. She (30) ________ one pair of glasses, and then another pair and another, but (31) ________ of them seemed to be right. The (32) ________ was a very (33) ________ man, and after some time he said to the old lady, "(34) ________________________________________________. Everything will be all right in the end. (35) ________________________________________________." "No, it isn't," answered the old lady. (36) ________________________________________________.

# Unit 2

## Section A

**Directions:** *In this section, you will hear 8 short conversations and 2 long conversations. At the end of each conversation, one or more questions will be asked about what was said. Both the conversation and the questions will be spoken only once. After each question there will be a pause. During the pause, you must read the four choices marked A, B, C and D, and decide which is the best answer. Then mark the corresponding letter on* ***Answer Sheet*** *2 with a single line through the centre.*

1. A. The man should stay a little longer. B. The man should leave at once.
   C. The man will miss the bus. D. The man must try to catch the last bus.
2. A. Fifty-five minutes. B. Thirty-five minutes.
   C. Twenty-five minutes. D. Twenty minutes.
3. A. Go and watch a volleyball match. B. Try to get some tickets.
   C. Go and buy a new dress. D. Do some washing at home.
4. A. He has a relative who once lived there. B. He is going to see his aunt there.
   C. He will visit the city soon. D. He used to have an apartment there.
5. A. 7:00. B. 8:30. C. 8:00. D. 7:30.
6. A. Husband and wife. B. Son and mother.
   C. Patient and doctor. D. Teacher and student.
7. A. The woman went to the theater, but the man didn't.
   B. The man went to the theater, but the woman didn't.
   C. The speakers did not go to the theater.

D. Both speakers went to the theater.

8. A. An English textbook. B. A Chinese textbook.
C. A chemistry book. D. A medical book.

**Questions 9 to 12 are based on the conversation you have just heard.**

9. A. It was roomy enough for him. B. It was more economical than the mid-sized one.
C. It had more features than the other vehicles. D. It had more protection policies.

10. A. He couldn't add an additional driver to the rental plan.
B. He was only limited to a certain number of miles per day.
C. The vehicle would probably consume a lot of gas.
D. It may be more comfortable.

11. A. It was a little larger than he expected. B. The car doesn't look very attractive.
C. The engine has problems and runs poorly. D. The car is good enough to rent.

12. A. You should call the police in case your car has mechanical difficulties.
B. Getting assistance might require some time and patience.
C. The company will compensate you for delays in your travel.
D. You can get some comfortable things from it.

**Questions 13 to 15 are based on the conversation you have just heard.**

13. A. Old friends. B. Brother and sister.
C. Colleagues from work. D. Common friends.

14. A. He feels medical treatment is still unproven for his condition.
B. He is worried about the side effects of the medication.
C. He thinks the treatment is too expensive.
D. He thinks the treatment is too cheap to believe.

15. A. He continues to look for other solutions to his problem.
B. He decides to visit a doctor at this girl's urging.
C. He finally accepts that state of his condition.
D. He decides to use some Chinese medicine.

## Section B

**Directions:** *In this section, you will hear 3 short passages. At the end of each passage, you will hear some questions. Both the passage and the questions will be spoken only once. After you hear a question, you must choose the best answer from the four choices marked A, B, C and D. Then mark the corresponding letter on* ***Answer Sheet*** *2 with a single line through the centre.*

### Passage One

**Questions 16 to 18 are based on the passage you have just heard.**

16. A. His travels. B. His short stories.
C. His finances. D. His family.

17. A. He wanted to be a journalist. B. His stories were inspired by his travels.
C. He wanted to get away from the army. D. He was sent there by his father.

18. A. His stories were inspired by his travels.
B. His travels prevented him from writing.
C. He traveled in order to relax from the pressure of writing.
D. He traveled around to publicize his writings.

### Passage Two

**Questions 19 to 21 are based on the passage you have just heard.**

19. A. The way you look at one person. B. The language you use.
C. The clothes you wear. D. The music you listen.

20. A. It means you want to protect yourself.
B. It means you want to take a rest.
C. It means you don't support the idea.
D. It means you want to keep away from others.

21. A. Drink up the tea.
B. Put his arms across his body.
C. His feet point towards the door.
D. Look directly at you.

**Passage Three**

**Questions 22 to 25 are based on the passage you have just heard.**

22. A. The shapes of water droplets.
B. The formation of fog.
C. The spread of air currents.
D. The characteristics of fog.

23. A. They both form in large spherical masses.
B. They are both made of tiny water droplets.
C. They are both common in cold climates.
D. They both change shape when temperatures vary.

24. A. The density of the water droplets in the air.
B. The size of the water droplets.
C. The temperature of the water droplets.
D. The purity of the water droplets.

25. A. Why frog freezes at low temperature.
B. How very small droplets can form.
C. Why the density of fog varies.
D. How water droplets stay suspended in air.

## Section C

**Directions:** *In this section, you will hear a passage three times. When the passage is read for the first time, you should listen carefully for its general idea. When the passage is read for the second time, you are required to fill in the blanks numbered from 26 to 33 with the exact words you have just heard. For blanks numbered from 34 to 36 you are required to fill in the missing information. For these blanks, you can either use the exact words you have just heard or write down the main points in your own words. Finally, when the passage is read for the third time, you should check what you have written.*

In police work, you can never predict the next crime or (26) ________. No working day is identical to any other, so there is no "typical" day for a police officer. Some days are (27) ________ slow, and the job is boring; other days are so busy that there is no time to eat. I think I can (28) ________ police work in one word: (29) ________. Sometimes it's dangerous. One day, for example, I was working (30) ________; that is, I was on the job, but I was wearing (31) ________ clothes, not my police (32) ________. I was trying to catch some (33) ________ who were stealing money from people as they walked down the street. (34) ________________________________________. One of them had a knife, and we got into a fight. Another policeman arrived, and together, we arrested three of the men; but the other four ran away. (35) ________________________________________. She was trying to get to the hospital, but there was a bad traffic jam. (36) ________________________________________. I thought she was going to have the baby right there in my car. But fortunately, the baby waited to arrive until we got to the hospital.

# Unit 3

## Section A

**Directions:** *In this section, you will hear 8 short conversations and 2 long conversations. At the end of each conversation, one or more questions will be asked about what was said. Both the conversation and the questions will be spoken only once. After each question there will be a pause. During the pause, you must read the four choices marked A, B, C and D, and decide which is the best answer. Then mark the corresponding letter on* ***Answer Sheet*** *2 with a single line through the centre.*

1. A. Someone fixed it.
B. Louise sold it.
C. Louise repaired it.
D. It's been thrown out.

2. A. $ 1.75.　　B. $ 2.50.　　C. $ 1.50.　　D. $ 1.05.

3. A. She felt sorry for the man.　　B. She had to pay the fine.
C. She couldn't accept the books.　　D. She had to ask the man to pay for the overdue.

4. A. He visited Kuwait last year.　　B. He was ill there.
C. He felt shameful.　　D. He is going to visit Kuwait and Barcelona this year.

5. A. Because of the yellow light.　　B. Because of the blue light.
C. Because of the piercing light.　　D. Because it was blue.

6. A. Yes, it's too far to walk.　　B. Yes, you must take a bus or a taxi.
C. No, it's within walking distance.　　D. No, but it's too far to walk.

7. A. The man is always absent in Mrs. Lee's class.　　B. The woman likes sleeping in the class.
C. Neither of them likes Mrs. Lee's class.　　D. They find Mrs. Lee's class is interesting.

8. A. The sun.　　B. Their children.
C. Right and wrong.　　D. The weather.

**Questions 9 to 12 are based on the conversation you have just heard.**

9. A. By talking to the salesclerk.　　B. By calling on the phone.
C. By going to the flower store to order himself.　　D. By asking others to do so for him.

10. A. 1.　　B. 2.　　C. 6.　　D. 12.

11. A. 4096239.　　B. 43.　　C. 47401.　　D. 452004.

12. A. All my love.　　B. All my love, Jim.
C. For my love, Jim.　　D. For my love.

**Questions 13 to 15 are based on the conversation you have just heard.**

13. A. Both the man and the woman's children like hamburgers.
B. Neither the woman's nor the man's children like hamburgers.
C. The man likes hamburgers.
D. The woman likes hamburgers.

14. A. Salt.　　B. Onion.　　C. Pepper.　　D. Sugar.

15. A. Fried balls made of meat.　　B. Flattened balls made of meat.
C. Fat balls made of meat.　　D. Lengthened balls made of meat.

## Section B

**Directions:** *In this section, you will hear 3 short passages. At the end of each passage, you will hear some questions. Both the passage and the questions will be spoken only once. After you hear a question, you must choose the best answer from the four choices marked A, B, C and D. Then mark the corresponding letter on **Answer Sheet** 2 with a single line through the centre.*

### Passage One

**Questions 16 to 18 are based on the passage you have just heard.**

16. A. Christmas is a public holiday of the world.
B. Christmas is now celebrated by Great Britain and the United States of America only.
C. Christmas is a holiday for churches.
D. Christmas is a public holiday in both Great Britain and the United States.

17. A. Rich Christmas cakes.　　B. Rich Christmas puddings.
C. Roast turkey.　　D. All the above.

18. A. About 156 years.　　B. About 129 years.
C. About more than 300 years.　　D. About 200 years.

### Passage Two

**Questions 19 to 21 are based on the passage you have just heard.**

19. A. Hawaii is a state of the USA.

B. Hawaii is an island of the USA.

C. Hawaii lies on a chain of more than 20 volcanic islands and atolls.

D. Hawaii lies in the central Pacific.

20. A. The capital of Hawaii is Honolulu.

B. The climate of Hawaii is tropical.

C. The main crops of Hawaii are sugar and pineapples.

D. The US navy is on the island of Oahu.

21. A. 228. B. 113.

C. 106. D. 47.

**Passage Three**

**Questions 22 to 25 are based on the passage you have just heard.**

22. A. Because Americans pay more attention to food value.

B. Because Americans are so rich that they like to have every kinds of food.

C. Because there are various groups of immigrants who brought over different varieties of cookery to America.

D. Because there are a lot of foreigners in America every day.

23. A. The turkey. B. Maine lobsters.

C. Californian oranges. D. Fried chicken.

24. A. Hamburgers and hot dogs. B. Sweet jellies.

C. Pickles with meat. D. Apple pie.

25. A. American Immigrants and Their Food B. American Food

C. The Native Food to America D. The Typically American Foods

## Section 

**Directions:** *In this section, you will hear a passage three times. When the passage is read for the first time, you should listen carefully for its general idea. When the passage is read for the second time, you are required to fill in the blanks numbered from 26 to 33 with the exact words you have just heard. For blanks numbered from 34 to 36 you are required to fill in the missing information. For these blanks, you can either use the exact words you have just heard or write down the main points in your own words. Finally, when the passage is read for the third time, you should check what you have written.*

Today I want to help you with a study reading method known as SQ3R. The letters stand for five steps in the reading (26) ________: Survey, Question, Read, Review, Recite. Each of the steps should be done carefully and in the order mentioned.

In all study reading, a survey should be the first step. Survey means to look quickly. In study reading, you need to look quickly at titles, words in darker or larger print, words with (27) ________ letters, (28) ________ and charts. Don't stop to read complete sentences. Just look at the important (29) ________ of the materials.

The second step is question. Try to form questions based on your survey. Use the question words who, what, when, where, why and how.

Now you are ready for the third step. Read. You will be reading the (30) ________ and important words that you looked at in the survey, but this time you will read the examples and (31) ________ as well. Sometimes it is useful to take notes while you read. I have had students who (32) ________ to underline important points, and it seemed to be just as useful as note-taking. What you should do, whether you take notes or underline, is to read (33)________. (34) ________________________________________.

The fourth step is review. Remember the questions that you wrote down before you read the material. You should be able to answer them now. (35) ________________________________________. Concentrate on those. Also review material that you did not consider in your questions.

The last step is recite. (36) ________________________________________.

SQ3R—Survey, question, read, review, and recite.

# Unit 4

## Section A

**Directions:** *In this section, you will hear 8 short conversations and 2 long conversations. At the end of each conversation, one or more questions will be asked about what was said. Both the conversation and the questions will be spoken only once. After each question there will be a pause. During the pause, you must read the four choices marked A, B, C and D, and decide which is the best answer. Then mark the corresponding letter on* ***Answer Sheet*** *2 with a single line through the centre.*

1. A. Tom is hard to find. B. Richard speaks with difficulty.
   C. Tom's classmate doesn't talk to him. D. Tom doesn't work very hard.
2. A. Catch a cold. B. Hurry to get the train.
   C. Sit next to the train stop. D. Fix his torn sleeve.
3. A. He's a boat builder. B. He smokes a pipe.
   C. He paints watercolors. D. He's a plumber.
4. A. Clean up her office. B. Get her report back.
   C. Not wait for him past noon. D. Not worry about her umbrella.
5. A. Getting another ticket at the door. B. Canceling the concert.
   C. Exchanging the ticket for a better one. D. Trying to sell the ticket.
6. A. Had finished last night. B. Fasted for ten days.
   C. Typed part of his paper. D. Paged his friend.
7. A. It will end on the fifth of May. B. The time was changed again.
   C. There are fifty students in it. D. It lasts longer than it is supposed to.
8. A. The noise in the laboratory. B. The heat inside.
   C. The late hour. D. The crowded room.

**Questions 9 to 12 are based on the conversation you have just heard.**

9. A. He saw the office on his way home from work.
   B. A friend referred him to Dr. Carter's office.
   C. He found Dr. Carter's number in the phone book.
   D. He found Dr. Carter's name on a newspaper.
10. A. Tuesday. B. Wednesday. C. Thursday. D. Monday.
11. A. He hurt his knee when a tall ladder fell on him.
    B. He injured his ankle when he fell from a ladder.
    C. He sprained his hand when he fell off the roof of his house.
    D. He injured his ankle when he was driving.
12. A. The man should put some ice on his injury.
    B. The man needs to come into the office right away.
    C. The man ought to take it easy for a few days.
    D. The man should put more ice into the paint can.

**Questions 13 to 15 are based on the conversation you have just heard.**

13. A. Linda's life and future. B. Linda's hobby and past working experience.
    C. Linda's family and marriage. D. Linda's work and marriage.
14. A. Went to a bar. B. Went to a Beauty salon.
    C. Went to a concert. D. Took a travel.
15. A. He's from TV station QRX. B. He's from radio station QRX.
    C. He's from radio station KLX. D. He's from TV station KLX.

## Section B

**Directions:** *In this section, you will hear 3 short passages. At the end of each passage, you will hear some questions. Both the passages and the questions will be spoken only once. After you hear a question, you must choose the answer from the four choices marked A, B, C and D. Then mark the corresponding letter on* ***Answer Sheet*** *2 with a single line through the centre.*

### Passage One

**Questions 16 to 18 are based on the passage you have just heard.**

16. A. All the old customs disappear together with the old nations.
    B. Many strange practices emerge with the new nations.
    C. The liberation of the nation doesn't mean the liberation of everything.
    D. A woman has freedom to choose anyone to be her husband.
17. A. He wrote a letter to set the girl free.
    B. He gave the girl enough money to repay the old man.
    C. He broadcasted the girl's story on a radio program.
    D. He changed the law that permitted women to be bought and sold.
18. A. 14 pounds. B. 40 pounds. C. 60 pounds. D. 2000 pounds.

### Passage Two

**Questions 19 to 22 are based on the passage you have just heard.**

19. A. The alcoholic can stop drinking easily.
    B. The alcoholic should be sent to jail.
    C. The alcoholic need medical attention.
    D. The alcoholic are healthy people.
20. A. By encouraging discussion of drinking problems.
    B. By finding work for its members.
    C. By providing money to its members.
    D. By giving its members sound advice.
21. A. To help drinkers find the meaning of life.
    B. To help drinkers get along well with others.
    C. To help drinkers keep away from any alcoholic drinking.
    D. To help drinkers realize the extent of their problems.
22. A. The law should be repealed.
    B. The law should be fair.
    C. The law should be strict.
    D. The law should be lenient.

### Passage Three

**Questions 23 to 25 are based on the passage you have just heard.**

23. A. Because they can use marks to judge students.
    B. Because they can control students' behavior.
    C. Because their words are usually respected by the children.
    D. Because they may judge a student from their own likes and dislikes.
24. A. Social studies.
    B. Science matters.
    C. The very atmosphere of the classroom.
    D. Criticism of children's behavior.
25. A. Because her personal attitudes may affect her students if she is prejudiced.
    B. Because she need to improve herself too.

C. Because she is also often influenced by her students.

D. Because she may not have a constant attitude towards some controversial issues.

## Section C

**Directions:** *In this section, you will hear a passage three times. When the passage is read for the first time, you should listen carefully for its general idea. When the passage is read for the second time, you are required to fill in the blanks numbered from 26 to 33 with the exact words you have just heard. For blanks numbered from 34 to 36 you are required to fill in the missing information. For these blanks, you can either use the exact words you have just heard or write down the main points in your own words. Finally, when the passage is read for the third time, you should check what you have written.*

The cost is going up for just about everything, and college (26)________ is no (27)________. According to a nationwide (28)________ the College Board's Scholarship Service, tuition at most American universities will be on an (29)________ of 9 percent higher this year over last year.

The biggest increase will occur at (30)________ colleges. Public colleges, heavily subsidized by tax founds, will also (31)________ their tuition, but the increase will be a few (32)________ points lower than their (33)________ sponsored neighbors. (34)________________________________________. To put that another way, the cost has climbed 150 percent in the last decade. (35)________________________________________ who must pay extra charges ranging from $ 200 to $ 2,000, (36)________________________________________.

# Unit 5

## Section A

**Directions:** *In this section, you will hear 8 short conversations and 2 long conversations. At the end of each conversation, one or more questions will be asked about what was said. Both the conversation and the questions will be spoken only once. After each question there will be a pause. During the pause, you must read the four choices marked A, B, C and D, and decide which is the best answer. Then mark the corresponding letter on* ***Answer Sheet*** *2 with a single line through the centre.*

1. A. The man wants to go to Washington.
   B. The man wants to go to San Francisco.
   C. There are no flights to Washington for the rest of the day.
   D. There are two direct flights to Washington within the next two hours.
2. A. Around 5:00. B. Around 3:00. C. At 2:00. D. At 1:00.
3. A. The train is crowded. B. The train is empty.
   C. The train is late. D. The train is on time.
4. A. She paid $ 40 for the necklace. B. Her husband presented it to her as a gift.
   C. She bought the coat on her fortieth birthday. D. Her friend sent it to her as a birthday gift.
5. A. Buying a motorcycle costs too much. B. Driving lessons are too expensive.
   C. Motorcycling is too dangerous. D. Taking the bus is more convenient.
6. A. By two o'clock. B. By four o'clock.
   C. By nine o'clock. D. By twelve o'clock.
7. A. The man is showing the woman round the way. B. The two persons are talking about washing.

C. The man is a stranger to the city.
D. The woman is asking the way.

18. A. One.
B. Two.
C. Three.
D. Four.

**Questions 9 to 12 are based on the conversation you have just heard.**

9. A. Practice the piano and soccer.
B. Practice soccer and take care of children.
C. Finish homework assignments and take care of children.
D. Practice the piano and take care of children.

10. A. She has to catch up on her French homework.
B. She needs to write a paper.
C. She must practice for a math test.
D. She has to practice the piano.

11. A. One hour.
B. An hour and a half.
C. Two hours.
D. Two hours and a half.

12. A. Clean the garage.
B. Pick up her room.
C. Finish her science project.
D. Pay a visit to her grandpa.

**Questions 13 to 15 are based on the conversation you have just heard.**

13. A. She says the matter is the owner's responsibility, not hers.
B. She is not on talking terms with her son who lives there.
C. She is afraid of what the man might say or do.
D. She is a little bit slow - minded.

14. A. There is an awful smell coming from the farm next door.
B. The property owners next door are illegally disposing of waste.
C. The neighbors are burning leaves which are drifting his way.
D. There is too much noise that he can't fall asleep at all.

15. A. The military is flying high - altitude jets overhead causing supersonic booms.
B. A coal company has resumed its mining operations using explosive devices.
C. The armed forces are carrying out artillery training exercises nearby.
D. There are too many battles around the hotel that he feels frightened.

## Section B

**Directions:** *In this section you will hear 3 short passages. At the end of each passage, you will hear some questions. Both the passage and the questions will be spoken only once. After you hear a question, you must choose the best answer from the four choices marked A, B, C and D. Then mark the corresponding letter on the* ***Answer Sheet*** *2 with a single line through the centre.*

### Passage One

**Questions 16 to 18 are based on the passage you have just heard.**

16. A. The earth's reaction to the sun.
B. The winds' blowing across the sea.
C. The sunshine.
D. The sea's reaction to the moon.

17. A. Because the water here is dirty.
B. Because the wind here is not strong enough.
C. Because the wind blows all the four seasons.
D. Because the sun heating is not strong.

18. A. 12 meters.
B. 24 meters.
C. 34 meters.
D. 42 meters.

### Passage Two

**Questions 19 to 21 are based on the passage you have just heard.**

19. A. Computers will be more powerful.
B. Pocket computer will be more popular.
C. All the schools and most families will own a computer in rich countries.
D. Computers will help people to be richer.

20. A. People can learn languages from computers.
B. Computers can be used to control a central heating.

C. Computer will bring more leisure.

D. Computers can compose nice music for you.

21. A. Computers will bring unemployment.

B. Computers will make people lazier than before.

C. Computers will affect children's health.

D. Computers are still very expensive.

**Passage Three**

**Questions 22 to 25 are based on the passage you have just heard.**

22. A. Its fast pace and rhythm. B. Its simple themes.

C. Its beautiful melodies. D. Both A and B.

23. A. South America. B. The countryside throughout America.

C. Urban areas in the Southern U. S. A. D. The countryside in the Southern U. S. A.

24. A. It originates from the American Indians.

B. It has simple themes and melodies.

C. It mainly expresses the miseries of the black people and their hard lives.

D. It describes the situations and feelings of the American people.

25. A. Five. B. Four. C. Six. D. More than six.

## Section C

**Directions:** *In this section, you will hear a passage three times. When the passage is read for the first time, you should listen carefully for its general idea. When the passage is read for the second time, you are required to fill in the blanks numbered from 26 to 33 with the exact words you have just heard. For blanks numbered from 34 to 36 you are required to fill in the missing information. For these blanks, you can either use the exact words you have just heard or write down the main points in your own words. Finally, when the passage is read for the third time, you should check what you have written.*

In the United States, people appear to be (26) ________ on the move. Think for a moment, how often do you see moving (27) ________ on the roads? They seem to be (28) ________. Are so many people (29) ________ changing their addresses? Yes, people in the United States are (30) ________ on the move. Within any five years (31) ________ about one third of the (32) ________ change their place of (33) ________. (34) ________________________________________. Some people may decide to move because of employment opportunities. (35) ________________________________________. And some have many other reasons. (36) ________________________________________.

# Unit 6

## Section A

**Directions:** *In this section, you will hear 8 short conversations and 2 long conversations. At the end of each conversation, one or more questions will be asked about what was said. Both the conversation and the questions will be spoken only once. After each question there will be a pause. During the pause, you must read the four choices marked A, B, C and D, and decide which is the best answer. Then mark the corresponding letter on **Answer Sheet** 2 with a single line through the centre.*

1. A. Yes. He will buy a house. B. No. He won't buy a house.

C. He will get a loan for buying a house.
D. He will have his house mortgaged.

2. A. 6:15. B. 6:40. C. 5:35. D. 5:15.

3. A. About twelve hours. B. About thirteen hours.
C. About one day. D. About one day and a half.

4. A. By boat. B. By airplane. C. By car. D. On foot.

5. A. Spanish. B. Arabic. C. Japanese. D. Chinese.

6. A. Forty-five pounds. B. Ninety pounds.
C. More than ninety pounds. D. Less than ninety pounds.

7. A. She's enjoying reading good books. B. She's going to earn more money.
C. She's writing a book. D. She's met some interesting persons.

8. A. It's not easy to recommend her a university. B. A large good university is good in every field.
C. Some universities have their own special fields. D. It's not good to go to only one university.

**Questions 9 to 12 are based on the conversation you have just heard.**

9. A. The minivan is three years old. B. The age of the minivan is around five.
C. The one she is dealing with is seven years old. D. What she is looking for is not three years old.

10. A. 55,000 miles. B. 65,000 miles. C. 75,000 miles. D. 45,000 miles.

11. A. It has several scratches in it. B. It will not open properly.
C. It is missing the door handle. D. The door is much too easy to be opened.

12. A. A faulty oil pump. B. A malfunctioning gage.
C. A worn out break drum. D. None of the above.

**Questions 13 to 15 are based on the conversation you have just heard.**

13. A. It's a horror film. B. It's a romance.
C. It's a science fiction. D. The film is about Swordsman.

14. A. She is getting a ride with her brother. B. Someone is coming to pick her up.
C. She is going by bus and will meet her date there. D. She will go there by subway.

15. A. The movie starts at 7:30 p. m. B. It is on show at 8:00 p. m.
C. The time for the film to start is 8:30 p. m. D. The starting time for the movie is 7:13 p. m.

## Section B

**Directions:** *In this section, you will hear 3 short passages. At the end of each passage, you will hear some questions. Both the passage and the questions will be spoken only once. After you hear a question, you must choose the best answer from the four choices marked A, B, C and D. Then mark the corresponding letter on **Answer Sheet** 2 with a single line through the centre.*

### Passage One

**Questions 16 to 18 are based on the passage you have just heard.**

16. A. The house where the President works and lives.
B. Mrs. Adams and her washing.
C. The President's Palace.
D. The President and his family.

17. A. President Washington lived in the White House.
B. All the Presidents of the U. S. have lived in the White House.
C. The East Room is no longer used for hanging up the wash.
D. President Lincoln lived in the White House.

18. A. President Adams had lived in the President's Palace before the building was finished.
B. British soldiers once set fire to the President's Palace.
C. The President's Palace was completely burned down by British soldiers.
D. The White House was formerly called the President's Palace.

### Passage Two

**Questions 19 to 21 are based on the passage you have just heard.**

19. A. To spray homes with insecticide every week.
    B. To tip leftovers and garbage at once.
    C. To keep the rooms clean and dry.
    D. To keep the lights brighter.
20. A. In all kinds of clothes.
    B. In anything made of wool or of hair.
    C. In cracks.
    D. Under floors.
21. A. Food or garbage. B. Fabric.
    C. Carpets. D. Wood.

### Passage Three

**Questions 22 to 25 are based on the passage you have just heard.**

22. A. A policeman. B. Guarding our home.
    C. Safety rules. D. The fuse.
23. A. When a fuse melts, we must examine the circuit carefully.
    B. Fuses can prevent the danger of fire.
    C. We can use electricity to toast bread.
    D. Fuse can stop cars.
24. A. Keeping lights shining for a long time. B. Making too many cars.
    C. Plugging too many lines into one outlet. D. Playing with fire.
25. A. Fuses guard our lives, so we can use electricity at will.
    B. Fuses work like traffic policeman.
    C. Proper-sized fuses should be used for circuits.
    D. We mustn't plug many lines into one outlet.

## Section C

**Directions:** *In this section, you will hear a passage three times. When the passage is read for the first time, you should listen carefully for its general idea. When the passage is read for the second time, you are required to fill in the blanks numbered from 26 to 33 with the exact words you have just heard. For blanks numbered from 34 to 36 you are required to fill in the missing information. For these blanks, you can either use the exact words you have just heard or write down the main points in your own words. Finally, when the passage is read for the third time, you should check what you have written.*

British people's (26)________ towards life is to (27)________ and enjoy themselves whenever they are free to do so. They (28)________ a lot on being (29)________ and having their (30)________ way during a (31)________ or party. I (32)________ it quite different from the Chinese way of (33)________ people.

At several parties I had been invited to in London, I observed that (34)________________________________________________. If the weather was nice, some guests might take their drinks or food to the garden, and some might just sit in the living room, chatting over the background music. (35)________________________________________________. It was the same for dinners. Guests were given things they asked for. If they said no the first time they were offered something, (36)________________________________________________.

# Unit 7

## Section A

**Directions:** *In this section, you will hear 8 short conversations and 2 long conversations. At the end of each conversation, one or more questions will be asked about what was said. Both the conversation and the questions will be spoken only once. After each question there will be a pause. During the pause, you must read the four choices marked A, B, C and D, and decide which is the best answer. Then mark the corresponding letter on* ***Answer Sheet*** *2 with a single line through the centre.*

1. A. Water the plants. B. Wash the car.
   C. Exercise in the sun. D. None of these.
2. A. He can't keep the children from making the noise.
   B. The children will make scratching noises with their pencils.
   C. He can keep the children very quiet.
   D. The children will draw pictures of mice.
3. A. He got angry with his boss. B. He made a mistake in his work.
   C. He was often late for work. D. He was frequently sick and absent.
4. A. $ 12.00. B. $ 7.50. C. $ 6.00. D. $ 9.00.
5. A. Pilots. B. Tim Johnson.
   C. How to become a pilot. D. How to get to know Tim Johnson.
6. A. Tuesday. B. Wednesday. C. Thursday. D. Friday.
7. A. He was speeding. B. He ran a red light.
   C. He drove in the wrong direction. D. He turned a corner too fast.
8. A. It is prettier. B. It is bigger.
   C. It has a prettier color. D. It has a bigger yard.

**Questions 9 to 12 are based on the conversation you have just heard.**

9. A. Fishing by the stream. B. Bird - watching in the park.
   C. Coming home from work. D. Going to work.
10. A. The thief was wearing a black striped dress. B. The thief was wearing a light red sweater.
    C. The thief was wearing a pair of tennis shoes. D. The thief was wearing a pair of black glasses.
11. A. He is about 170 cm tall. B. He is about 180 cm tall.
    C. He is about 190 cm tall. D. He is around 199 cm tall.
12. A. A man who dresses up like a woman. B. A woman who robs men in the park.
    C. A man who lives in the park. D. A woman who lives in the park.

**Questions 13 to 15 are based on the conversation you have just heard.**

13. A. They went to watch the flick at a friend's house.
    B. The place where they watch the flick is downtown.
    C. They enjoyed the flick at a local bar.
    D. They watched it in a charming countryside.
14. A. He thought it was outstanding. B. He thought it was ridiculous.
    C. He thought it was weird. D. He thought it was awful.
15. A. He had a basketball game the next morning.
    B. He wasn't feeling well because he drank too much.
    C. He needed plenty of rest for his test the following day.
    D. He usually slept during that period.

## Section B

**Directions:** *In this section, you will hear 3 short passages. At the end of each passage, you will hear some questions. Both the passage and the questions will be spoken only once. After you hear a question, you must choose the best answer from the four choices marked A, B, C and D. Then mark the corresponding letter on* ***Answer Sheet*** *2 with a single line through the centre.*

### Passage One

**Questions 16 to 18 are based on the passage you have just heard.**

16. A. Blood and white work clothes. B. Blood and bandage.
C. Blood and basin. D. Enthusiasm and purity.

17. A. Bleeding is really a good cure.
B. Barbers serve mostly men.
C. Signs sometimes outlast their meanings.
D. Barbers no longer cure the sick.

18. A. Barbers. B. Hair.
C. Bleeding. D. Beard.

### Passage Two

**Questions 19 to 21 are based on the passage you have just heard.**

19. A. Simple plants. B. Algae and sand.
C. Red dust. D. Colored snow.

20. A. Algae all live in air.
B. Algae have different colors.
C. Some algae live in water or on land.
D. All algae have neither roots nor stems.

21. A. Sunlight. B. Dust. C. Insects. D. Smoke.

### Passage Three

**Questions 22 to 25 are based on the passage you have just heard.**

22. A. Russia. B. The United States.
C. Alaska. D. Both A and B.

23. A. Because there were no natural resources there.
B. Because they didn't like to do business with Russians.
C. Because they knew too much about Alaska.
D. Because they knew little about Alaska.

24. A. Gold. B. Fish. C. Natural gas. D. Timber.

25. A. It was a good buy.
B. It was a folly.
C. It was just so-so.
D. It was neither too good nor too bad.

## Section C

**Directions:** *In this section, you will hear a passage three times. When the passage is read for the first time, you should listen carefully for its general idea. When the passage is read for the second time, you are required to fill in the blanks numbered from 26 to 33 with the exact words you have just heard. For blanks numbered from 34 to 36 you are required to fill in the missing information. For these blanks, you can either use the exact words you have just heard or write down the main points in your own words. Finally, when the passage is read for the third time, you should check what you have written.*

Any mistake made in the (26)________ of a stamp raises its value to stamp (27)________. A mistake on a (28)________ stamp has made it worth a million and a half times its face value. Do you think it (29)________? Well, it is true. And this is how it (30)

______.

The mistake was made more than a hundred years ago in the former British (31)______ of Mauritius, a small island in the India Ocean. In 1847, an (32)______ for stamps was sent to London. Mauritius was about to become the (33)______ country in the world to put out stamps.

Before the order was filled and the stamps arrived from England, a big dance was planned by the commander – in – chief of all the armed forces on the island. (34)________________________________________. Stamps were badly needed to post the letters.

Therefore, an islander, who was a good printer, was told to copy the pattern of the stamps. (35)________________________________________.

Today, there are only twenty – six of these misprinted stamps left—fourteen One – penny Reds and twelve Two – penny Blues. Because there are so few Two – penny Blues and because of their age, (36)________________________________________.

# Unit 8

## Section A

**Directions:** *In this section, you will hear 8 short conversations and 2 long conversations. At the end of each conversation, one or more questions will be asked about what was said. Both the conversation and the questions will be spoken only once. After each question there will be a pause. During the pause, you must read the four choices marked A, B, C and D, and decide which is the best answer. Then mark the corresponding letter on* ***Answer Sheet*** *2 with a single line through the centre.*

1. A. She can do the job. B. She should phone a friend.
C. She's just switched off the light. D. She's already replaced the shelf.
2. A. Gas station. B. Police station.
C. Lost and found department. D. Bar.
3. A. He grew up in Chicago. B. He was born in Chicago.
C. He was born in Boston. D. He grew up in Boston.
4. A. Lend James some money. B. Use some of James' money.
C. Help James save his money. D. Help James win some money.
5. A. The woman. B. The man and Fred.
C. The man and the woman. D. Fred's parents.
6. A. Worried. B. Indifferent. C. Unhappy. D. Sorry.
7. A. He wishes he were her. B. He thinks her going abroad will be good for him.
C. He thinks he is delighted to go with her. D. He thinks it's a good chance for her.
8. A. It's not a good place. B. It's as good as usual.
C. She doesn't want a dinner. D. She likes the cafeteria.

**Questions 9 to 12 are based on the conversation you have just heard.**

9. A. A digital camera. B. A TV. C. A stereo. D. A MP3 player.
10. A. People generally have a difficult time getting out of debt.
B. Students often apply for more credit cards than they need.
C. The interest rates on student cards are very high.
D. Students cannot apply for a credit card.
11. A. She hopes that someone will give her the money.
B. She plans on getting rid of her student credit cards.

C. She is going to return the items she purchased on the card.

D. She's going to borrow money from her parents.

12. A. To help her find a better paying job to cover her expenses.

B. To teach her how to prepare a financial management plan.

C. To show her how she can apply for low – interest student credit cards.

D. To persuade their parents to lend money to her.

**Questions 13 to 15 are based on the conversation you have just heard.**

13. A. He's 7 years old. B. He's 10 years old. C. He's 17 years old. D. He's 20 years old.

14. A. He's studying physics. B. He's studying biology.

C. He's studying psychology. D. He's studying chemistry.

15. A. She works as a sales representative. B. She is a computer programmer.

C. She is a receptionist in a hotel. D. She works as a typist in a company.

## Section B

**Directions:** *In this section, you will hear 3 short passages. At the end of each passage, you will hear some questions. Both the passage and the questions will be spoken only once. After you hear a question, you must choose the best answer from the four choices marked A, B, C and D. Then mark the corresponding letter on* ***Answer Sheet*** *2 with a single line through the centre.*

### Passage One

**Questions 16 to 18 are based on the passage you have just heard.**

16. A. The queen ant. B. Caring for baby ants. C. An ant nest. D. A nursery.

17. A. Ants eat sugar. B. The queen does not care for her babies.

C. The ants' nurses are their mothers. D. The queen takes care of her babies.

18. A. Ants hatch from eggs. B. Ants live in groups.

C. Ant babies sleep all the time. D. The queen ant lays eggs.

### Passage Two

**Questions 19 to 21 are based on the passage you have just heard.**

19. A. To make his house more beautiful.

B. Because the picture was worth more money.

C. To show to his son.

D. Because he liked the artist's painting.

20. A. He felt very sad. B. He felt very glad.

C. He felt very angry. D. He felt very disappointed.

21. A. The artist's painting was very good.

B. The artist's painting was not good.

C. The farmer's son would like the artist's painting very much.

D. The artist's painting would make the farmer's son more determined to be an artist.

### Passage Three

**Questions 22 to 25 are based on the passage you have just heard.**

22. A. Paper enables people to receive education more easily.

B. More and more paper is being used nowadays.

C. The invention of paper is of great significance to man.

D. Paper contributes a great deal to the human being.

23. A. More people could be educated than before.

B. More books could be printed than before.

C. More ways could be used to exchange knowledge.

D. More workers were needed than before.

24. A. Around 400.　　B. Around 1400.　　C. Around 1800.　　D. Around 1900.
25. A. More than fifty kilograms.　　B. About one kilogram.
C. About forty kilograms.　　D. About fifteen kilograms.

## Section C

**Directions:** *In this section, you will hear a passage three times. When the passage is read for the first time, you should listen carefully for its general idea. When the passage is read for the second time, you are required to fill in the blanks numbered from 26 to 33 with the exact words you have just heard. For blanks numbered from 34 to 36 you are required to fill in the missing information. For these blanks, you can either use the exact words you have just heard or write down the main points in your own words. Finally, when the passage is read for the third time, you should check what you have written.*

Visual aids offer several advantages. The (26)________ advantage is clarity. If you are discussing an object, you can make your (27)________ clearer by showing the object or some (28)________ of it. If you are citing statistics, showing how something works, or demonstrating a (29)________, a visual aid will make your (30)________ more vivid to your (31)______. After all, we live in a visual age. Television and movies have conditioned us to (32)______ a visual image. By using visual aids in our speeches, you often will make it easier for listeners to understand exactly what you are trying to (33)______.

Another advantage of visual aids is interest. (34)____________________________________________, not just speechmaking. Still another advantage of visual aids id retention. (35)____________________________________________. We've all heard that works can "go in one ear and out the other." Visual images tend to last.

(36)____________________________________________. Let us look first at the kinds of usual aids you are likely to use, then at guidelines for preparing visual aids, and finally at some tips for using visual aids effectively.

# Unit 9

## Section A

**Directions:** *In this section, you will hear 8 short conversations and 2 long conversations. At the end of each conversation, one or more questions will be asked about what was said. Both the conversation and the questions will be spoken only once. After each question there will be a pause. During the pause, you must read the four choices marked A, B, C and D, and decide which is the best answer. Then mark the corresponding letter on **Answer Sheet** 2 with a single line through the centre.*

1. A. The pear.　　B. The seafood.　　C. The weather.　　D. The fever.
2. A. Hans' neighbor.　　B. Hans' father.
C. Hans' father-in-law.　　D. Hans' brother.
3. A. She can use his car.　　B. She can borrow someone else's car.
C. She must get her car fixed.　　D. She can't borrow his car.
4. A. At 7:35.　　B. At 7:45.　　C. At 8:00.　　D. At 7:15.
5. A. To the bank.　　B. To a book store.
C. To a shoe store.　　D. To the grocer's.
6. A. Near her work place.　　B. In the countryside.
C. Near the station.　　D. In the city.
7. A. At a bus station.　　B. At Uncle Tom's.
C. At a gas station.　　D. At a cigarette store.
8. A. Policeman and driver.　　B. Policeman and thief.

C. Teacher and pupil. D. Director and actress.

**Questions 9 to 12 are based on the conversation you have just heard.**

9. A. He wants to play outside. B. He wants to watch a football match.
   C. He plans to play video games. D. He wants to buy something for drink.
10. A. He is required to clean the bathroom. B. He has to vacuum the floors.
    C. He is assigned to wash the walls. D. He must dust everything.
11. A. He has to put away his books. B. He has to make his bed.
    C. He must pick up his dirty clothes. D. He must tide his books.
12. A. Wash the car. B. Paint the house.
    C. Wash clothes. D. Go to office.

**Questions 13 to 15 are based on the conversation you have just heard.**

13. A. She's in a meeting. B. She's out of the office.
    C. She's talking with another customer. D. She's in the home.
14. A. Information on after - sales service. B. A picture of the newest computers.
    C. A list of software products. D. A list of computer names.
15. A. Cordell. B. Kordel. C. Kordell. D. Cordel.

## Section B

**Directions:** *In this section, you will hear 3 short passages. At the end of each passage, you will hear some questions. Both the passage and the questions will be spoken only once. After you hear a question, you must choose the best answer from the four choices marked A, B, C and D. Then mark the corresponding letter on* ***Answer Sheet*** *2 with a single line through the centre.*

### Passage One

**Questions 16 to 19 are based on the passage you have just heard.**

16. A. Demanding. B. Genuine. C. Indifferent. D. Friendly.
17. A. The farmer. B. The farmer's wife.
    C. A pastor. D. A mule.
18. A. The farmer's wife. B. The pastor.
    C. The female folks. D. The male folks.
19. A. Because the pastor took the mule away.
    B. Because he was unwilling to lend his mule to them.
    C. Because someone else had already borrowed the mule.
    D. Because the mule died in a tragedy.

### Passage Two

**Questions 20 to 22 are based on the passage you have just heard.**

20. A. Psychology and Sociology.
    B. Human Behavior Study.
    C. Main Concerns of Social Sciences.
    D. Individual and Group.
21. A. The role of religion or art in a society.
    B. The main reason for revolution in a society.
    C. The reasons why society progress more rapidly than another.
    D. The causes of antisocial behavior.
22. A. Psychology has no influence on Sociology.
    B. Psychology determines the subjects studied in Sociology.
    C. Sociology determines the subjects studied in Psychology.
    D. They work together in their study of human behavior.

**Passage Three**

**Questions 23 to 25 are based on the passage you have just heard.**

23. A. 1760. B. 1776. C. 1716. D. 1850.
24. A. No more than 10. B. Over 10. C. About 10. D. More than 20.
25. A. A light. B. Living quarters. C. A dog. D. A keeper.

## Section C

**Directions:** *In this section, you will hear a passage three times. When the passage is read for the first time, you should listen carefully for its general idea. When the passage is read for the second time, you are required to fill in the blanks numbered from 26 to 33 with the exact words you have just heard. For blanks numbered from 34 to 36 you are required to fill in the missing information. For these blanks, you can either use the exact words you have just heard or write down the main points in your own words. Finally, when the passage is read for the third time, you should check what you have written.*

Every person writes a different handwriting. A (26) ________ can determine a person's (27) ________ from his or her handwriting. (28) ________ handwriting is often the sign of a person who really doesn't want to (29) ________ with others. Such a person may have a hard time getting along with people. Sometimes, though, handwriting is not easy to read because the person writes too fast. To a graphologist, this means that the writer has a lot of (30) ________.

People who are feeling happy are likely to write (31) ________. With generally happy people, the dots of their i's are likely to be the right of the letter (32) ________. If one's t-bars are wavy and i-dots are more like (33) ________ than like dots, or the endings of words are up-curved, this person often has a sense of humor. (34) ________________________________________________. (35) ________________________________________________. (36) ________________________________________________.

# Unit 10

## Section A

**Directions:** *In this section, you will hear 8 short conversations and 2 long conversations. At the end of each conversation, one or more questions will be asked about what was said. Both the conversation and the questions will be spoken only once. After each question there will be a pause. During the pause, you must read the four choices marked A, B, C and D, and decide which is the best answer. Then mark the corresponding letter on* ***Answer Sheet*** *2 with a single line through the centre.*

1. A. At a lawyer's office. B. At a library.
   C. At a post office. D. At an airport.
2. A. She agreed. B. She disagreed.
   C. She was impatient. D. She was worried.
3. A. Fiji. B. Switzerland. C. Singapore. D. Italy.
4. A. Father and daughter. B. Teacher and student.
   C. Patient and doctor. D. Athlete and coach.
5. A. They are wife and husband. B. They are waitress and customer.
   C. They are friends. D. They are a young couple.
6. A. Two. B. Three. C. Four. D. Five.
7. A. The dean was very warm-hearted. B. The man had troubled the dean a lot.

C. The woman went to bed at twelve o'clock. D. The woman was worried about the dean.

8. A. In a garden. B. In the woods.
C. At a florist's shop. D. At a post office.

**Questions 9 to 12 are based on the conversation you have just heard.**

9. A. She can use it to check her email.
B. She can call family in case of an emergency.
C. She can make cheaper long - distance calls with it.
D. Her old cellphone is stolen.

10. A. 3 months. B. 6 months. C. 9 months. D. 4 months.

11. A. Their current car is in bad shape, and it doesn't look good.
B. The cellphone's power supply won't work in an older car.
C. Their car isn't big enough to ride in with all her friends.
D. Their car is broken in an accident.

12. A. They talk about where to buy a new car. B. The girl will take a ride with her friends.
C. They read newspapers together. D. The man gives the girl a lesson.

**Questions 13 to 15 are based on the conversation you have just heard.**

13. A. Play basketball with friends from work. B. Try out for the company baseball team.
C. Get in shape and compete in a cycling race. D. Cycle with his colleagues in the company.

14. A. She is worried her husband will spend too much time away from home.
B. She is afraid her husband will become a fitness freak.
C. She is concerned about her husband's health.
D. She cares about her husband's performance.

15. A. It is good for improving muscle tone. B. It helps strengthen the heart.
C. It helps develop mental toughness. D. It will be helpful for improving the figure.

## Section B

**Directions:** *In this section, you will hear 3 short passages. At the end of each passage, you will hear some questions. Both the passage and the questions will be spoken only once. After you hear a question, you must choose the best answer from the four choices marked A, B, C and D. Then mark the corresponding letter on* ***Answer Sheet*** *2 with a single line through the centre.*

### Passage One

**Questions 16 to 18 are based on the passage you have just heard.**

16. A. The first President of the USA.
B. The capital city of the USA.
C. A state of the USA, in the extreme north-west of the continent.
D. A state of the northwest USA.

17. A. More than 200 years old. B. More than 100 years old.
C. More than 110 years old. D. More than 147 years old.

18. A. The fix of the border. B. Lumbering and fishing.
C. The discovery of gold. D. The arrival of the railways.

### Passage Two

**Questions 19 to 21 are based on the passage you have just heard.**

19. A. The US law is derived from English law.
B. The US law is based on common law, statute law and the Constitution.
C. The USA has two separate sets of courts, state and federal.
D. The US law is based on English law.

20. A. A federal district with judge.

B. The Courts of Appeal.

C. The F. B. I.

D. The state systems in local, district and country courts.

21. A. Common law, statute law and the Constitution of the US.

B. The system of the US law.

C. The US two separate sets of courts, state and federal.

D. The US federal legal system.

**Passage Three**

**Questions 22 to 25 are based on the passage you have just heard.**

22. A. Both radio and TV are privately controlled in the US.

B. Both radio and TV are run by commercial companies.

C. America has only three TV networks: ABC, CBS and NBC.

D. There are a lot of local TV stations in the US.

23. A. 400 minutes. B. 260 minutes. C. 240 minutes. D. 120 minutes.

24. A. Americans prefer radios to TV sets.

B. Many Americans listen to the radio while driving their motor cars.

C. Americans prefer TV sets to radios.

D. Americans like motor cars only.

25. A. The American TV and Radio Programs. B. The American TV networks.

C. The American Newspapers. D. The Broadcasting of the USA.

## Section C

**Directions:** *In this section, you will hear a passage three times. When the passage is read for the first time, you should listen carefully for its general idea. When the passage is read for the second time, you are required to fill in the blanks numbered from 26 to 33 with the exact words you have just heard. For blanks numbered from 34 to 36 you are required to fill in the missing information. For these blanks, you can either use the exact words you have just heard or write down the main points in your own words. Finally, when the passage is read for the third time, you should check what you have written.*

In a (26)________ economy, the consumer usually has the choice of several different (27) ________ of the same product. Yet underneath their labels, the products are often nearly (28) ________. One manufacturer's toothpaste (29) ________ to differ very little from another manufacturer's. Thus, manufacturers are (30) ________ with a problem—how to keep sales high enough to stay in business. Manufacturers solve this problem by advertising. They try to appeal to consumers in various ways. In fact, advertisements may be (31) ________ into three types according to the kind of appeals they use.

One type of advertisement tries to (32) ________to the consumer's reasoning mind. It may offer a claim that seems scientific. For example it may say the dentists (33) ________ flash toothpaste. (34) ____________________________________________. A scientific approach gives the appearance of truth.

Another type of advertisement tries to amuse the potential buyer. (35) ____________________________________________.

One way of doing this is to make the products appear alive. For example, the advertisers may personify cans of insecticide, and show them attacking mean-faced bugs. Ads of this sort are silly, but they also tend to be amusing. (36) ____________________________________________.

# Unit 11

## Section A

**Directions:** *In this section, you will hear 8 short conversations and 2 long conversations. At the end of each conversation, one or more questions will be asked about what was said. Both the conversation and the questions will be spoken only once. After each question there will be a pause. During the pause, you must read the four choices marked A, B, C and D, and decide which is the best answer. Then mark the corresponding letter on* ***Answer Sheet*** *2 with a single line through the centre.*

1. A. Mostly English. B. Only sometimes the foreign language.
   C. Mostly the students' language. D. Each language about half the time.
2. A. He must have some money in a bank. B. He must use the utilities.
   C. He must pay the security deposit. D. He must clean his own room.
3. A. A tailor. B. A cook.
   C. A dentist. D. A singer.
4. A. 45 minutes. B. 50 minutes.
   C. 55 minutes. D. 5 minutes.
5. A. In a grocery store. B. Over the telephone.
   C. At a party. D. In Mary's house.
6. A. 2:00 p.m. B. 8:00 p.m.
   C. 6:00 p.m. D. 4:00 p.m.
7. A. Drive the woman home. B. Work together with the woman.
   C. Do the work for the woman. D. Leave the work to others.
8. A. He ran a red light. B. He was speeding.
   C. He went through a stop sign. D. He turned a corner too fast.

**Questions 9 to 12 are based on the conversation you have just heard.**

9. A. He was returning home from a party. B. He just got off work when he saw the UFO.
   C. He was driving home from a restaurant. D. He just took some medicine.
10. A. It was a giant deer. B. A strange man appeared.
    C. A hairy alien jumped out. D. It was an angry alien.
11. A. He walked to a flying saucer. B. He followed the animal to a plane.
    C. He was carried to a spaceship. D. He flied to a flying saucer.
12. A. The man should call the fire department.
    B. The man should seek counseling.
    C. The man should contact the newspaper.
    D. The man should not drink beer when he needs to drive a car.

**Questions 13 to 15 are based on the conversation you have just heard.**

13. A. The presence of life-forms far below the Earth's surface.
    B. The risk of infection from rare strains of bacteria.
    C. Fictional representations of a hidden underground world.
    D. The reliability of evidence collected by new drilling methods.
14. A. Its texture. B. Its size. C. Its preservation. D. Its shape.
15. A. The bacteria would be killed by the human immune system.
    B. The bacteria would die if brought to the surface.
    C. Many antidotes and remedies are available.
    D. Drilling operations are always closely monitored.

## Section B

**Directions:** *In this section, you will hear 3 short passages. At the end of each passage, you will hear some questions. Both the passages and the questions will be spoken only once. After you hear a question, you must choose the answer from the four choices marked A, B, C and D. Then mark the corresponding letter on* ***Answer Sheet*** *2 with a single line through the centre.*

### Passage One

**Questions 16 to 18 are based on the passage you have just heard.**

16. A. The age between one and three.
    B. The age between four and six.
    C. The age between six and eight.
    D. The age between eight and ten.
17. A. Because they are curious.
    B. Because they are not taken good care of.
    C. Because they want to help.
    D. Because they are unsupervised.
18. A. Poison.
    B. For External Use Only.
    C. Keep Out of Reach of Children.
    D. Harmful if Swallowed.

### Passage Two

**Questions 19 to 21 are based on the passage you have just heard.**

19. A. On the morning of April 18th, 1906.
    B. On the evening of October 17th, 1989.
    C. On the evening of April 18th, 1906.
    D. On the morning of October 17th, 1989.

30. A. 250,000. B. 5,000. C. 700. D. 100.

21. A. Because the Earth plates are in movement.
    B. Because the Pacific plate is moving slowly.
    C. Because the Pacific plate meets the North American plate.
    D. Because the Pacific plate jumps high.

### Passage Three

**Questions 22 to 25 are based on the passage you have just heard.**

22. A. Because they lost their houses in a flood.
    B. Because they planned to plant corn on the hills.
    C. Because they didn't have enough room to live in.
    D. Because they wanted to avoid a war.
23. A. Because corn didn't need as much as water as rice.
    B. Because corn was more delicious than rice.
    C. Because corn was more expensive than rice.
    D. Because corn was more nutritious than rice.
24. A. People in the west boil corn, and eat it with salt and butter.
    B. People in the west toast corn and eat it with sugar.
    C. People in the west fry corn and eat it with water.
    D. People in the west eat corn without cooking.
25. A. Potato. B. Tomato. C. Corn. D. Bean.

## Section C

**Directions:** *In this section, you will hear a passage three times. When the passage is read for the first time, you should listen carefully for its general idea. When the passage is read for the second time, you are required to fill in the blanks numbered from 26 to 33 with the exact words you have just heard. For blanks numbered from 34 to 36 you are required to fill in the missing information. For these blanks, you can either use the exact words you have just heard or write down the main points in your own words. Finally, when the passage is read for the third time, you should check what you have written.*

Editing and (26)________ benefit richly from (27)________. Instead of crossing out mistakes, or (28)________ an entire paper to correct (29)________ errors, you can make all (30)________ changes within the most recent draft. If you find (31)________ or proofreading on the screen hard on your eyes, print out a copy. Mark any (32)________ on that copy, and then transfer them to the final draft.

If the word-processing (33)________ you're using includes spelling and grammar checks, by all means use them. (34)________________________________________________. Keep in mind, however, that the spell-check can't tell you how to spell a name correctly or when you have mistakenly used. (35)________________________________________________. Any errors it doesn't uncover are still your responsibility.

A word-processed paper, with its clean appearance and handsome formatting, looks so good that you may feel it is in better shape than it really is. Do not be fooled. (36)________________________________________________.

# Unit 12

## Section A

**Directions:** *In this section, you will hear 8 short conversations and 2 long conversations. At the end of each conversation, one or more questions will be asked about what was said. Both the conversation and the questions will be spoken only once. After each question there will be a pause. During the pause, you must read the four choices marked A, B, C and D, and decide which is the best answer. Then mark the corresponding letter on **Answer Sheet** 2 with a single line through the centre.*

1. A. A movie makes her sad. B. A movie makes her laugh.
   C. A movie makes her cry. D. A movie makes her feel frightened.
2. A. He has no time to visit the outside London. B. He can see only a few places.
   C. He visits a lot of places. D. He forgets what places he had seen.
3. A. In a car. B. In an office.
   C. In a restaurant. D. In a supermarket.
4. A. To solve his problem. B. To pay attention to her house.
   C. To go to the seaside with her. D. To have his eye checked.
5. A. She doesn't know the statistics about the new machine.
   B. What's the value of the paper?
   C. She knows how to operate the machine without reading the instructions.
   D. She doesn't understand the instructions, either.
6. A. In a plane. B. In a car. C. At a shop. D. On a farm.
7. A. At a bus station. B. At a weather station.
   C. At an airport. D. On the campus.
8. A. The Best of Jazz. B. Christmas Carols.

C. Rock Music Collection. D. Classical Favorites.

**Questions 9 to 12 are based on the conversation you have just heard.**

9. A. He doesn't think it is very special. B. Someone else bought it before him.
 C. He has no need for a ring. D. The ring is too small.
10. A. The buttons are scratched. B. The CD casing is chipped.
 C. The handle is damaged. D. The display is loose.
11. A. Because it is stained. B. Because he already has one.
 C. Because its too expensive. D. Because the seams are coming undone.
12. A. Only records. B. Only a vase.
 C. Some records and a vase. D. Nothing.

**Questions 13 to 15 are based on the conversation you have just heard.**

13. A. Problems with living in an apartment. B. A search for a new apartment.
 C. The cost of rent near universities. D. The distance to the university.
14. A. Somewhere that is within a short walking distance of campus.
 B. An apartment without furniture in it.
 C. A place where she can live alone.
 D. A place with a good sight.
15. A. He is planning on calling a friend who owns an apartment building.
 B. He will check the newspapers to see if he can find an apartment for rent.
 C. He is going to visit an apartment building near his place.
 D. He will ask his sister to share the house with her.

## Section B

**Directions:** *In this section, you will hear 3 short passages. At the end of each passage, you will hear some questions. Both the passage and the questions will be spoken only once. After you hear a question, you must choose the best answer from the four choices marked A, B, C and D. Then mark the corresponding letter on **Answer Sheet** 2 with a single line through the centre.*

### Passage One

**Questions 16 to 18 are based on the passage you have just heard.**

16. A. Three. B. Four. C. Five. D. More than four.
17. A. He should show a correct signal. B. He should let other cars pass.
 C. He should keep a proper distance. D. He should guess other drivers' direction.
18. A. To show drivers around in the country.
 B. To teach drivers how to avoid a car accident.
 C. To require drivers to learn some safety regulations
 D. To warn drivers of the dangers on the road.

### Passage Two

**Questions 19 to 21 are based on the passage you have just heard.**

19. A. It was stolen. B. It was robbed. C. It was hurt. D. It was eaten.
20. A. Two. B. Four. C. Three. D. Five.
21. A. From what the neighbor had told him. B. From what he had seen and heard.
 C. From what he was shown. D. From what kind of horse it was.

### Passage Three

**Questions 22 to 25 are based on the passage you have just heard.**

22. A. Computers have become part of our daily lives.
 B. Computers have disadvantages as well as advantages.
 C. People have different attitudes to computers.

D. More and more families will own computers.

23. A. Computers can bring financial problems.

B. Computers can bring unemployment.

C. Computers can be very useful in families.

D. Computerised robots can take over some unpleasant jobs.

24. A. Computers may change the life they have been accustomed to.

B. Spending too much time on computers may spoil people's relationship.

C. Buying computers may cost a lot of money.

D. Computers may take over from human beings altogether.

25. A. Affectionate. B. Disapproving. C. Approving. D. Neutral.

## Section C

**Directions:** *In this section, you will hear a passage three times. When the passage is read for the first time, you should listen carefully for its general idea. When the passage is read for the second time, you are required to fill in the blanks numbered from 26 to 33 with the exact words you have just heard. For blanks numbered from 34 to 36 you are required to fill in the missing information. For these blanks, you can either use the exact words you have just heard or write down the main points in your own words. Finally, when the passage is read for the third time, you should check what you have written.*

Let us (26) __________ that you are in the position of a parent. Would you allow your children to read any book they wanted without first checking its (27) __________? Would you take your children to see any film without first finding out whether it is (28) __________ for them? If your answer to these questions is "yes", then you are just plain (29) __________. If your answer is "no", then you are exercising your right as a parent to (30) __________ your children from what you consider to be (31) __________ influences. In other words, by acting as an (32) __________ yourself, you are (33) __________ that there is a strong case for censorship.

Now, of course, you will say that it is one thing to exercise censorship where children are concerned and quite another to do the same for adults. Children need protection and it is the parents' responsibility to provide it. But what about adults? Aren't they old enough to decide what is good for them? (34) ______________________________________________________________. Censorship is for the good of society as a whole. Like the law, it contributes to the common good.

Some people think that it is a shame that a censor should interfere with works of art. (35) ______________________________________________________________.

When censorship laws are relaxed, dishonest people are given a chance to produce virtually anything in the name of "art". One of the great things that censorship does is (36) ______________________________________________________________. To argue in favour of absolute freedom is to argue in favour of anarchy. Society would really be the better if it were protected by correct censorship.

# Unit 13

## Section A

**Directions:** *In this section, you will hear 8 short conversations and 2 long conversations. At the end of each conversation, one or more questions will be asked about what was said. Both the conversation and the questions will be spoken only once. After each question there will be a pause. During the pause, you must read the four choices marked A, B, C and D, and decide which is the best answer. Then mark the corresponding letter on* ***Answer Sheet*** *2 with a single line through the centre.*

1. A. At 6:30. B. At 7:00. C. At 7:30. D. At 8:00.

2. A. Mary didn't eat those sweets because of her toothache.
   B. Mary had a fever last night.
   C. Mary had a toothache because she ate those sweets.
   D. Mary could not speak because she had sweets in her mouth.
3. A. To her Aunt Mary. B. To her Aunt Agate. C. To her parents. D. To her uncles.
4. A. Bill doubted the value of the car. B. Bill thought the car had a good value.
   C. The price of the car was too high. D. The car was valuable.
5. A. She will fail. B. She will get zero.
   C. She will get a high score. D. She will pass away.
6. A. In a mill. B. In a bank. C. In an airport. D. In a post office.
7. A. By plane. B. By bus. C. By car. D. By train.
8. A. Four. B. Five. C. Six. D. Seven.

**Questions 9 to 12 are based on the conversation you have just heard.**

9. A. A vacation trip to Yellowstone Park. B. A lecture by a visiting professor.
   C. Her biology thesis. D. A research project.
10. A. More buffalo are surviving the winter. B. Fewer buffalo are dying of disease.
    C. More buffalo are being born. D. Fewer buffalo are being killed by hunters.
11. A. She is from Wyoming. B. She needs the money.
    C. She has been studying animal diseases. D. Her thesis adviser is heading the project.
12. A. Collecting information about the bacteria. B. Working on a cattle ranch.
    C. Writing a paper about extinct animals. D. Analyzing buffalo behavior.

**Questions 13 to 15 are based on the conversation you have just heard.**

13. A. It shrank. B. The colour faded.
    C. The fabric is coming apart. D. It is stained.
14. A. The customer didn't follow the instructions for using the item.
    B. The item was on clearance.
    C. The man no longer has the store receipt.
    D. The receipt is out of date.
15. A. The store clerk eventually gives the customer a refund.
    B. The customer is able to exchange the item.
    C. The customer leaves the store without the item.
    D. The customer got part of the refund.

## Section B

**Directions:** *In this section, you will hear 3 short passages. At the end of each passage, you will hear some questions. Both the passage and the questions will be spoken only once. After you hear a question, you must choose the best answer from the four choices marked A, B, C and D. Then mark the corresponding letter on **Answer Sheet** 2 with a single line through the centre.*

### Passage One

**Questions 16 to 18 are based on the passage you have just heard.**

16. A. The car was pulled away by a police. B. The car was parked in an unusual place.
    C. The car was missing. D. The car was driven away by one of his neighbors.
17. A. He found that his car was damaged. B. He found two theater tickets in the car.
    C. He found that the seats in the car were stolen. D. He found a letter on the car's window.
18. A. Because he wanted to thank them for their lending him the car.
    B. Because he knew that they liked drama.
    C. Because he could steal when they were out.

D. Because he wanted to play a joke with them.

**Passage Two**

**Questions 19 to 21 are based on the passage you have just heard.**

19. A. In the nineteen century. B. In the early twentieth century.
C. In the eighteenth century. D. In the middle of twentieth century.

20. A. It didn't work because of quality problem.
B. It was broken by the violent waves in the English Channel.
C. A fisherman mistook the cable for a long fish and caught it.
D. A fisherman mistook the cooper wire in the cable for gold and cut it.

21. A. 2. B. 3. C. 4. D. 5.

**Passage Three**

**Questions 22 to 25 are based on the passage you have just heard.**

22. A. The earth is a magnet with two magnetic poles.
B. Scientists understand completely about magnetism.
C. The Chinese knew about magnets in the eleventh century.
D. The earth's magnet has little impact on its beings.

23. A. Science has proved that the center of the earth is an enormous magnet.
B. The center of the earth gives out a very weak magnetic force.
C. The center of the earth has the strongest magnetic force than elsewhere.
D. The magnetic force at the center of the earth is unstable.

24. A. Doves will not be able to find their way back. B. People will commit suicide.
C. People will become mad and abnormal. D. Rats will reproduce very rapidly.

25. A. The magnetic force is controlling more and more people.
B. The magnetic force is disappearing little by little.
C. The magnetic force is becoming weaker and weaker.
D. The magnetic force is hard to predict.

## Section C

**Directions:** *In this section, you will hear a passage three times. When the passage is read for the first time, you should listen carefully for its general idea. When the passage is read for the second time, you are required to fill in the blanks numbered from 26 to 33 with the exact words you have just heard. For blanks numbered from 34 to 36 you are required to fill in the missing information. For these blanks, you can either use the exact words you have just heard or write down the main points in your own words. Finally, when the passage is read for the third time, you should check what you have written.*

Most people think that the older you get, the (26) ________ it is to learn a new language. That is, they (27) ________ that children learn more easily than (28) ________. Thus, at some (29) ________ in our lives, maybe around age twelve or thirteen, we lose the (30) ________ to learn language well. Is it true that children learn a foreign language more (31) ________ than adults? One report showed that the (32) ________ learned more, in less time, than the younger children. Another report showed that the ability to learn a language increases as the age increases, from (33) ________ to adulthood. (34) ______________________________ ______________________________. For one thing, adults know more about the world and therefore are able to understand meanings more easily than children. (35) ______________________________. Finally, adults are more self-controlled than children. (36) ______________________________.

# Unit 14

## Section A

**Directions:** *In this section, you will hear 8 short conversations and 2 long conversations. At the end of each conversation, one or more questions will be asked about what was said. Both the conversation and the questions will be spoken only once. After each question there will be a pause. During the pause, you must read the four choices marked A, B, C and D, and decide which is the best answer. Then mark the corresponding letter on* ***Answer Sheet*** *2 with a single line through the centre.*

1. A. On the 16th. B. On the 17th.
   C. On the 18th. D. Next week.
2. A. She has lost her coat. B. She feels very cold.
   C. She is going to have another coat. D. She is catching another cold.
3. A. This morning. B. Very soon.
   C. Maybe later in the day. D. Perhaps tomorrow morning.
4. A. The test consisted of one page. B. The exam was difficult for the woman.
   C. The woman found the exam easy. D. The woman completed the exam in one hour.
5. A. A bus station. B. A superhighway.
   C. An airport. D. A train station.
6. A. That Sally is serious about Bob. B. That Bob is serious about Sally.
   C. That Sally is not serious about Bob. D. That Bob is not serious about Sally.
7. A. 8. B. 12. C. 24. D. 36.
8. A. Half an hour. B. Longer than half an hour.
   C. Longer than an hour. D. It depends on different people.

**Questions 9 to 12 are based on the conversation you have just heard.**

9. A. They first met at a party. B. They first met at school.
   C. They got to know each other at a friend's house. D. They knew each other at a club.
10. A. Sharon. B. Susan. C. Sherry. D. Chalon.
11. A. The woman is majoring in engineering. B. Computer science is the woman's major.
    C. The woman is majoring in education. D. She is majoring in English.
12. A. At this time he is considering the international business.
    B. What he thinks over a lot is accounting.
    C. Marketing is what the man cares.
    D. Education is what he is thinking about most now.

**Questions 13 to 15 are based on the conversation you have just heard.**

13. A. Business. B. Pleasure.
    C. Business and pleasure. D. None of the above.
14. A. Clothing, a computer, and books. B. CD player, clothing, and books.
    C. Books, gifts and a computer. D. Books, clothing and gifts.
15. A. Her parents are on the same trip.
    B. She enjoys traveling to different countries.
    C. She was born in that country.
    D. She plans to pay a visit to one of her old friends during the trip.

## Section B

**Directions:** *In this section, you will hear 3 short passages. At the end of each passage, you will heat some questions. Both the passage and the questions will be spoken only once. After you hear a question, you must choose the best answer from the four choices marked A, B, C and D. Then mark the corresponding letter on* ***Answer Sheet*** *2 with a single line through the centre.*

### Passage One

**Questions 16 to 18 are based on the passage you have just heard.**

16. A. It's an international, non-commercial exhibition.
    B. It's another name for Olympic Games.
    C. It's a world economy organization.
    D. It's a global organization for architectures.
17. A. To entertain the Queen's guests.
    B. To show the country's power and pride.
    C. To celebrate the Queen's birthday.
    D. To provide rooms for foreign visitors.
18. A. China. B. America. C. Belgium. D. Russia.

### Passage Two

**Questions 19 to 22 are based on the passage you have just heard.**

19. A. How well the listeners understand the speeches.
    B. What effects the speeches had on listeners' attitude to speakers.
    C. How well the speeches were organized.
    D. What parts of a speech listeners paid most attention to.
20. A. How to write a wonderful speech.
    B. How to use their voices appropriately.
    C. How to organize their ideas effectively.
    D. How to attract their audiences' attention.
21. A. Introduction, body, and conclusion.
    B. The speaker, the words and the audience.
    C. The speaker's voice, his dress and his feeling.
    D. The time, the place and the people.
22. A. The findings of two experiments.
    B. Students' requirements in a speech class.
    C. The basic parts of a speech.
    D. The importance of organization in speechmaking.

### Passage Three

**Questions 23 to 25 are based on the passage you have just heard.**

23. A. Associate. B. Bachelor's. C. Master's. D. Doctor's.
24. A. A technical associate degree.
    B. A degree designed for transfer.
    C. The last degree one can get.
    D. Part of a bachelor's degree.
25. A. Associate. B. Bachelor's. C. Master's. D. Doctor's.

**Directions:** *In this section, you will hear a passage three times. When the passage is read for the first time, you should listen carefully for its general idea. When the passage is read for the second time, you are required to fill in the blanks numbered from 26 to 33 with the exact words*

*you have just heard. For blanks numbered from 34 to 36 you are required to fill in the missing information. For these blanks, you can either use the exact words you have just heard or write down the main points in your own words. Finally, when the passage is read for the third time, you should check what you have written.*

The new Victoria Line was (26)________ in 1969. This new line is very (27)________ from the others. The stations on the other lines need a lot of workers to sell (28)________, and to check and collect them when people (29)________ the trains.

This is all different on the Victoria Line. Here a machine checks and (30)________ the tickets, and there are no (31)________ on the (32)________. On the train, there is only one worker. If (33)________, the man can drive the train. (34)________________________________________________. The trains are controlled by electrical signals which are sent by the so - called "command spots".

(35)________________________________________________. Each sends a certain signal. (36)________________________________________________. If the command spots send no signals, the train will stop.

# Unit 15

## Section A

**Directions:** *In this section, you will hear 8 short conversations and 2 long conversations. At the end of each conversation, one or more questions will be asked about what was said. Both the conversation and the questions will be spoken only once. After each question there will be a pause. During the pause, you must read the four choices marked A, B, C and D, and decide which is the best answer. Then mark the corresponding letter on* ***Answer Sheet*** *2 with a single line through the centre.*

1. A. It's not as good as it was.　　B. It's better than it used to be.
   C. It's better than people say.　　D. It's worse than people say.
2. A. 8:15.　　B. 8:30.　　C. 9:15.　　D. 9:30.
3. A. Lawyer—Client.　　B. Teacher—Student.
   C. Customer—Waitress.　　D. Boss—Secretary.
4. A. He wants to make a long-distance call.
   B. He wants to save money.
   C. He wants to talk to the operator.
   D. He wants to know the best time to make a long-distance call.
5. A. The lady is wearing a new hat.　　B. The lady is wearing a hair piece.
   C. The lady has a new hair style.　　D. The lady's hair is combed nicely.
6. A. She wants to buy a new skirt.
   B. She wants to return the skirt her husband bought.
   C. She wants to change the smaller skirt for a bigger one.
   D. She wants to change the bigger skirt for a smaller one.
7. A. His wife.　　B. A saleslady.
   C. A store detective.　　D. A customs official.
8. A. No one knows how Mary gets it to work.
   B. She doesn't like the record player.
   C. She threw the old records away.
   D. It's surprising that Mary can get it work.

**Questions 9 to 12 are based on the conversation you have just heard.**

9. A. He runs. B. He plays tennis. C. He does aerobics. D. He plays football.

10. A. He swims before it. B. He does push – ups.
C. He stretches his muscle. D. He runs for a period.

11. A. In order to strengthen his muscles. B. With a view to improving his endurance.
C. To increase his flexibility. D. Just to lose his weight.

12. A. It helps him get rid of his worries from the week.
B. Hiking allows him to burn off weight from overeating.
C. Walking with his dog provides him opportunities to enjoy nature.
D. Meeting someone he wants to meet during the hiking.

**Questions 13 to 15 are based on the conversation you have just heard.**

13. A. In Scotland. B. In France.
C. In England. D. In the United States.

14. A. He is a travel agent. B. He is a professor.
C. He is a computer programmer. D. He is a French teacher.

15. A. Why her parents lived in England for several years.
B. Where her mother works at the present time.
C. Where the girl grew up.
D. How old the girl was when she moved to her present location.

## Section B

**Directions:** *In this section, you will hear 3 short passages. At the end of each passage, you will hear some questions. Both the passage and the questions will be spoken only once. After you hear a question, you must choose the best answer from the four choices marked A, B, C and D. Then mark the corresponding letter on **Answer Sheet** 2 with a single line through the centre.*

### Passage One

**Questions 16 to 18 are based on the passage you have just heard.**

16. A. A person is doing a job which he likes very much.
B. A person is doing a job which he doesn' t like.
C. A person is doing a job which he is not suited for.
D. A person is doing a job which he thinks very important.

17. A. Because it will make a person earn a lot of money.
B. Because many people in the world don' t want to be square pegs.
C. Because good jobs make them happy.
D. Because it will make full use of their talents.

18. A. Businessmen, managers and accountants.
B. Chemists, physicists and biologists.
C. Governors, doctors and teachers.
D. Engineers, public servants and news reporters.

### Passage Two

**Questions 19 to 22 are based on the passage you have just heard.**

19. A. Television' s bad effects on the young.
B. Television' s bad effects on society in general.
C. The history of television.
D. The good sides of television.

20. A. Children do not know whether TV shows are true or not.
B. Children can tell that some programs are real and some are not.
C. Children know clearly that TV shows present an unreal world.

D. Children don't care whether or not TV shows are real.

21. A. About 625 days. B. About 700 days.
C. About 750 days. D. About 500 days.

22. A. Production of TV sets will be stopped due to its bad effects.
B. The number of TV sets will remain the same in the future.
C. Television will still be in use for all its shortcomings.
D. Television must stay where it is now.

**Passage Three**

**Questions 23 to 25 are based on the passage you have just heard.**

23. A. Eating out all the time. B. Sitting with his friends.
C. Playing games with words. D. Taking bread with meat.

24. A. Lord Sandwich. B. Scientists.
C. The friends of Lord Sandwich. D. Inventors.

25. A. He dismissed the agent. B. He praised the agent.
C. He removed the poor tenants. D. He increased the rents and taxes.

## Section C

**Directions:** *In this section, you will hear a passage three times. When the passage is read for the first time, you should listen carefully for its general idea. When the passage is read for the second time, you are required to fill in the blanks numbered from 26 to 33 with the exact words you have just heard. For blanks numbered from 34 to 36 you are required to fill in the missing information. For these blanks, you can either use the exact words you have just heard or write down the main points in your own words. Finally, when the passage is read for the third time, you should check what you have written.*

The history of the United States is quite (26) ________. It began a little more than 200 years ago. In 1776, 13 (27) ________ located on the eastern (28) ________ of North America (29) ________ independence and fought a (30) ________ against the British. In 1783 the (31) ________ won the revolution. After the revolution, the United States (32) ________ a large (33) ________ of country from Napoleon of France.

(34) ________________________________________. Later, the United States and Mexico went to war. (35) ________________________________________.

In 1861 one-half of the United States went to war with the other half. (36) ________________________________________.

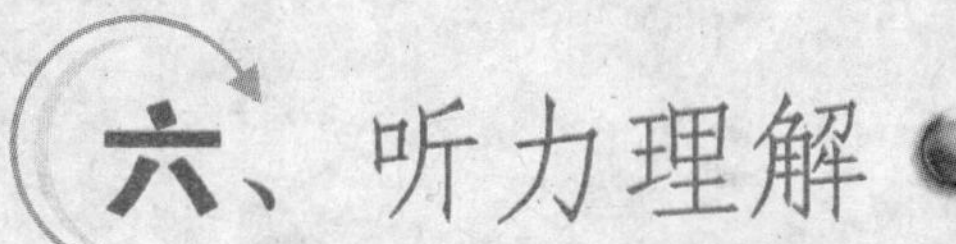

# 六、听力理解

# 答案详解

## Unit 1

**录音文字材料**

**Section A**

1. W: I haven't heard from my mother since the month before last.
   M: Don't worry Mary. Overseas mail is often slow.
   Q: What can you learn about the woman from the conversation?
2. W: Tom, how are you doing with your composition?
   M: I've written and rewritten it so many times that I wonder if I'll ever finish it.
   Q: What can we learn about the man?
3. M: Has Lucy returned from Australia yet?
   W: Yes, but she was only here for 3 days before her company sent her to Europe.
   Q: Where is Lucy now?
4. M: Excuse me, do you know when the train departs?
   W: I'm sorry. The train is behind schedule. You'll have to wait another twenty-five minutes.
   Q: What do we learn from this conversation?
5. W: I have a sore throat and I feel a bit dizzy.
   M: Well, let me see, Hmm. Inflammation of the tonsils. I'll give you some pills. Take one big one and two small ones, three times a day.
   Q: How many pills will the woman take a day?
6. W: How often did you write home?
   M: I used to write home once a week.
   Q: What do you understand from the man's answer?
7. M: Shall we have a foreign language test this weekend?
   W: No, it is postponed because the teacher has to attend a meeting.

**答案与详解**

1.【答案】B
【详解】解答本题时,考生在看到选项时可初步判断该题是围绕女士的说话来展开的,在听音时"heard from"表明是来信,时间定语是"the month before last",地点隐含在男士说话中的"overseas",综合判断答案为B。

2.【答案】D
【详解】本题关键要理解男士说话的意思,"written and rewritten it so many times"和"I wonder if I'll ever finish it",反复写了很多遍,虚拟语气"真希望已经完成了",事实正好相反,因此答案选D。

3.【答案】B
【详解】本题通过关键词"returned from Australia"可排除C、D选项。然后根据女士的回答可知Lucy已离开这儿去了欧洲,答案选B。

4.【答案】D
【详解】本题通过女士的回答可知"The train is behind schedule."火车比时刻表上的时间要晚到,因此表示火车晚点了,答案选D。

5.【答案】C
【详解】这是一道数字计算题,关键要听懂"Take one big one and two small ones, three times a day."这句话就不难计算出每天吃9片药,答案为C。

6.【答案】B
【详解】本题的关键词是"used to"过去常做某事,言下之意是现在不常做了,答案很容易选B。

7.【答案】B
【详解】这题女士的回答否定了本周将会考试的这一说法,因此排除选项D。另外说明了考试推迟的原因是老师不得不参加一

Q: What do we learn from this conversation?

8. W: Have you seen my sister?

M: No, I haven't seen her since the year before last.

Q: When did he see the woman's sister?

**Conversation One**

W: Dr. Jones, how exactly would you define eccentricity?

M: Well, we all have our own particular habits which others find irritating or amusing, but an eccentric is someone who behaves in a totally different manner from those in the society in which he lives.

W: When you talk about eccentricity, are you referring mainly to matters of appearance?

M: Not specifically, no. There are many other ways in which eccentricity is displayed. For instance, some individuals like to leave their mark on this earth with strange buildings. Others have the craziest desires which influence their whole way of life.

W: Can you give me an example?

M: Certainly. One that immediately springs to mind was a Victorian surgeon by the name of Buckland. Being a great animal lover he used to share his house openly with the strangest creatures, including snakes, bears, rats, monkeys and eagles.

W: That must've been quite dangerous at times. Does one of these stand out in your mind at all?

M: Yes, I suppose this century has produced one of the most famous ones: the American billionaire, Howard Hughes.

W: But he wasn't a recluse all his life, was he?

M: That's correct. In fact, he was just the opposite in his younger days. He was a rich young man who loved the Hollywood society of his day. But he began to disappear for long periods when he grew tired of high living. Finally, nobody was allowed to touch his food and he would wrap his hand in a tissue before picking anything up. He didn't even allow a barber to go near him too often and his hair and beard grew down to his waist.

W: Did he live completely alone?

M: No, that was the strangest thing. He always stayed in luxury hotels with a group of servants to take care of him. He used to spend his days locked up in a penthouse suite watching adventure films over and over again and often eating nothing but ice cream and chocolate bars.

W: It sounds a very sad story.

M: It does. But, as you said earlier, life wouldn't be the same without characters like him, would it?

**Conversation Two**

W: Dad, I need a few supplies for school, and I was wondering if...

M: Yeah. There are a couple of pencils and an eraser in the kitchen drawer, I think.

W: Dad, I'm in eight grade now, and I need REAL supplies for

个会议，而不是学生要参加会议，可以排除选项 A、C，因此答案为 B。

8.【答案】C

【详解】本题要理解"the year before last"的含义为前年，因此答案选 C。

**Questions 9 to 12 are based on the conversation you have just heard.**

9. What did the woman want to talk to Dr. Jones about?

【答案】A

【详解】这段对话一开头就提到了这个主题 define eccentricity。因此选 A。其余的几项都是不准确的。大家要在开头特别注意，往往主题会出现在开始的部分。

10. According to the dialogue, what's the meaning of eccentricity?

【答案】A

【详解】对于对话中关于喂养奇怪的动物，吃些奇怪的东西的描述是对主题的反映，从而得知此题应该是选 A。

11. Who is Howard Hughes?

【答案】B

【详解】the American billionaire, Howard Hughes 这是细节考查题。做题前要先掌握题干的意思，有的放矢。

12. Which following statement is correct according to your apprehension of this dialogue?

【答案】D

【详解】这种题目难度较大，既要了解细节，又需要把握全局，最好能先阅读题目，边听边记。这种能力需要经常锻炼方能培养，此题选 D。

**Questions 13 to 15 are based on the conversation you have just heard.**

13. Why doesn't the father want to buy his daughter some of these supplies?

【答案】B

【详解】爸爸的论据就是 I didn't have any of that when I was in

my demanding classes.

M: Oh, so you need a ruler too?

W: Dad, I need some high - tech stuff like a calculator, a Palm Pilot, and a laptop computer.

M: Uh. I didn't have any of that when I was in middle school, and I did just fine.

W: Yeah, and they weren't any cars either, were there? And things are just more progressive now.

M: Well, we can rule out the hand pilot. Whatever, and the computer... unless mom lets you sell the car. And as for the adding machine. Yeah, I think mine from college is kicking around here somewhere.

W: Dad, I need a calculator for geometry, and I have heard you can download free software from the Internet.

M: Great. My daughter will be playing video games in geometry class.

W: Dad.

M: Okay. How much is this thing going to cost me?

W: Well, I saw it at the store for only $99, with a $10 mail - in rebate, or you could buy it online.

M: Oh. Do they throw in a few aspirin so your father can recover from sticker shock?

W: Dad. Please!!! Everyone has one and you always say you want me to excel in school, and I'll chip in $10 of my own, and I'll even pick up my room. Dad, Hmmm, 100 bucks.

M: Well, you'll be supporting me in my old age, so, I guess so. When do you need it?

W: Now, right now. Mom's already waiting in the car for us. She said she would buy me an ice cream if I could talk you into to buying it for me today.

**Section B**

**Passage One**

Every human being, no matter what he is doing, gives off body heat. The designers of the Johnstown campus of the University of Pittsburgh set their hearts on how to collect body heat and utilize it. They have designed a collection system which utilizes not only body heat, but the heat given off by light bulbs and refrigerators as well. The system works so well that no conventional fuel is needed to make the campus's six buildings comfortable.

Some parts of most modern buildings, theatres and offices as well as classrooms are heated by people and lights. The technique of saving heat and redistributing it is called "heat recovery". A few modern buildings recover heat from some buildings and reuse it in others.

Along the way, Pitt has learned a great deal about some of its heat producers. The harder a student studies, the more heat his body gives off. Male students emit more heat than female students, and the larger a student, the more heat he produces.

**Passage Two**

A lot of people don't like to give waiters extra money - a tip, but maybe those people don't understand about waitresses and waiters. We count on the tips as part of our salary. If waiters

middle school, and I did just fine,向女儿表明不要这些东西一样可以把学习搞好,所以不支持买这些用品,他们不是必需的。答案是B。

14. What specific argument does the girl give her father to persuade him to buy these things?

【答案】C

【详解】and you always say you want me to excel in school, and I'll chip in $10 of my own, and I'll even pick up my room,这就是他女儿的许诺。选项只有C符合要求。

15. Why does the father eventually give in to his daughter?

【答案】B

【详解】最后,爸爸从长远打算:Well, you'll be supporting me in my old age, so, I guess so,也就被女儿劝服了。答案是B。

**Questions 16 to 18 are based on the passage you have just heard.**

16. What did the designers of the Johnstown campus of the University of Pittsburgh set their hearts on?

【答案】A

【详解】根据短文第一段第二句话"The designers of the Johnstown... and utilize it."可知答案几乎就是句中原话,可直接选择答案A。

17. At the Johnstown campus, how many of the buildings are heated entirely by the heat collection system?

【答案】D

【详解】本题也是文中原话,在第一段的最后一句"The system works... the campus's six buildings comfortable."可知答案为D。

18. According to the passage, which of the following would produce the LEAST amount of heat?

【答案】B

【详解】根据短文第三段列举的几种比较,再比较选项得出答案为B。

**Questions 19 to 21 are based on the passage you have just heard.**

19. What is the aim of the writer?

【答案】D

【详解】这是一道主旨题,问作者的意图是什么。综合全篇分

and waitresses didn't get tips, they wouldn't get enough money to live.

People ask me, "What's a good tip?" I like to get 15% of the bill. So if a customer has to pay $20.00 for her dinner, I like to get about $3.00 for a tip. Sometimes I expect 20% if I did a lot of work for the customer. For example, if I got her a special kind of food or recipe from the chef. But do you know something? Very often it's the person you work the most for who gives you the smallest tips.

Once I looked up "tipping" in a dictionary. It said that the letters in the word "tip" stand for "To Insure Promptness." In other words, to make sure that we do things right away. The dictionary said that no one knows if that is the real meaning of "tip", but it makes sense to me. If we know a regular customer is a good tipper, then we make sure he gets good service. But if someone gives small tips, we aren't in a hurry to bring him food or forget his drinks. So remember, be nice to your waitress and she'll be nice to you.

### Passage Three

Everybody wastes time. Instead of doing his homework, the schoolboy watches television. Instead of writing her essay, the student goes out with her friends. They all had good intentions, but they keep putting off the moment when they must start work. Consequently, they begin to feel guilty, and then waste even more time wishing they had not allowed themselves to be distracted.

I know two writers who seem to work in quite different ways. Bob is extremely methodical. He arrives at his office at 9 a.m. and is creative until 12:30. At 2 p.m. he returns to his desk and is creative until 5 p.m., when he goes home and switches off until the following morning. Alan, on the other hand, works in inspired periods, often missing meals and sleep in order to get his ideas down on paper. Such periods of intense activity are usually followed by days when he stays in his flat, listening to Mozart and flicking through magazines.

Their places of work also reflect their styles. Bob's books are neatly arranged on the shelves; he can always find the books he wants. Alan, on the other hand, has books and magazines all over the place. They are about every subject mostly unconnected with his work but he has a special talent of making use of the unlikely information to enrich his books.

### Section C

An old lady who lived in a (26) village went into town one Saturday. After she had (27) bought fruit and (28) vegetables in the market for herself and for a friend who was ill, she went into a shop which (29) sold glasses. She (30) tried one pair of glasses, and then another pair and another, but (31) none of them seemed to be right. The (32) shopkeeper was a very (33) patient man and after some time he said to the old lady, (34) "Now, don't worry, madam. Everything will be all right in the

析,作者以一个 waiter 的口吻来叙述,主要想说明的是一个人如何付小费才能得到更好的服务。

20. Who very often gives the waiter or waitress the smallest tips according to the writer?

【答案】B

【详解】根据短文的第二段最后一句"Very often it's the person you work the most for who gives you the smallest tips."可直接得出答案是 B。

21. What do the letters in the word "tip" mean?

【答案】A

【详解】文章中提到了"looked up 'tipping' in a dictionary",解释了其含义为"To Insure Promptness",虽然这样解释了,意义还是不明朗,我们再往下听"In other words",后面紧着的就是答案关键句了,答案为 A。

**Questions 22 to 25 are based on the passage you have just heard.**

22. When do people start to do when they found they were wasting time instead of getting down to work?

【答案】A

【详解】文章中有"Consequently, the begin to feel guilty...",从这里我们就要注意下面将要听到的内容了,抓住了关键词"waste even more time"就不难选出正确答案为 A。

23. Which of the following would describe Bob's way of working?

【答案】D

【详解】题目问如何描述 Bob 的工作方式,从文章中我们可以听到"extremely methodical",非常有条理,理解了这个词组的意思,很容易得出正确答案为 D。

24. Why does Alan have a lot of books and magazines?

【答案】C

【详解】文章中 Bob 和 Alan 来做比较,"Alan, on the other hand, has books and magazines all over the place."但关键句在后一句"They are about... but he has a special talent of making use of the unlikely information to enrich his books."。

25. Which of the following can be used to describe Alan's working schedule?

【答案】B

【详解】如何描述 Alan 的工作时间表,关键句在文章第二段"Alan,..., works in inspired period, often missing meals and sleep in order to get his ideas down on paper."可以看出他工作时间不固定且没有规律的。

### Section C

26. village　27. bought　28. vegetables　29. sold
30. tried　31. none　32. shopkeeper　33. patient
34. Now, don't worry, madam
35. It isn't easy to get just the right glasses, you know
36. And it is even more difficult when you are shopping for a friend

end. (35) It isn't easy to get just the right glasses, you know."
"No, it isn't," answered the old lady. (36) "And it is even more difficult when you are shopping for a friend."

# Unit 2

## 录音文字材料

**Section A**

1. M: I've got to go now. I have to catch the last bus.
   W: But it's too late. You should have said so earlier.
   Q: What does the woman mean?
2. W: How long does it take you to ride home when there is not much traffic?
   M: Only twenty-five minutes. But if I can't leave my school before 5: 30 p. m., it sometimes takes me thirty-five minutes.
   Q: How long does it take the man to ride home when it isn't rush hour?
3. M: Would you like to come with me to the volleyball match?
   W: No, not today. I've got some washing to do now, but you can get some tickets for next Saturday.
   Q: What does the woman want to do next weekend?
4. W: I've just moved here from Houston.
   M: Houston? My aunt used to live there.
   Q: What does the man say about Houston?
5. W: You'd better hurry, Tom. There isn't much time left. The lecture is to begin at 8 o'clock.
   M: Don't worry. We still have half an hour.
   Q: What time was it when the conversation took place?
6. M: Have you found anything wrong with my liver?
   W: Not yet. I'm still examining. I'll let you know the result the day after tomorrow.
   Q: What's the probable relationship between the man and the woman?
7. W: You didn't go to the theater last night either, did you?
   M: No, I had a slight pain in my stomach.
   Q: What did you learn from this conversation?
8. W: Has your brother bought his books yet?
   M: He bought a medical book. The Chinese and English textbooks were sold out.
   Q: Which book has the man's brother got?

**Conversation One**

W: Hi. How can I help you?
M: Yeah. I'd like to rent a mid - size car for three days.
W: Well, the economy car would work. We have one right out front.
M: Where? That one? It looks more like a shoebox to me. I'm really tall and trying squeeze into that thing... I don't think so.
W: Well, if you need more room or comfort, I recommend the full - size car. It also has a nice stereo system, CD player, safety rear door locks, and cruise control, and power locks and windows.

## 答案与详解

1.【答案】C
【详解】女士说"You should have said so earlier."说明男士走得太晚,赶不上末班车了。

2.【答案】C
【详解】非高峰时期需25分钟。

3.【答案】A
【详解】女士建议男士可以拿一些下个星期六的票。可见她会去看排球。

4.【答案】A
【详解】男士的姨妈曾经住在休斯顿。A项意思符合。

5.【答案】D
【详解】离8点还有半个小时即7点半。答案是D。

6.【答案】C
【详解】女士在检查男士的肝脏。可见是病人与医生的关系。

7.【答案】C
【详解】关键词是either,表示否定的"也"。推知两人都没去。

8.【答案】D
【详解】他买了一本医学书籍。

**Questions 9 to 12 are based on the conversation you have just heard.**

9. Why did the man settle on renting the full - size car?
【答案】A
【详解】Well, if you need more room or comfort, I recommend the full - size car. 可以推断出 full - size car 能够提供更多的空间,roomy 意思就是宽敞的。
10. What was one of his major concerns about renting the car?

M：Well, I'm not so concerned about how it's equipped. I just want to make sure it is comfortable to drive. And what is the daily rate for that anyway?

W：It'll come to fifty - seven ninety - five a day. Hey, all of our cars have unlimited miles, but of course, that doesn't include gas.

M：Yeah, right. I bet that car probably eats up gas, and now that were in the middle of the vacation season, gas stations are gouging consumers with astronomical prices.

W：Well, as they say, it comes down to the law of supply and demand.

M：I'll go with the full - size car. Wait, uh... what does it look like?

W：Uh, it's right out there in the parking lot. The one over there next to the sidewalk.

M：Do you mean that old lemon with the missing hubcap?

W：Sir, It's just that it's one of the last cars on our lot, but it runs like a dream. And would you like to purchase our daily car protection plan?

M：What's that exactly?

W：Well, the car protection plan is a complete insurance package covering damage to the vehicle, injury or loss of life to you or your passengers.

M：But wouldn't my own car insurance cover those problems?

W：It might, but each insurance policy is different. With our car protection plan, however, you deal directly with us in case there is a problem, and we handle everything quickly, and you don't have to contact your own insurance company. Okay. Let me just confirm this. A full - size car for three days, plus the car protection package. Is that right? Okay, I'll have our mechanic, Louie, check the car over and pull it up to the door.

M：Push it up to the door? I hope this car really runs.

W：Well, in case it does break down on some out - of - the - way, deserted road, just call the toll - free number for assistance. They'll come to assist you within two business days. Enjoy your trip.

**Conversation Two**

W：Ron, what are you doing?

M：Ah, nothing. I'm just looking up some information on the Internet.

W：Like what? Let me see.

M：No, no, it's okay. I mean, you know:..

W：Baldness? What are you looking that up for? I... I mean, you're not that bad off.

M：Ah, there you go. Bringing it up again!

W：No. I mean it. You look great! Honestly, it's not that bad.

M：Hey, I get enough of it from friends, and the people at work, and now from you!

W：Well, maybe you could wear a toupee? I think you'd look great.

M：Oh no. And have it slip off my head on to my date's dinner plate as I lean over to kiss her? Uh - uh.

W：Well, have you ever thought about seeking medical advice? There are new advances in medicines that not only retard hair loss, but help regenerate new growth.

M：Ah, I still don't give much credibility to medical treatment to prevent permanent hair loss.

W：Well, what about accepting the fact that you're just losing your hair?

M：I just can't give up hope. I know appearances shouldn't matter, but I don't know. I just feel that women just avoid me.

【答案】C

【详解】Well, I'm not so concerned about how it's equipped. I just want to make sure it is comfortable to drive. And what is the daily rate for that anyway? 他所关心的不是车子是如何装备的，而是每天驾车所需要费用，后来对话有很大的篇幅讨论了油价和耗油量的问题，可见他最关心的就是耗油量的问题。

11. How would you describe his rental car?

【答案】B

【详解】that old lemon with the missing hubcap, Push it up to the door? I hope this car really runs. 这些描述都说明了该车是存在许多缺陷的，所以答案选 B。

12. What can we infer from the closing statement about roadside assistance?

【答案】B

【详解】They'll come to assist you within... two business days. 援助一般需要两天才可以到达，这对顾客当然是一种耐心的考验，答案为 B。

**Questions 13 to 15 are based on the conversation you have just heard.**

13. Based on the conversation, what is the most probable relationship between the speakers?

【答案】B

【详解】这个题是考查对英美国家文化的理解。一般说来，他们都是很注意保护个人的隐私的，而对话中女孩却可以随时干涉男孩的隐私，轻易地就知道了他的隐私，而且男孩还并不生气，可见除了 brother 和 sister 的关系外，就很难解释通了。所以应该是选 B。

14. Why is the boy hesitant about seeking medical treatment for his condition?

【答案】A

【详解】男孩在这里说：I still don't give much credibility to medical treatment to prevent permanent hair loss. 说明他对于这种治疗并不抱很大的希望，他觉得可信度不高。答案是 A。

15. What does the boy probably do at the end of the conversation?

【答案】A

【详解】I just can't give up hope. I know appearances shouldn't matter, but I don't know. I just feel that

W: Come on. You can't be serious.

M: No really. I've seen it many times. It just, I don't know...

**Section B**

## Passage One

Mark Twain, who wrote many famous novels, traveled quite a lot often because circumstances usually financial circumstances forced him to. He was born in Florida, Missouri in 1835 and moved to Hannibal, Missouri with his family when he was about 4 years old. Most people think he was born in Hannibal but that isn't true. After his father died when he was about 12, Twain worked in Hannibal for a while and then left, so he could earn more money. He worked for a while as a river pilot on the Mississippi. Twain loved this job and many of his books show it. And his name also came from this job. The river job didn't last however, because of the outbreak of the Civil War. Twain was in the Confederate Army for just 2 weeks and he and his whole company went West to get away from the war and the army. In Nevada and California, Twain prospected for silver and gold without much luck, but did succeed as a writer. Once that happened Twain traveled around the country, giving lectures and trying to earn enough money to go to Europe. Twain didn't travel much the last 10 years of his life and he didn't publish much either. Somehow his travels even inspired his writings. Like many other popular writers Twain derived much of the materials for his writing from the wealth and diversity of his own personal experiences.

## Passage Two

People often show their feelings by the body positions they adopt. These can contradict what you are saying, especially when you are trying to disguise the way you feel. For example, a very common defensive position, assumed when people feel threatened in some way, is to put your arm or arms across your body. This is a way of protecting yourself from a threatening situation. Leaning back in your chair especially with your arms folded is not only defensive, it's also a way of showing your disapproval, of a need to distance yourself from the rest of the company.

A position which betrays an aggressive attitude is to avoid looking directly at the person you are speaking to. On the other hand, approval and desire to cooperate are shown by coping the position of the person you are speaking to. This shows that you agree or are willing to agree with someone. The position of one's feet also often shows the direction of people's thoughts, for example, feet or a foot pointing towards the door can indicate that a person wishes to leave the room. The direction in which your foot points can also show which of the people in the room you feel most sympathetic towards, even when you are not speaking directly to that person.

## Passage Three

You may not realize that fog is simply a cloud that touches the ground. Like any cloud, it is composed of tiny droplets of water, or, in rare cases, of icy crystals that form an ice fog. Ice fogs usually occur only in extremely cold climates, because water droplets are so tiny they don't solidify until the air temperature is far below freezing, generally thirty degrees below zero or lower. The droplets of fogs are nearly spherical, the ball - shape. The transparency of fog depends mainly on the concentration of droplets. The

women just avoid me. 根据他的这段心声的表露,可见他想要改变目前的状况,心中仍抱希望的去追求新的解决方法。答案 A 符合题目意思。

**Questions 16 to 18 are based on the passage you have just heard.**

16. The speaker focuses on which aspect of Mark Twain's life?

【答案】A

【详解】听力材料讲了 Mark Twain 一生的游记。

17. Why did Twain go West?

【答案】C

【详解】材料中说到"he and his whole company went west to get away from the war and the army."答案为 C。

18. What connection does the passage suggest between Twain's travels and his writings?

【答案】A

【详解】Somehow his travels even inspired his writings. A 项意思符合。

**Questions 19 to 21 are based on the passage you have just heard.**

19. Which of the following one can reveal your true feeling according to the passage?

【答案】A

【详解】一种暴露挑衅态度的姿势是不正视与你说话的人。可见看人的方式会泄露你的真情。B、C、D 选项内容均未在材料中提及。所以选 A。

20. What does leaning back in your chair with your arms folded NOT mean according to the passage?

【答案】B

【详解】材料中说背靠在你的椅子里尤其是环抱着你的双臂不仅是自卫的,而且是一种表示你不想和其他人员接近的方式。A、C、D 符合材料意思,所以选 B。

21. Which body position can make you know your guest want to leave according to the passage?

【答案】C

【详解】材料中提到"feet or a foot pointing towards the door can indicate that a person wishes to leave the room",因此 C 项符合题目意思。

**Questions 22 to 25 are based on the passage you have just heard.**

22. What is the main topic of talk?

【答案】B

【详解】听力材料讲了雾的组成,雾滴的形状,以及雾中水微粒不落地的原因。可见主题是雾的形成。答案是 B。

23. According to the passage, how are fog and clouds relat-

more droplets, the denser the fog is. Since water is eight hundred times denser than air, investigators were puzzled for a long time as to why the water particles in fog didn't fall to the ground, making the fog disappear. It turns out that droplets do fall at a predictable rate but in fog - creating conditions, they are either supported by rising air currents or continually replaced by new droplets condensing from water vapor in the air. And now we have a new word "smog", it's the unhealthy dark mixture of gases, smoke and the vehicle waste in the air, which make the people can hardly see clearly the road. It's the man - made fog.

**Section C**

In police work, you can never predict the next crime or (26) problem. No working day is identical to any other, so there is no "typical" day for a police officer. Some days are (27) relatively slow, and the job is boring; other days are so busy that there is no time to eat. I think I can (28) describe police work in one word: (29) variety. Sometimes it's dangerous. One day, for example, I was working (30) undercover; that is, I was on the job, but I was wearing (31) normal clothes, not my police (32) uniform. I was trying to catch some (33) robbers who were stealing money from people as they walked down the street. (34) Suddenly, seven bad men jumped out at me. One of them had a knife, and we got into a fight. Another policeman arrived, and together, we arrested three of the men; but the other four ran away. (35) Another day, I helped a woman who was going to have a baby. She was trying to get to the hospital, but there was a bad traffic jam. (36) I put her in my police car to get her there faster. I thought she was going to have the baby right there in my car. But fortunately, the baby waited to arrive until we got to the hospital.

ed?

【答案】B

【详解】材料谈到"Like any cloud, it is composed of tiny droplets of water",所以应该是选 B。

24. What is the main factor that determines the transparency of the fog?

【答案】A

【详解】雾的透明度主要由雾滴的密度决定。雾滴越多,雾越厚。所以选 A。

25. What used to puzzle investigators about fog?

【答案】D

【详解】材料中讲到既然水比空气稠密 800 倍,研究者们曾长期为雾中水微粒不落地的原因而困惑。D 项意为水微粒是如何悬浮在空中的。与材料意思一致,所以选 D。

**Section C**

26. problem 27. relatively 28. describe
29. variety 30. undercover 31. normal
32. uniform 33. robbers
34. Suddenly, seven bad men jumped out at me
35. Another day, I helped a woman who was going to have a baby
36. I put her in my police car to get her there faster

# Unit 3

## 录音文字材料

**Section A**

1. M: Louise, do you want me to try to fix that broken camera of yours?
   W: Thanks, but I already had it taken care of.
   Q: What has happened to the camera?
2. M: I'll take these three books. Are they fifty cents each?
   W: These two books are, but this one is 75 cents.
   Q: How much will the three books cost?
3. M: I'd like to return these books.
   W: Let me see. These books are one week overdue. I'm sorry, but you'll have to pay a fine.
   Q: What does the woman mean?
4. M: I've heard about Kuwait, but I haven't been there yet. Actually, I wanted to go there last year, but I couldn't. I was ill.

## 答案与详解

1.【答案】A
【详解】注意"had it taken care of "意为让别人修了。

2.【答案】A
【详解】计算题。三本书中 2 本每本 50 美分,另一本为 75 美分。总共 $0.5\times2+0.75=1.75$ 美元。

3.【答案】D
【详解】女士说"but you'll have to pay a fine."可见答案为选项 D。

4.【答案】D
【详解】女士说"But let's go this year, and then we can

W: Oh, that's a shame. But let's go this year, and then we can visit Barcelona, too.

Q: What do we know about the man?

5. W: At first I thought this rug was yellow, but now it looks green to me.

M: You were right the first time. It's this blue light in the store that makes everything look different.

Q: Why did the rug look green?

6. W: How can I get to the shopping center from here?

M: You can take a bus or a taxi, but it isn't too far. Maybe you'd like to walk.

Q: Is the shopping center far away?

7. W: I'm always absent-minded in Mrs. Lee's class.

M: It was noisy, as usual.

Q: What do they think of Mrs. Lee's class?

8. M: The newspaper says it'll be cloudy and rainy today. What do you think?

W: I don't believe it. Look! The sun is out. There's not a cloud in the sky.

Q: What are they talking about?

**Conversation One**

W: Hello, Worldwide Flowers. Mrs. Green is speaking.

M: This is Jim Kelly. I'd like to order some flowers and have them sent to my home.

W: Fine, Mr. Kelly. What kind of flowers did you have in mind?

M: I'd like to send a dozen red roses.

W: OK. I'll need your complete address with the zip code, Mr. Kelly.

M: The address is: 43 Pennsylvania Avenue, Bloomington, Indiana, 47401.

W: What would you like us to put on the card?

M: Hmm, just something simple. How about: All my love, Jim.

W: OK. Now, when should they arrive?

M: They should be there before six in the evening on September 12.

W: That should be no problem. And how does your name appear on the card?

M: James William Kelly.

W: Is that K-E-L-L-Y?

M: That's right.

W: OK. Thank you for calling Worldwide Flowers. We'll have those dozen red roses delivered before six p. m. on September 12th.

M: Thank you very much.

W: You're welcome. Good-bye.

M: Bye.

**Conversation Two**

M: Hamburgers, hamburgers! My children keep asking for hamburgers. I get so tired of hearing about hamburgers!

W: I know what you mean. Nowadays, teenagers don't like anything but hamburgers. I get tired of making them.

visit Barcelona, too."可知答案为 D。

5.【答案】B

【详解】关键句是"It's this blue light in the store that makes everything look different."答案为选项 B。

6.【答案】C

【详解】由男士说也许你愿意走着去,可推知答案为选项 C。

7.【答案】C

【详解】一个说总走神,一个说和平常一样吵,可见二人都不喜欢 Mrs. Lee 的课堂。

8.【答案】D

【详解】两人谈论的都是天气。

**Questions 9 to 12 are based on the conversation you have just heard.**

9. How is the man ordering flowers?

【答案】B

【详解】这篇对话中共有两个地方让我们可判断出 Jim 是通过电话来定购鲜花。一处是在开头的对话中:W: Hello, Worldwide Flowers. Mrs. Green speaking. 和 M: This is Jim Kelly. 是打电话典型的开场白,其中 Worldwide Flowers 是花店的名称。再就是结尾处 Thank you for calling Worldwide Flowers. 如果抓住了这句话不难找到正解。答案 A,C 项迷惑性最强,在解析中提到的句子如果未注意的话很容易会以为这就是一般购物对话。D 项作为干扰项,很容易排除。

10. How many roses does the man want?

【答案】D

【详解】对话中的关键句是 I'd like to send a dozen red roses, a dozen 即 12 支,故答案选 D。

11. What is the man's zip code?

【答案】C

【详解】47401 这个数字出现在地名 Bloomington, Indiana 之后,且按照常理邮政区码为 5,6 位数字左右,不难作出正确选择。

12. What does the man want to be put on the card?

【答案】B

【详解】to be put on the card 指写在卡片上,这里的卡片指附在花上表明送花人身份和心意的卡片。明白这个,可知 Jim 在此处是落款。C, D 选项中的 For 和 All 读音相近,意思上也说得过去,为干扰项。

**Questions 13 to 15 are based on the conversation you have just heard.**

13. Which one is TRUE in the following statements?

【答案】A

【详解】对话一开头男子说他的孩子一直向他要汉堡

M：How do you make them? I suppose I should learn how.
W：They're very simple.
M：What kind of meat do you buy?
W：I usually ask for ground round steak. It's more expensive than ground chuck, but it has less fat.
M：What do you mix with the ground beef?
W：I just add a little salt, some pepper and sometimes a little chopped onion. Some people don't like it with onions, but I do.
M：So you mix the salt and pepper and onions with the ground beef.
W：Yes, and divide the mixture into balls. Then make the balls into patties.
M：What is a patty?
W：You just press down on a ball and make it flat. Patties are flattened balls, about the size of the palm of your hand. If you want to get more than four patties out of your pound of meat, just make the patties thinner. But don't handle the meat too much.
M：How long do you cook them?
W：That depends on how you like them. Some people like them rare – red inside.
M：I don't like them rare!
W：If you like them medium or well – done, cook them longer. Then you put the hamburger inside some big rolls, and serve it with ketchup and pickles. I let my family spread ketchup on their own hamburgers, if they want it.
M：Thank you for your helpful ideas.

**Section B**

Passage One

In both Great Britain and the USA Christmas is a public holiday, and is celebrated with religious services and various forms of merrymaking. Traditional foods are eaten, especially rich Christmas cakes and Christmas puddings, which date from the Middle Ages, and roast turkey. Homes and churches are decorated with holly and mistletoe: this is a survival of pagan customs. Christmas trees were introduced to Great Britain and the USA from Germany. Santa Claus (sometimes called Father Christmas) is a corrupt form of St. Nicholas, patron saint of children, whose feast day is in fact 6 December. The first Christmas card was designed in England in 1843, and the habit of sending them became popular in about 1870; the custom then spread to the USA and other countries.

Passage Two

Hawaii is a state of the USA in the central Pacific. Hawaii is a chain of more than 20 volcanic islands and atolls. Its area is 6,424 square miles and its population more than 770,000. The capital is Honolulu. The climate is tropical and the vegetation rich and varied; the main crops are sugar and pineapples, and the chief industry is tourism. The US naval base, Pearl Harbor, is on the island of Oahu. Captain Cook discovered the islands in 1778. Hawaii was a kingdom until 1893, then it was taken over by the USA in 1898, became a territory in 1900 and a state of the Union in 1959.

包，让他心烦。而女士则说如今孩子们什么都不喜欢，除了汉堡包，且还说出关键一句：她开始厌倦做汉堡包，由此可推断两者的孩子都喜欢汉堡包。故选A。此题要求听出 implied meaning，即暗含之意。对话中并没有直接说孩子们喜欢汉堡包，但却明确传达了这个意思。在此题中，需要大家注意的是 nothing but 这个结构，表示什么也不……，除了……的意思。懂得这个，此题不难解。

14. What is NOT added in the meat by the woman?

【答案】D

【详解】文中女士说她将盐、胡椒和洋葱与肉混合，未提到糖。本题难度不大，只要听出这几个词就能做出，唯一需要小心的是题目要求选出没有加入肉的东西。文中提到有人不喜欢洋葱，可能这会迷惑人，但这位女士喜欢。(注意文中有：but I do)

15. What are patties?

【答案】B

【详解】文中对 patties 明确下了定义：Patties are flattened balls. 故选 B。该题难度在于 fried，flattened，fat 和 lengthened 这四个词的读音。

**Questions 16 to 18 are based on the passage you have just heard.**

16. According to the passage, which of the following statements is true?

【答案】D

【详解】听力材料开头便有提到。

17. What are the main traditional foods for Christmas?

【答案】D

【详解】传统食物有"rich christmas cakes and Christmas puddings, and roast turely"。

18. According to the passage, how long has the popular habit of sending Christmas cards existed for?

【答案】B

【详解】计算题，发送贺卡的习惯于 1870 年开始流行，得出 B 项最接近。

**Questions 19 to 21 are based on the passage you have just heard.**

19. According to the passage, which of the following statements is NOT true?

【答案】B

【详解】材料开始讲到夏威夷是美国的一个州，位于太平洋中心，它是一系列的 20 多个火山岛和珊瑚岛。所以 B 项是错误的。

20. Which of the following is not mentioned in the passage?

【答案】D

【详解】此题容易错选 D。注意材料只提到 The US naval base, Pearl Harbor, is on the island of Oahu，并没

## Passage Three

It is difficult to generalize about American food, since so many different varieties of cookery were brought over by the various groups of immigrants. Some foods were taken from the American Indians, e. g. sweet corn. The turkey, too, is native to America. In such a highly industrialized country food tends to be rather uniform: the products of particular regions, such as Maine lobsters or Californian oranges, can be found all over the country. But there are some regional specialities, such as the Creole cooking of New Orleans, and Southern fried chicken, hominy grits and pecan pie. Other typically American foods are ham and eggs, corn on the cob, apple pie, ice cream, and snacks like Hamburgers and Hot Dogs, sweet jellies and pickles with meat. Most American food is mass-produced, and on the whole more attention is paid to food value than taste. In big towns, every kind of food can be found in the USA, from Japanese to Mexican.

## Section C

Today I want to help you with a study reading method known as SQ3R. The letters stand for five steps in the reading (26) process: Survey, Question, Read, Review, Recite. Each of the steps should be done carefully and in the order mentioned.

In all study reading, a survey should be the first step. Survey means to look quickly. In study reading, you need to look quickly at titles, words in darker or larger print, words with (27) capital letters, (28) illustrations, and charts. Don't stop to read complete sentences. Just look at the important (29) divisions of the materials.

The second step is question. Try to form questions based on your survey. Use the question words who, what, when, where, why and how.

Now you are ready for the third step. Read. You will be reading the (30) titles and important words that you looked at in the survey, but this time you will read the examples and (31) details as well. Sometimes it is useful to take notes while you read. I have had students who (32) preferred to underline important points, and it seemed to be just as useful as note-taking. What you should do, whether you take notes or underline, is to read (33) actively. (34) Think about what you are reading as a series of ideas, not just a sequence of words.

The fourth step is review. Remember the questions that you wrote down before you read the material. You should be able to answer them now. (35) You will notice that some of the questions were treated in more detail in the reading. Concentrate on those. Also review material that you did not consider in your questions.

The last step is recite. (36) Try to put the reading into your own words. Summarize it either in writing or orally.

SQ3R – Survey, question, read, review, and recite.

有提到 the US navy 的所在地。

21. According to the passage, how old is Hawaii as a state of the USA?

【答案】D

【详解】夏威夷为一个州。2006 - 1959 = 47。所以选 D。

**Questions 22 to 25 are based on the passage you have just heard.**

22. According to the passage, why does the USA have every kind of food?

【答案】C

【详解】听力材料一开始便提到。

23. According to the passage, which of the following foods is native to America?

【答案】A

【详解】材料中说"The turkey, too, is native to America."

24. According to the passage, which of the following foods doesn't belong to snacks?

【答案】D

【详解】材料中提及的零食有"Hamburgers and Hot dogs, sweet jellies and pickles with meat."。

25. What would be the best title for this passage?

【答案】B

【详解】材料谈到美国食品,既有本土的,也有引进的,还列举了典型的美国食品。所以选 B。

## Section C

26. process　27. capital　28. illustrations
29. divisions　30. titles　31. details
32. preferred　33. actively
34. Think about what you are reading as a series of ideas, not just a sequence of words
35. You will notice that some of the questions were treated in more detail in the reading
36. Try to put the reading into your own words. Summarize it either in writing or orally

# Unit 4

## 录音文字材料

### Section A

1. W: Tom studies as hard as his classmates.
   M: That's not saying very much.
   Q: What conclusion can be drawn from the man's statement?
2. M: Could you tell me when the next train leaves for Shanghai?
   W: The next train leaves in three minutes; if you run you might catch it.
   Q: What will the man probably do?
3. W: The pipe is leaking and there is water all over the floor.
   M: Why don't you call Mr. Smith?
   Q: What does Mr. Smith do?
4. W: I left my umbrella in the office. Wait while I go back to get it.
   M: Don't bother, the weather report said it would clear up by noon.
   Q: What does the man advise the woman to do?
5. M: I have an extra ticket to the concert tonight. Would you like to come along?
   W: Thanks, but I already have my own ticket. Perhaps you can sell the other one at the door.
   Q: What does the woman suggest?
6. W: Did you finish your paper last night?
   M: Almost, I still have ten more pages to type.
   Q: What did the man do?
7. W: When is your seminar over?
   M: It's supposed to end at 5:50, but he never lets us out on time.
   Q: What does the man say about the seminar?
8. W: It's so hot today I can't work. I wish there were a fan in the laboratory.
   M: So do I. I'll fall asleep if I don't get out of this stuffy room soon.
   Q: What are these people complaining about?

### Conversation One

W: Dr. Carter's Office.
M: Yes, I'd like to make an appointment to see Dr. Carter, please.
W: Is this your first visit?
M: Yes it is.
W: Okay. Could I have your name please?
M: Yes. My name is Ronald Schuller.
W: And may I ask who referred you to our office?
M: Uh, I drove past your office yesterday.
W: Okay. How about the day after tomorrow on Wednesday at 4:00 O'clock?
M: Uh. Do you happen to have an opening in the morning? I usually pick up my kids from school around that time.

## 答案与详解

1.【答案】D
【详解】对话中男子说话中的关键词"not",可推断他不同意女子的观点,即他认为 Tom 学习不认真。

2.【答案】B
【详解】这段对话只要听到"next train"这个词就可以很容易排除 A、D。再根据女子说的"... you might catch it."可知答案为 B。

3.【答案】D
【详解】考生很容易在听到"pipe"这个词后,误认为是烟斗的意思,而错选答案 B。但只要听完整后面半句"there is water all over the floor"就可知"pipe"指的是水管,所以应选 D。

4.【答案】D
【详解】对话中"clear up"指的是天气变晴朗。弄懂这个词组的意思后,不难发现这段对话是和天气有关,所以应选答案 D。

5.【答案】D
【详解】从对话中女子说"... sell the other one...",可知她建议卖掉另一张票。

6.【答案】C
【详解】根据男子说的"... the more pages to type."很容易选择答案 C。

7.【答案】D
【详解】对话中"on time"指的是按时、准时,根据男子说的"... never lets us out on time."可知应选答案 D。

8.【答案】B
【详解】根据对话中关键词"hot, a fan",可知他们是在抱怨天气太热,heat 也是指高温、热,所以应选答案 B。

**Questions 9 to 12 are based on the conversation you have just heard.**

9. From the conversation, how did the man probably find out about Dr. Carter?
【答案】A
【详解】对话中,the man 在被问到是怎么知道 Dr. Carter 的,他回答说"I drove past your office yesterday",因此只有 A 符合答案。

10. What time does he schedule an appointment to see Dr. Carter?
【答案】C
【详解】这里考的是一个关于时间的细节,对话中出现了几个时间,最后预约见医生的时间定在星期四,注意不能

W: Okay. Um... how about Tuesday at 8:00 A. M. or Thursday at 8:15 A. M.?

M: Uh, do you have anything earlier, like 7:30?

W: No. I'm sorry.

M: Well, in that case, Thursday would be fine.

W: Okay. Could I have your phone number please?

M: It's 643 - 0547.

W: Alright. And what's the nature of your visit?

M: Uh...

W: Yes sir.

M: Well, to tell the truth, I fell from a ladder two days ago while painting my house, and I sprained my ankle when my foot landed in a paint can. I suffered a few scratches on my hands and knees, but I'm most concerned that the swelling in my ankle hasn't gone down yet.

W: Well, did you put ice on it immediately after this happened?

M: Well yeah. I just filled the paint can with ice and...

W: And so after you removed the paint can... Sir, sir, Mr. Schuller, are you still there?

M: Well that's part of the problem. Uh, the paint can is still on my foot.

W: Look, Mr. Schuller. Please come in today. I don't think your case can wait.

**Conversation Two**

M: Good morning, Miss. I'm from radio station QRX, and I wonder if you could answer a few questions for our survey.

W: Sure.

M: What's your name?

W: Linda Montgomery.

M: Linda, what do you do for a living?

W: Uh, well, right now I'm going to beauty school.

M: Beauty school?

W: Yeah.

M: Uh - huh. And what do you do for fun?

W: Oh, what for fun, I hang out with my friends—you know, go for pizza, stuff like that.

M: I understand. What's the most exciting thing that's happened to you recently?

W: Oh, this was so great! (Yeah?) Four of my friends and I, we went to a Bruce Springsteen concert. We actually—we got tickets.

M: Wonderful.

W: It was the best.

M: Who do you admire most in the world?

W: Who do I admire—I guess (Mm - hmm.) my dad, (Uh - huh.) probably my dad. Yeah.

M: And what do you want to be doing five years from now?

W: I would love it if I could have my own beauty salon.

M: Uh - huh.

W: That would be great.

M: Thanks very much for talking to us today.

W: Okay.

混淆,答案是C。

11. Why does the man want to see the doctor?

【答案】B

【详解】注意题目中问的是Ronald为什么要见医生,虽然他在粉刷房子的时候也划伤了手和膝盖,但是并不严重,他要见医生的原因只有一个,那就是他的脚踝伤的比较严重,因此答案选B。

12. What does the receptionist suggest at the end of the conversation?

【答案】B

【详解】在对话的最后,the receptionist发现Ronald的脚还卡在油漆桶里面,所以她建议Ronald今天就到医院来,因为他的情况紧急,不能拖延,因此答案选B。

**Questions 13 to 15 are based on the conversation you have just heard.**

13. What is the main topic of the interview?

【答案】A

【详解】记者的问题包括姓名,现在的工作,学习,爱好,最近高兴的事和未来的打算等等。符合题意的只有A。该题考查考生的综合分析能力,要学会从总体上把握大意。

14. What's the most exciting thing that's happened to Linda recently?

【答案】C

【详解】"four of my friends and I, we went to a Bruce Springsteen concert."答案很明显是C。该题是考查对细节信息的捕捉能力,听力过程中要适当的做笔记。

15. Where is the reporter from?

【答案】B

【详解】"I'm from radio station QRX."只要集中精力,此题是很容易答对的。但要学会区分几个字母的发音,即Q与K,R和L。

**Section B**

## Passage One

In the twentieth century numerous new nations have been formed. Though their people often enjoy full political liberty, there exist at the same time a great many strange practices. Native populations may be free to vote and to elect whom they please to govern them, but popular prejudices, unusual and harmful customs take a long time to die out. However, now that people are better educated, they need not suffer in silence, for they are able to express their views. In this way many unpleasant customs disappear rapidly.

There was a good example of this recently in a newly formed republic when a girl of fourteen refused to marry a sixty-year-old man who had "bought" her for 40 pounds. Just before the marriage ceremony, the girl ran away and wrote to the president of republic. In her letter she pointed out that although her country was independent, its people were still not truly free. Some human beings were like slaves, she said, and woman could be bought and sold like cattle. She asked the president if he felt that this was right. This letter caused the president a great deal of concern and he immediately changed the cruel law which permitted women to be bought and sold.

The girl had won a considerable victory but she still had a big problem. She had to find 40 pounds to repay the man who might have become her husband. There seemed to be no way of raising so much money. Fortunately, however, the girl's story was broadcast on a radio program in Europe and nearly 2000 pounds poured in from listeners. The buyer got his money back and the girl was free to marry anyone she chose. She had won true freedom for herself and for others like her.

## Passage Two

Like all diseases, alcoholism responds to treatment best when it is found early. If alcohol is in any way changing your life, you need help now. Perhaps you are one of those few lucky people who can really stop drinking. If so, you are truly blessed. One of the most common comments of alcoholic is that they can stop their drinking anytime they choose to. Very few can manage this on their own feet.

Alcoholics Anonymous is a group of recovering alcoholics who seek to help one another through the sharing of mutual problems. Group members can be reached easily by anyone wishing to discuss an alcohol-related problem. They make no moral speeches, and they do not provide welfare or jobs. AA has done more to help the alcoholic see his problem than any other group has done.

Society, after years of trying to sweep the alcoholic beneath the public opinion of apathy and loathing, is at last facing up to the responsibility of dealing openly and fairly with the alcoholic. While laws against drunken driving must be as severe as possible to stem the death on our highways, the alcoholic should not be quenched for the disease itself. Public drunkenness has been taken off the books of many states. Jail is not a place in which to cure the sick. The problem drinker needs medical attention and sound advice.

**Questions 16 to 18 are based on the passage you have just heard.**

16. What do we learn from the passage?

【答案】C

【详解】本题要求考生能听懂全文大意,并进行概括。四个选项在原文中都提到了。A 项与原文相反,B、D 项概述过于片面,因此答案应选 C。

17. What did the President do to help the girl?

【答案】D

【详解】本题是一道细节题,考生听清"... he immediately changed the cruel law which permitted women to be bought and sold."这句话,就可以得出答案 D。

18. How much did the girl have to repay the old man?

【答案】B

【详解】本题是道关于数字的细节题,文中出现了"40 pounds"和"2000 pounds",但考生只要听到"She had to find 40 pounds to repay the man..."便不难选出答案。

**Questions 19 to 22 are based on the passage you have just heard.**

19. What does the passage tell us about the alcoholic?

【答案】C

【详解】本题要求考生综合理解全篇大意。A 项文中并未提及,B 和 C 项与文义相反。文中最后提到"... needs medical attention and sound advice."可知应选 C 项。

20. By what means does Alcoholics Anonymous help alcoholics?

【答案】A

【详解】本题是道细节题,要求考生正确理解"... through the sharing of mutual problems... by any one wishing to discuss alcohol – related problem."在理解此句基础上,不难得出正确答案为 A。

21. What is the main goal of Alcoholics Anonymous?

【答案】D

【详解】本题考查考生对篇章大意的理解能力。A、B 项文中并未提及,而 C 项 keep away from any alcoholic drinking 的说法过于绝对,因此只有 D 项为正确答案。

22. What do we know about the laws against drunken driving?

【答案】C

【详解】本题考查考生对文中细节的掌握,考生听清关键句"While laws against drunken driving must be as severe as possible...",就不难得出正确答案 C。

## Passage Three

The classroom teacher in the elementary school is in strategic position to influence attitudes. This is true partly because children acquire attitudes from those whose word they respect.

Another reason it is true is that pupils often search somewhat deeply into a subject in school that has only been touched upon at home or has possibly never occurred to them before. To a child who had previously acquired little knowledge of China, his teacher's method of handling such a unit would greatly affect his attitude toward Chinese.

The teacher can develop proper attitudes through social studies, science matters, the very atmosphere of the classroom, etc. However, when children come to school with undesirable attitudes, it is unwise to attempt to change their feeling by criticizing them. The teacher can achieve the proper effect by helping them obtain constructive experience.

Finally, a teacher must constantly evaluate her own attitudes, because her influence can be harmful if she has personal prejudices. This is especially true in respect to controversial issues and questions on which children should be encouraged to reach their own decisions as a result of objective analysis of all the facts.

**Section C**

The cost is going up for just about everything, and college (26) tuition is no (27) exception. According to a nationwide (28) survey the College Board's Scholarship Service, tuition at most American universities will be on an (29) average of 9 percent higher this year over last year.

The biggest increase will occur at (30) private colleges. Public colleges, heavily subsidized by tax funds, will also (31) increase their tuition, but the increase will be a few (32) percentage points lower than their (33) privately sponsored neighbors. (34) Ten years ago, the tuition was only $ 2,150. To put that another way, the cost has climbed 150 percent in the last decade. (35) An additional burden is placed on out-of-state students who must pay extra charges ranging from $ 200 to $ 2,000, (36) and foreign students who are eligible for scholarships at state-funded universities.

**Questions 23 to 25 are based on the passage you have just heard.**

23. Why may teachers have great influence on children's attitudes?

【答案】C

【详解】本题考查考生对细节的理解能力,根据原文"...from those whose word they respect."可知C为正确答案。

24. Through which of the following factors a teacher can NOT develop proper attitudes of students?

【答案】D

【详解】本题需要考生抓住关键词"unwise",可知"...change their feeling by criticizing them."是不可取的方法,因此D为正确答案。

25. Why must a teacher constantly evaluate her own attitudes?

【答案】A

【详解】本题考查考生对文中细节的推断能力。根据"a teacher must constantly evaluate her own attitudes,..."可知A最符合文义。

**Section C**

26. tuition 27. exception 28. survey 29. average
30. private 31. increase 32. percentage 33. privately
34. Ten years ago, the tuition was only $ 2,150
35. An additional burden is placed on out-of-state students
36. and foreign students who are eligible for scholarships at state-funded universities

# Unit 5

## 录音文字材料

**Section A**

1. M: Excuse me, would you please tell me when the next flight to Washington is?
   W: Certainly. The next direct flight to Washington is two hours from now, but if you do not mind transferring at San Francisco you can board now.
   Q: What can you learn from the conversation?
2. W: Bob, are you going straight home after work today?
   M: No. I have a meeting until one o'clock and after that I'm go-

## 答案与详解

1.【答案】A

【详解】由对话可知男士想去 Washington。

2.【答案】B

【详解】计算题。注意 a couple of hours 指 2 个小时,所以选

ing to spend a couple of hours at the laboratory before going home.

Q: When is Bob going home this afternoon?

B。

3. M: I'm sorry, madam. The train is somewhat behind schedule. Take a seat and I'll tell you as soon as we know something definite.

W: Thank you. I'll just sit here and read the newspaper in the meantime.

Q: What can you conclude about the train from the conversation?

3.【答案】C

【详解】关键词是 behind schedule,意为落后于预定计划(或时间)。

4. M: That's a lovely necklace you are wearing.

W: Oh, thank you. My husband gave it to me for my fortieth birthday.

Q: What did the woman say about the necklace?

4.【答案】B

【详解】女士说项链是她的丈夫送给她的 40 岁生日礼物。

5. W: Are you going to learn to ride a motorcycle, Tom? You only have to pay fifteen dollars for the driving license.

M: No, I'm not. I'm afraid of breaking my leg. So I always prefer to take the bus.

Q: Why doesn't the man like to motorcycle?

5.【答案】C

【详解】男士不愿学骑摩托车而宁愿乘公车是因为怕弄伤腿,可见他认为骑摩托车很危险。

6. W: Have you finished your term-paper? I handed mine in last Friday.

M: I finished typing mine at four o'clock this morning. I have to submit it by noon today.

Q: When must the boy turn in his paper?

6.【答案】D

【详解】关键词是 noon,中午之前即 12 点之前。

7. W: Excuse me, sir, could you tell me where the laundry is, please?

M: Walk down the street for a block. The laundry is right on the left-hand corner.

Q: What do you learn from this conversation?

7.【答案】D

【详解】此题易错选 A 项。男士只是告诉女士洗衣房的地点,但并没有给她带路。

8. W: I haven't seen you for such a long time. Where have you been?

M: I have had a couple of meetings, one in Australia and one in Switzerland.

Q: How many meetings did the man attend?

8.【答案】B

【详解】一个在澳大利亚,一个在瑞士。

**Conversation One**

W: Dad, can I go to a movie this week with Shannon?

M: Here. Try this. It's called a book. Moby Dick. An Amercian classic. Okay. Let me look at the calendar here. Hmm. When are you thinking about going to a movie?

W: Uh, we're thinking about seeing a movie on Wednesday after school.

M: Well, that's not going to work. You have piano lessons after school and then you have to babysit for the neighbors until 9:00.

W: What about Monday?

M: Monday's out. You haven't practiced your clarinet at all... for an entire month, so you have to catch up on that. And, don't you have an essay due in your English class on Tuesday?

W: Oh, I forgot about that, and anyway, I was going to finish that during first period at school. So, what about Tuesday?

M: Uh, you have soccer practice from 4:00 until 5:30, and after that, you have to do your homework.

W: Ah, you can help me with that. Oh, I forgot you don't know how to do geometry. So, can I see the movie on Thursday?

**Questions 9 to 12 are based on the conversation you have just heard.**

9. What does the girl have to do on Wednesday after school?

【答案】D

【详解】星期三放学后女孩要上钢琴课,还要为邻居照顾小孩一直到 9 点钟,所以没有时间去看电影,因此答案是 D。

10. Why can't the girl go to a movie on Monday?

【答案】B

【详解】女孩星期一也不能去看电影是因为她要练竖琴,还要完成一篇文章在星期二的英语课上要交,因此答案是 B。

11. How long is her soccer practice on Tuesday?

【答案】B

【详解】女孩星期二练足球时间是 4:00 – 5:30,计算后得出她总共要练一个半小时的球,因此选 B。

12. What chore does the girl have to do on Saturday?

【答案】A

【详解】星期六女孩的安排是上午做家务,下午帮爸爸整理车库,符合答案的只有 A。

M：Well, remember the science fair at school is on Friday, right? Is, is your project finished yet?

W：Umm, what about Friday night? I checked the paper, and there's a midnight showing.

M：Uh – uh. Forget that idea.

W：And Saturday?

M：Well, you have to do your chores in the morning before noon. And then, we have to clean out the garage. Well, that should only take a couple of hours. And then, after that, we can go to the movie.

W：We?

M：Yeah, We. Mom and I and you and Shannon. Now, let me check the paper for showtimes. You already checked, I see.

**Conversation Two**

W：Well, hi Mr. Brown. How's your apartment working out for you?

M：Well, Mrs. Nelson. That's what I would like to talk to you about. Well, I want to talk to you about that noise! You see. Would you mind talking to the tenant in 4B and ask him to keep his music down, especialily after 10:00?

W：Ohhh. Who me?

M：Why yes. The music is blaring almost every night, and it should be your job as manager to take care of things.

W：Hey, I just collect the rent. Besides, the man living there is the owner's son, and he's a walking refrigerator. Hey, I'll see what I can do. Anything else?

M：Well, yes. Could you talk to the owners of the property next door about the pungent odor drifting this way.

W：Well, the area is zoned for agricultural and livestock use, so there's nothing much I can do about that.

M：Well, what about the... That, that noise.

W：What noise? I don't hear anything.

M：There, there it is again.

W：What noise?

M：That noise.

W：Oh, that noise. I guess the military has resumed its exercises on the artillery range.

M：You have to be kidding. Can't anything be done about it?

W：Why certainly. I've protested this activity, and these weekly activities should cease... within the next three to five years.

M：Hey, you never told me about these problems before I signed the rental agreement.

**Section B**

**Passage One**

Waves are beautiful to look at, but they can destroy ships at sea, as well as houses and buildings near the shore. What causes waves? Most waves are caused by winds blowing over the surface of the water. The sun heats the earth, causing the air to rise and the winds to blow. The winds blow across the sea, pushing little waves into bigger and bigger ones.

The size of a wave depends on how strong the wind is, how long it blows, and how large the body of the water is. In a small bay big waves will never build up. But at sea the wind can build up giant, powerful waves.

**Questions 13 to 15 are based on the conversation you have just heard.**

13. Why is the woman hesitant about carrying out the first request?

【答案】C

【详解】根据她对所有反映问题的态度,我们可以推断她之所以这样说,是怕顾客会让她处理这些事情,以至于难于解决,还造成与老板间的矛盾。所以应该选择 C。

14. What is the man's second complaint?

【答案】A

【详解】第二个抱怨:Could you talk to the owners of the property next door about the pungent odor drifting this way? 这里 pungent odor 就是很难闻的刺激性气味。

15. What is the source of the man's third complaint?

【答案】C

【详解】I guess the military has resumed its exercises on the artillery range. resume its exercises 就是 drill 演习的意思,也就是 training,答案为 C。

**Questions 16 to 18 are based on the passage you have just heard.**

16. What cause the formation of the waves?

【答案】B

【详解】材料说"Most waves are caused by winds blowing over the surface of the water."B 项符合。

17. Why can't the big waves be built up in a small bay?

【答案】B

【详解】材料中提到"The size of a wave depends on how strong the wind is, how long it blows, and how large the body of the water is."所以选 B。

18. What's the height of the largest measured wave in history?

A rule says that the height of a wave (in meters) will usually be no more than one - tenth of the wind's speed (in kilometers). In other words, when the wind is blowing at 120 kilometers per hour, most waves will be about twelve meters. Of course, some waves may combine to form giant waves that are much higher. In 1993 the United States Navy reported the largest measured wave in history. It rose in the Pacific Ocean to a height of thirty-four meters.

### Passage Two

Computers are now part of our daily lives. With the price of a small home computer now as low as $150, experts predict that before long all schools and businesses and most families in the richer parts of the world will own a computer of some kind. Among the general public, computers arouse strong feelings - people either love them or hate them.

The computer lovers talk about how useful computers can be in business, in education and in the home. Apart from all the games, you can do your accounts on them, learn languages from them, write letters on them, use them to control your central heating, and in some places even do your shopping with them. Computers, they say, will also bring some more leisure, as more and more unpleasant jobs are taken over by computerized robots.

The haters, on the other hand, argue that computers bring not leisure but unemployment. They worry, too, that people who spend all their time talking to computers will forget how to talk to each other. And anyway, they ask, "What's wrong with going shopping, using pens and typewriters?" But their biggest fear is that computers may eventually take over human beings together.

### Passage Three

One of American's most important exports is her modern music. American popular music is played all over the world. It is enjoyed by people of all ages in all countries. The reasons for its popularity are its fast pace and rhythmic beat. The music has many origins in the United States. Country music coming from the rural areas in the Southern United States is one source. Country music features simple themes and melodies, describing day-to-day situations and feelings of country people. Many people appreciate this music because of the emotions expressed by country music songs. A second origin of American popular music is the blues. It describes mostly sad feelings reflecting the difficult lives of American blacks. It is usually played and sung by black musicians but it is popular with all Americans. Rock music is a newer form of music. This music style featuring fast and repetitious rhythm was influenced by the blues and country music. It was first known as rock - and - roll in the 1950s. Since then, there have been many forms of rock music: hard rock, soft rock, pop rock, disco music and others. Many performers of popular rock music are young musicians. New popular songs are heard on the radio several times a day.

**Section C**

In the United States, people appear to be (26) constant on the move. Think for a moment, how often do you see moving (27) vans

【答案】C

【详解】1993年美国海军部报道的史上测出的最大海浪在太平洋海域升至34米高。

**Questions 19 to 21 are based on the passage you have just heard.**

19. According to the passage, what is predicted by experts for computers in the near future?

【答案】C

【详解】根据文中"expert predict that before long all schools and businesses and most families in the richer parts of the world will own a computer of some kind."可知答案为C。

20. Which of the following is not mentioned by computer - lovers in the passage?

【答案】D

【详解】材料并未谈及电脑的作曲功能。

21. Which is one of the main reasons that computer - haters have against computers?

【答案】A

【详解】从文中"The haters, on the other hand, argue that computers bring not leisure but unemployment."可知选择答案A。

**Questions 22 to 25 are based on the passage you have just heard.**

22. What makes American popular music so popular throughout the world?

【答案】A

【详解】文章提到"The reasons for its popularity are its fast pace and rhythmic beat."可知答案为A。

23. Where does country music come from?

【答案】D

【详解】根据文章中提到的"Country music coming from the rural areas in the Southern United States is one source."可知选择D项。

24. What do you know about the blues?

【答案】C

【详解】蓝调主要描述美国黑人的艰难生活的忧伤情绪，通常是黑人音乐家演奏并演唱,但它受到全美国人的欢迎。

25. How many forms of rock music are especially mentioned in the passage?

【答案】B

【详解】材料谈到 hard rock, soft rock, pop rock, disco music,可知答案为B。

**Section C**

| 26. constant | 27. vans | 28. everywhere |
|---|---|---|
| 29. actually | 30. indeed | 31. period |

on the roads? They seem to be (28) everywhere. Are so many people (29) actually changing their addresses? Yes, people in the United States are (30) indeed on the move. Within any five years (31) period about one third of the (32) population change their place of (33) residence.

(34) Every person who moves has his or her own personal reasons for making such a decision. Some people may decide to move because of employment opportunities. (35) Some may wish to live in a warmer or a colder climate. And some have many other reasons. (36) Regardless of the specific causes the amount of movement in this country is substantial.

32. population 33. residence
34. Every person who moves has his or her own personal reasons for making such a decision
35. Some may wish to live in a warmer or a colder climate
36. Regardless of the specific causes the amount of movement in this country is substantial

# Unit 6

## 录音文字材料

### Section A

1. W: Can you afford to buy a house?
   M: We'll have to get a loan, of course.
   Q: Do you think the man has enough money to buy a house?
2. W: The plane leaves at 6:15. Do we have time to eat first?
   M: No, we've only got 40 minutes before departure time.
   Q: What time is it now?
3. W: When did you arrive in Hong Kong?
   M: I left London on the morning of the 12th and landed in Hong Kong on the evening of the 13th.
   Q: How long did the man spend on his traveling?
4. M: It's getting late and the sea looks wavy. We'd better head home.
   W: The winds are against us just now. It'll take hours.
   M: That's all right because if it gets dark, the harbor is well-lighted.
   Q: How are the man and the woman traveling?
5. M: Mr. Black is fluent in Spanish and now he's beginning to study Arabic.
   W: He also knows a few words in Japanese and Chinese.
   Q: Which language does Mr. Black speak well?
6. M: I have a single and a double room for two nights.
   W: Forty-five pounds a night, plus VAT.
   Q: What is the total price for two nights?

## 答案与详解

1.【答案】C
【详解】此题主要考查对对话含义的理解。男士回答"We'll have to get a loan."可知他需要贷款才能买房,所以C为正确答案。

2.【答案】C
【详解】此题主要考查时间理解能力。文中出现了两个时间6:15和40分钟,男士回答"We're only got 40 minutes before departure time."句意为他们离出发时间还有40分钟,那么现在的时间就是5:35。

3.【答案】D
【详解】此题主要考查对信息细节的理解,对话中出现了两个时间12号的早上和13号的晚上,从男士的回答中"I left London on the morning of the 12th and landed in Hongkong on the evening of 13th."可知他12号早上登机,13号晚上到达,这样只有D项符合答案。

4.【答案】A
【详解】此题主要考查对对话中关键词的理解,例如"the sea look wavy""the wind""the harbor is well - lighted",从以上词中可判断他们是出海旅行,所以他们乘坐的是船,A为正确选项。

5.【答案】A
【详解】此题主要考查对句子关键词的理解能力,对话中出现了Spanish, Arabic, Japanese and Chinese几种语言,这就给学生造成了选择障碍,听的时候一定要抓住关键的词,对话中男士说"Mr Black is fluent is Spanish...",fluent意为"流利",这样Spanish就是Mr Black说得好的语言,所以A为正确选项。

6.【答案】C
【详解】此题主要考查对信息细节的理解,对话中女士说"Forty - five pounds a night, plus VAT." VAT意为"增值税",两晚上的房费为2×45 = 90,再加上增值税就超过了90元,所以C为正确选项。

7. M: You really seem to enjoy literature class.
   W: I sure do! It's opened new worlds for me. I'm exposed to the thoughts of some of the world's best writers. I've never read so much in my life.
   Q: Why is she so excited?
8. W: Can you recommend some universities with good graduate schools?
   M: Well, generally in the USA, each university has its own specialty which is particularly outstanding. A large university is not necessarily good in every field.
   Q: What does the man mean?

**Conversation One**

M: Oh. Hi there. A beauty, isn't she?

W: Well...

M: Do you want to take her a test ride?

W: Well... Um. How old is it?

M: Well, it's only three years old.

W: And what's the mileage?

M: Uh, let me check. Oh yes, 75,000 miles.

W: 75,000 miles? That's quite a bit for a car that's only three years old.

M: Well, once you're in the driver's seat, you'll fall in love with her. Get in.

W: Ugh... Uh, I can't seem to get the door open. [Ah, it's okay.] It could be broken.

M: Ah, just give her a little tap. Ugh. Now she's opened.

W: Great. A door I have to beat up to open.

M: Hey. Get in and start her up.

(Woman tries to start the car...)

M: Well, it's probably the battery. I know she has enough gas in her, and I had our mechanic check her out just yesterday. Try it again.

W: Uh. It sounds a little rough to me. [Well...] How much is this minivan anyway?

M: Oh. It's a real bargain today and tomorrow only at $15,775, plus you get the extended warranty covering defects, wear, and tear beyond the normal maintenance on the vehicle for an extra $500 for the next 30,000 miles. [Oh...] with a few minor exclusions.

W: Like...?

M: Well, I mean, it covers everything except for the battery, and light bulbs, and brake drums, exhaust system, trim and moldings, upholstery and carpet, paint, tires... Well, a short list, you know.

W: Uh. Well, almost $16,000 is a little out of my price range, plus the seats covers are torn a little.

M: Well, hey, I might be able to talk the manager into lowering the price another two hundred dollars, but that's about all.

W: No thanks. I think I'll just keep looking.

**Conversation Two**

W: He's here. Bye Dad.

M: Wait, wait, wait... Where are you going?

7.【答案】A

【详解】此题主要考查对话内容的理解。男士问她是否真的对文学课感兴趣,女士说"I sure do!... I've never read so much in my life."可知她对阅读很感兴趣,所以A为问题的正确答案。

8.【答案】A

【详解】此题主要考查对说话人意见的判断,从男士的回答"each university has its own speciality... A large university is not necessarily good in every field."他没有正面回答女士的问题,但从他的回答可知,每个学校都有自己独特的优势,给她推介一个好学校不容易,B、C、D三项明显与对话内容不符,所以A为正确答案。

**Questions 9 to 12 are based on the conversation you have just heard.**

9. How old is the minivan the woman is looking at?

【答案】A

【详解】这个题目考的是细节,顾客问How old is it? 男士回答到:It's only three years old. 所以选A。

10. What is the vehicle's mileage?

【答案】C

【详解】这是细节加数字区分的问题。是75,000 miles。

11. What is the problem with the minivan's door?

【答案】B

【详解】这是个隐含意思推断的问题。Beat up to open 就是说很难打开。

12. What is one thing the extended warranty would NOT cover on the vehicle according to the conversation?

【答案】C

【详解】这是个综合细节的考查题目,要学会辨别有用信息,找到关键点 not cover…所以要听出没有涉及的部分。

**Questions 13 to 15 are based on the conversation you have just heard.**

13. What kind of movie is the girl going to see on her date?

W: Dad. I've already told mom. I'm going out tonight.

M: Who with? You mean you're going on a date?

W: Yeah. Mom met Dirk yesterday. He's so cool. We're going on a double-date with Cindy and Evan.

M: Dirk?

W: I have to go.

M: Wait, wait. I want to meet this guy.

W: He's waiting for me.

M: Well, so what are you going to do tonight? Going to the library?

W: Dad! We're going out to eat, and then we're going to catch a movie.

M: What movie and what is it rated?

W: It's a science fiction thriller called... well, I don't know what it is called, but it's rated PG.

M: And where's the movie showing?

W: Down at the Campus Plaza Movie Theater.

M: Hey, I was thinking about seeing a movie down there tonight, too.

W: Ah, Dad.

M: Hey, let me meet that guy.

(Father looks out the living room window...)

M: Hey, that guy has a moustache!

W: Dad. That's not Dirk. That's his older brother. He's taking us there! Can I go now?

M: Well...

W: Mom said I could, and mom knows his parents.

M: Well...

W: Dad.

M: Okay, but be home by 8:00.

W: Eight!? The movie doesn't start until 7:30. Come on, Dad.

M: Okay. Be back by 11:00.

W: Love you, Dad.

M: Love you, too.

W: Bye.

M: Bye.

**Section B**

Passage One

The President's helpers work in office buildings. But the President does his work in the White House. This is where he and his family live. A few lucky children have called the White House "home" for a while. One of President Theodore Roosevelt's boys once took a pony for a ride in the White House elevator!

All of our Presidents except George Washington have lived in the White House. Americans are proud of this fine building.

At first the White House was not white. It was made of gray stone and called the President's Palace. President Adams made his home there in 1800, even though the building was not finished. Mrs. Adams used to hang her washing in the East Room to dry! Today the white-and-gold East Room is quite different. It is used for great occasions, not for hanging up the wash!

During the War of 1812, the President's Palace was burned by British soldiers. Afterwards, it was rebuilt. The walls were painted white to cover up marks of the fire. People then began calling the Pres-

【答案】C

【详解】It's a science fiction thriller... 证明了该电影题材为C——科幻片。

14. How is the girl getting to the movie?

【答案】B

【详解】Dirk 的哥哥将会来接她，所以应该是 Someone is coming to pick her up. 选 B。

15. What time does the movie begin?

【答案】A

【详解】The movie doesn't start until 7:30 说明了电影7:30才开始上演，答案为 A。

**Questions 16 to 18 are based on the passage you have just heard.**

16. What is the passage about?

【答案】A

【详解】此题主要考查全文大意。从全文看，主要是围绕着白宫叙述，写了几位总统在白宫里的生活，综合比较，A选项比较全面和准确，所以 A 为正确答案。

17. Which of the following is implied but not stated?

【答案】D

【详解】此题考查对文章信息的理解能力，从文中第二段"All of our Presidents except George Washington have lived in the White House."可知 A、B 两项错误，在文中"Today the white-and-gold East Room is quite different. It is used for great occasions, not for hanging up the wash."已说明 C 项内容，D 项 Lincoln 为美国总统，所以他也居住过白宫，只是文中没有叙述，所以 D 为正确选项。

18. Which of the following is not true?

【答案】C

ident's home the "White House". The name caught on and has remained in use ever since.

Passage Two

The cockroach is a common house pest. This insect lives in cracks, under floor, behind baseboards, or in dark, damp crevices. The cockroach does not like bright light. Because of this, cockroach usually comes out only at night to eat whatever food or garbage they can find. Homes can usually be kept free of cockroaches by keeping the rooms clean and dry.

The clothes moth is another household pest. The female clothes moth lays her eggs in anything made of wool or of hair. These eggs hatch into caterpillars that eat the fabric around them. Since moths do not like sunlight or fresh air, airing clothing and carpets in spring helps reduce the damage done by moths.

Termites are pests that eat wood. These creatures often get into the wooden parts of a house and do a great deal of damage. To avoid termites, the wooden parts of houses should be built well above the ground. Also, the parts of houses that do touch the ground should be made of stone, brick, or concrete.

Passage Three

Why is the fuse often called the twenty-four-hour policeman? Fuse is an electrical safety device. When too much electric current is flowing, a piece of metal in the fuse melts. This breaks the circuit. Without a fuse to break the circuit, very strong current could cause a fire.

Just as the policeman stops traffic to make sure that the cars don't crash into one another, so a fuse in our home stops electrons from overcrowding. For when too many lines are plugged into one outlet, the electrons, like the cars begin crashing into each other. When that happens, the extra movement makes the wires warm. The wires may get so hot that the walls of the house could catch fire.

The fuse stops these electrons just as they begin to get hot. "Stop", says the policeman. "Blow," says the fuse. The lights go out. The toaster stops toasting and the broiler stops broiling.

The little fuse box is our policeman. It is guarding our homes and our lives day and night.

**Section C**

British people's (26) attitude towards life is to (27) relax and enjoy themselves whenever they are free to do so. They (28) emphasize a lot on being (29) independent and having their (30) own way during a (31) dinner or party. I (32) found it quite different from the Chinese way of (33) entertaining people.

【详解】此题主要考查对文章信息的理解能力。文中"Presidest Adoms... even though the building was not finished."可知A项正确,文中"the President... by British soldiers"可知B项正确。文中"It was... called the President's Palace."可知D项正确,文中"it was rebuilt"可知白宫只是重建,并没有完全烧毁,所以C项错误。

**Questions 19 to 21 are based on the passage you have just heard.**

19. According to the passage, what's the usual way to keep homes free of cockroaches?

【答案】C

【详解】此题主要考查抓住文中的关键句。从文中"Home can usually be kept free from cockroaches by keeping the rooms clean and dry."可知C为正确选项。

20. Where do female moths lay their eggs?

【答案】B

【详解】此题主要考查抓住文中的关键句。从文中"the female clothes moth lays her eggs in anything made of wool or of hair."可知B为正确答案。

21. What do termites eat?

【答案】D

【详解】此题主要考查抓住文中的关键句,从文中"Termites are prests that eat wood."可知D为正确答案。

**Questions 22 to 25 are based on the passage you have just heard.**

22. What is this passage about?

【答案】D

【详解】此题主要考查文章大意。从文章第一句就可找到全文的主旨句,全文主要是讲述的是"Fuse",所以D为正确答案。

23. What is implied but not stated in the passage?

【答案】B

【详解】此题主要考查文章信息理解能力。文章第二段"The wires may get so hot that the walls of the house could catch fire."可知如果没有保险丝盒的保护,就可能会发生火灾,侧面说明了B项是正确的选项。

24. In order to prevent the danger of fire, what should we avoid?

【答案】C

【详解】此题主要抓住文章关键句,从文中"For when too many lines are plugged into... the house could catch fire."可知C为正确答案。

25. Which of the following is not true?

【答案】A

【详解】此题主要考查文章信息理解能力,从文章第一、二段可判断B、C、D三项均符合原文内容,从原理上判断A不正确,所以A为正确答案。

**Section C**

26. attitude　　27. relax
28. emphasize　　29. independent
30. own　　31. dinner
32. found　　33. entertaining
34. people ate whatever they liked and even could ask for their

At several parties I had been invited to in London, I observed that (34) people ate whatever they liked and even could ask for their favourite drinks or food. If the weather was nice, some guests might take their drinks or food to the garden, and some might just sit in the living room, chatting over the background music. (35) The host seldom asked people what they preferred or that they should eat more. It was the same for dinners. Guests were given things they asked for. If they said no the first time they were offered something, (36) they would never expect the host to repeat the request.

favourite drinks or food

35. The host seldom asked people what they preferred or that they should eat more

36. they would never expect the host to repeat the request

## Unit 7

### 录音文字材料

**Section A**

1. W: It's so hot today. I was going to wash the car and water the plants, but now I'm not sure I should.
   M: I agree. You should relax in heat like this. Too much hard work could cause you to faint.
   Q: What will she probably do?
2. W: The baby has just fallen asleep. Do you think you can keep the child quiet for a while?
   M: As quiet as mice. I'll have them draw some pictures.
   Q: What does the man mean?
3. M: I was sorry to hear about Bill's being fired. I know that he was sick a lot and that he usually went to work late.
   W: Oh, it wasn't that. Bill made a big error in last month's accounting. Even though it wasn't really his fault, his boss was very angry.
   Q: Why did Bill lose his job?
4. M: How much are the tickets, please?
   W: They are $ 3 each for adults, but students can get in at half price.
   Q: How much will the man pay if he gets two adult tickets and two student tickets?
5. M: He's Tim Johnson, an airplane pilot. Airplane pilots spend a lot of time away from home and they have a lot of responsibility.
   W: I think it's an interesting job. They see a lot of interesting places. And they earn a good salary.
   Q: What are they talking about?
6. M: May I see Professor Kent, please?
   W: I'm sorry, he's been ill since last Friday. I think he'll be back for his regular office hours next Thursday, but you might telephone Tuesday or Wednesday to make sure.
   Q: On what day will the man be able to see Professor Kent?
7. W: I don't understand how you got a ticket. I always thought you were a careful driver.
   M: I usually am, but I thought I could make it before the light turned.

### 答案与详解

1.【答案】D

【详解】此题主要考查对人物行为的判断。从女士说"I'm nost sure I should"可知她现在不想做了。从男士的回答也可看出"You should relax... cause you faint."综上可判断女士什么也不会做,所以D为正确答案。

2.【答案】C

【详解】此题主要考查对对话含义的理解。从男士的回答"As quiet as a mice",可知他能使孩子们很安静,所以C为正确答案。

3.【答案】B

【详解】此题主要考查对对话内容的理解,从女士的回答"Bill made a big error... his boss was very angry."可知Bill是因为工作失误而失去了工作。所以B为正确答案。

4.【答案】D

【详解】此题主要考查对数字的理解。two adult: $3\times2=6$, two children: $1.5\times2=3$,这个人要付的钱数为: $6+3=9$,所以A为正确答案。

5.【答案】A

【详解】此题主要考查对对话内容的理解。从对话中可知两人主要谈到飞行员这项工作,所以A为正确答案。

6.【答案】C

【详解】此题主要考查对时间的判断。从女士回答"I think he will be back for his regular office hours next Tursday...",其中A、B、D三项可能会造成混淆,听的时候要加以辨别,所以C为正确答案。

7.【答案】B

【详解】此题主要考查对对话含义的理解。从男士的回答"I though I could make it before the light turned."这里是个虚拟语气,他原以为在红灯亮之前可以通过,可知他是闯了红

Q: Why did the man get a ticket?

8. M: Of the two houses we saw today, which do you prefer?
 W: I think the white one is better, but the brick one has a bigger yard, so I like it better.
 Q: Why does the woman like the brick house better?

**Conversation One**

M: Yes, I'd like to report a theft.

W: Okay. Can you tell me exactly what happened?

M: Well, I was walking home from work two days ago, enjoying the nature all around me... the birds, the frogs, the flowing stream... [Okay, Okay] when this woman knocked me right off my feet, grabbed my stuff, and ran off through the trees. [Hmm]. I was so surprised by the ordeal that I didn't go after her.

W: Yeah. Can you describe the woman for me?

M: Yeah. She was about a hundred and ninety centimeters tall...

W: Wait. You said a woman robbed you.

M: Well, I'm not really sure. [Hmm] You see, the person was wearing a white and black polka dot dress, a light red sweater over it, and she... or he... was wearing a pair of basketball shoes.

W: Humm. What else can you tell me?

M: Okay. Like I said, the person was about 190 centimeters tall, heavy build, with long wavy hair. She... or he... was probably in his or her late 30's. I didn't get a good look at the person's face, but well... uh...

W: What? Was there something else?

M: Well, the person... had a beard.

W: Ah! What was, uh, taken... exactly?

M: Well, just my left shoe. Bizarre, isn't it?

W: Ah. The "bearded woman" has struck again!

M: The "bearded woman"?

W: Yeah. It's this man who dresses up like a woman and, for some unknown reason, removes the left shoe from his victims. He's really quite harmless, though, and he usually returns the shoe to the crime scene a couple of days later.

M: Hey, he can keep my shoe, and I'll just take off my left shoe every time I walk through the park.

**Conversation Two**

W: Hey Dave. How was your weekend?

M: Not bad. I went downtown to watch a flick with my roommate.

W: How was it?

M: Oh, the movie was awesome, it's really great. But the company wasn't.

W: What do you mean?

M: Well, I liked the movie, but my roommate is a real airhead. What a stupid person he is. We started talking about this and that, and he thought the Titanic was a Japanese boat which was sunk during World War II.

W: Oh, oh. Not a very bright guy.

M: Yeah, and he's a real couch potato. He just spends a lot of time in watching TV, especially some soap opera. He invited me over

灯,所以B为正确答案。

8.【答案】D

【详解】此题主要是对人物行为的判断,要注意女士的回答,通常"but"的转折部分才是主要部分,对话中"but the brick one has a big yard, so I like it better."可知女士更喜欢砖房的原因是它有一个更大的花园。所以D为正确选项。

**Questions 9 to 12 are based on the conversation you have just heard.**

9. What was the man doing when he was robbed?
 【答案】C
 【详解】I was walking home from work two days ago,他正享受大自然的鸟语花香,但不是在公园里,更不是在小溪边,是在下班的途中。

10. What was the thief wearing?
 【答案】B
 【详解】通过被抢者对贼的描述,只有B项符合,其余项和他描述的不一致。

11. How tall was the thief?
 【答案】C
 【详解】这是个数字辨听题,注意区分190和199的发音。答案为C。

12. Who is the "bearded woman"?
 【答案】A
 【详解】根据警方的最后判断:Yeah. It's this man who dresses up like a woman and, for some unknown reason, removes the left shoe from his victims. 可以推出答案为A。

**Questions 13 to 15 are based on the conversation you have just heard.**

13. Where did the man go to watch the flick?
 【答案】B
 【详解】I went downtown to watch a flick with my roommate,信息很直接就可以获得,是downtown,选B。

14. How did the man like the movie?
 【答案】A
 【详解】the movie was awesome, awesome的意思是great, fantastic and outstanding,还要区分awful和awesome的意思。答案是A。

15. Why did Dave decide to hit the sack?
 【答案】C
 【详解】首先要了解hit the sack的意思,这是个美国俚语,

to his parent's house to watch TV. Wow. I bet watching TV is his only hobby, but he sure doesn't know much.

W: Too bad. Hey, do you wanna go out and get something to drink? It's pretty early.

M: Nah. I'm gonna hit the sack. Usually I just go to bed around 12. But I have a test tomorrow morning, and I wanna be ready for it.

**Section B**

### Passage One

Do you know how the red and white striped pole became the sign of a barbershop? Many years ago, barbers not only cut hair but also served as doctors. Their main work as doctors was bleeding people. At that time it was believed that bleeding helped cure the sick. The white stripes on the pole represent the bandage with which the barber wrapped the patient after bleeding him. The red strips represent the blood. Long ago, a basin hung beneath the striped pole. This basin stood for the real basin used by barbers to catch the blood. Although barbers no longer do this kind of work, they have kept the sign.

Our word barber comes from "barba", a very old word meaning "beard". Barbers still shave and trim men's beards. Their chief work, however, is cutting hair. When women first began having their long hair cut, they went to barber shops. After shops that specialized in cutting women's hair appeared in the U. S., barbers once again served mostly men.

### Passage Two

Would you be surprised to see colored snow? Although snow itself is always white, it sometimes appears to be pink, brown, blue, or green! The most common cause of snow that seems colored is very small plants called algae, which have varying colors.

Algae are the simplest of all plants. Many of them have neither roots nor stems. Some of these simple plants live in the air. When snow falls, algae in the air may be carried down with it. The plants are too minute to be seen separately. Only their color is visible. Because of this, it seems that the snow has changed color.

Another cause of snow that appears colored is the red dust from sandstorms. This dust is sometimes carried hundreds of miles through the air. High in the air, it mixes with snow. When the snow falls, there is sometimes enough red dust mixed in with it to give it a pink color. In 1933, for example, a dust storm took place. Later in the year, dust from the storm caused snow falling in New England to look pink.

### Passage Three

Alaska, which became the forty-ninth state of the United States in 1959, was bought from Russia in 1867. The price paid to the Russian government for this huge piece of land was $7,200,000. Secretary of State Seward arranged the treaty and the purchase. Because people in the United States at that time knew little about Alaska, many of them did not approve of purchasing it. Some jokingly called Alaska "Se-

单独看你不清楚它的意思,但通过后面的一句,平常在12点睡,我们可以得知这个词语的意思是睡觉的意思。But I have a test tomorrow morning, and I wanna be ready for it. 就解释了原因,是因为明天有考试。

**Questions 16 to 18 are based on the passage you have just heard.**

16. What do the red strips and white stripes stand for respectively?

【答案】B

【详解】此题主要考查对文章信息细节的理解。文中"The white stripes on the pole represent the bandage... The red strips represent the blood."所以B为正确答案。

17. What is implied but not stated in the passage?

【答案】C

【详解】此题主要考查对文章信息细节的理解。A、B、D三项文中都已提及,从第一段"Although barbers no longer do this kind of work, they have kept the sign."可知B项内容符合原文,虽然文中并没直接说明此点。

18. What did the old word "barba" mean?

【答案】D

【详解】此题主要考查对文章信息的的把握。从原文"a very old word meaing beard"可知D为正确答案。

**Questions 19 to 21 are based on the passage you have just heard.**

19. Which of the following might be the best title for the passage?

【答案】D

【详解】此题主要考查对文章主旨的理解。文章第一句就点明主旨"Would you be surprised to see colored snow?"所以文章标题应与此有关,所以D为正确选项。

20. What is implied but not stated in the passage?

【答案】C

【详解】此题主要考查对文章信息的理解。A项从文中第二段可知明显错误。B项可从文中第一段最后一句找到答案。D项从原文"Many of them neither roots nor stems"可知此项错误,C项从原文"When snow falls, alage in the air may be carried down with it."可知alage可存在水里和地表,所以C为正确答案。

21. According to the passage, besides algae, what may change the color of snow?

【答案】B

【详解】此题主要考查对文章信息的题解。从文中"Another cause of snow that appears colored is the red dust from sandstorms."可知B为正确答案。

**Questions 22 to 25 are based on the passage you have just heard.**

22. Which of the following was the buyer?

【答案】B

【详解】此题主要考查对文章信息的理解。从文中第一句"Alaska, which became... state of the United States in 1959, was bought from Russia 1867."可知美国从俄罗斯购买了

ward's Folly".

However, Alaska proved to be a wonderful buy. Over $450,000,000 in gold has been taken from Alaska since it was bought. Alaskan streams and rivers are rich in fish, and so many salmon are caught each year in Alaska that it has developed the largest salmon canning industry in the world. Alaskan hills have thousands of acres of valuable timber. In addition, the area has many fur-bearing animals, such as seals, sea otters, minks, foxes, and beavers.

If Mr. Seward were alive today, he would be proud of his purchase.

Alaska,所以 B 为正确选项。

23. Why didn't many people approve of purchasing the huge piece of land?

【答案】D

【详解】此题主要考查对文章信息的理解。从文中"Because people in the United States at that time knew little about Alask, many of them didn't approve of purchasing it."可知 D 为正确选项。

24. Which of the following resources is not mentioned in the passage?

【答案】C

【详解】此题主要考查对文章信息的掌握。文中第二段提到了"gold, fish, timber"并没有提到 natural gas, 所以应选 C。

25. What did the author think of the buy?

【答案】A

【详解】此题主要考查对作者态度的判断。从文章最后一句可看出作者对 Seward's buy 是赞赏的态度,所以他应觉得这是一次好的购买,A 为正确答案。

**Section C**

Any mistake made in the (26) painting of a stamp raises its value to stamp (27) collectors. A mistake on a (28) two-penny stamp has made it worth a million and a half times its face value. Do you think it (29) impossible? Well, it is true. And this is how it (30) happened.

The mistake was made more than a hundred years ago in the former British (31) colony of Mauritius, a small island in the India Ocean. In 1847, an (32) order for stamps was sent to London. Mauritius was about to become the (33) fourth country in the world to put out stamps.

Before the order was filled and the stamps arrived from England, a big dance was planned by the commander-in-chief of all the armed forces on the island. (34) The dance would be held in his house and letters of invitation would be sent to all the important people in Mauritius. Stamps were badly needed to post the letters.

Therefore, an islander, who was a good printer, was told to copy the pattern of the stamps. (35) He carelessly put the words "Post Office" instead of "Post Paid", two words seen on stamps at that time, on the several hundred that he printed.

Today, there are only twenty-six of these misprinted stamps left—fourteen One-penny Reds and twelve Two-penny Blues. Because there are so few Two-penny Blues and because of their age, (36) collectors have paid as much as $16,800 for one of them.

**Section C**

26. painting 27. collectors 28. two-penny
29. impossible 30. happened 31. colony
32. order 33. fourth
34. The dance would be held in his house and letters of invitation would be sent to all the important people in Mauritius
35. He carelessly put the words "Post Office" instead of "Post Paid", two words seen on stamps at that time, on the several hundred that he printed
36. collectors have paid as much as $16,800 for one of them

# Unit 8

## 录音文字材料

**Section A**

1. M: Are you going to replace the light switch yourself?
   W: Why should I call an electrician?
   Q: What does the woman imply?
2. W: This doesn't look at all familiar. We must be lost. We'd better get some directions.
   M: Let's pull in here. While I'm filling the tank, you can ask about the directions and get me a soft drink.
   Q: Where will the man and woman go for assistance?
3. M: Professor Jones was born in Boston, wasn't he?
   W: I don't know. But I've heard him say he grew up in Chicago.
   Q: What does the woman know about Professor Jones?
4. W: Wasn't James lucky winning all the money?
   M: Yes, I wish I could help him spend it.
   Q: What would the man like to do?
5. W: How did you go to Canada? Did you fly?
   M: I was planning to, because it's such a long trip by bus or train. But Fred decided to drive and invited me to join him. It took us 2 days and a night.
   Q: Who went to Canada?
6. W: It's too bad you miss the class today. Professor Smith gave us a review session. And he said most of the exam will come from that.
   M: That means I'll have to do twice as much work for the exam.
   Q: How does man feel about the class he has missed?
7. W: Do you think it's good for me to go abroad for further study?
   M: If I were you, I'd be delighted with such a chance.
   Q: What does the man mean?
8. M: How was the dinner at the cafeteria tonight?
   W: It was noisy, as usual.
   Q: What does the woman mean?

**Conversation One**

M: Hi, Sis. I just came over to give you the DVDs you wanted, and... Hey, wow!? Where did you get all of this stuff?
W: I bought it. So, what do you think of my new entertainment center? And the widescreen TV...
M: Bought it?
W: ... and my new DVD player. Here, let me show you my stereo. You can really rock the house with this one.
M: But where did you get the dough to buy all this? You didn't borrow money from mom and dad again, did you?
W: Of course not. I got it with this!
M: This? Let me see that... Have you been using Dad's credit card again?
W: No, silly. It's mine. It's a student credit card.
M: A student credit card? How in the world did you get one of these?
W: I got an application in the mail.

## 答案与详解

1.【答案】A
【详解】女士回答"Why should I...",是反问的口吻。"为什么我要找电工呢?"可知她认为自己可以做好这件事。

2.【答案】A
【详解】由男士的话可知他说:当我去加油时,你可以去询问方向,同时买饮料。可判断他们将去加油站。

3.【答案】A
【详解】此题易错选为C,但由女士首先说:不知道他是否出生于 Boston,因此应选 A。

4.【答案】B
【详解】由男士的话"I wish I could help him spend it."可知他很希望花一部分 James 的钱。

5.【答案】B
【详解】由男士的话可知不仅他去 Canada,Fred 也要去。

6.【答案】D
【详解】由女士的话可知那节课很有价值,而男士说他不得不花费更多的功夫去弥补那节课的损失。

7.【答案】D
【详解】男士说,"假如我是你,我将会为有此机会而高兴。"可知 D 对。

8.【答案】A
【详解】由女士所说:It was noisy, as usual,像平常一样吵闹。可知她认为不是个好地方。

**Questions 9 to 12 are based on the conversation you have just heard.**

9. According to the conversation, which item did the woman NOT purchase with her credit card?
【答案】A
【详解】注意题干是什么东西没有买,采取排除法选 A。

10. According to the man, what is one reason for NOT having a credit card?
【答案】C
【详解】在对话中,the man 不赞成申请信用卡,因为会导致 impulse spending,而且"the interest rates of student credit cards are usually sky high",因此只有 C 符合答案。

11. What does the woman imply about how she plans on resolving her credit card problems?
【答案】A
【详解】当被问到怎样来支付账单时,the woman 提到她的

M: Well, why did you get one in the first place?

W: Listen. Times are changing, and having a credit card helps you build a credit rating, control spending, and even buy things that you can't pay with cash... like the plane ticket I got recently.

M: What plane ticket?

W: Oh yeah, my roommate and I are going to Hawaii over the school break, and of course, I need some new clothes for that so...

M: I don't want to hear it. How does having a student credit card control spending? It sounds you've spent yourself in a hole. Anyway, student credit cards just lead to impulse spending... as I can see here. And the interest rates of student credit cards are usually sky-high, and if you miss a payment, the rates, well, just jump!

W: Ah. The credit card has a credit limit...

M: ...of $20,000?

W: No, no quite that high. Anyway,...

M: I've heard enough.

W: Did I tell you we now get a digital cable with over 100 channels? Oh, and here's your birthday present. A new MP3 player...

M: Yeah. Oh, don't tell me you purchased them on the credit card. Listen. Hey, I don't think having a student credit card is a bad idea, but this is ridiculous. And how in the world are you going to pay your credit card bill?

W: Um, with my birthday money. It's coming up in a week.

M: Hey, let's sit down and talk about how you're going to pay things back, and maybe we can come up with a budget that will help you get out of this mess. That's the least I can do.

**Conversation Two**

M: Hi. I don't think we've met. My name's Tom.

W: Hi, Tom. Nice to meet you. My name is Juanita, but everybody calls me Jenny.

M: Nice to meet you, Jenny. So, where are you from?

W: Well, originally I'm from Argentina, but we moved to the United States when I was about five years old. My parents now live in Chile. That's where they first met. How about you, Tom?

M: I was born in Fresno, California, and we lived there until I was seven. Then, since my father worked for the military, we moved all over the place.

W: Oh yeah? Where are some of the places you've lived?

M: Mostly, we were overseas. We spent a total of ten years in Korea, Germany, and Okinawa, Japan. We were transferred back to the States three years ago, but I think my parents would have liked to live overseas for at least 20 more years.

W: Wow. It sounds like you've had an interesting life. So, what do you do now?

M: I'm a student at Purdue University.

W: Oh really? What are you studying?

M: I'm majoring in psychology. How about you? What do you do?

W: Well, I'm working as a sales representative for Vega Computers downtown.

M: No kidding! My brother works there too.

生日快到了,这样就可以拿她的 birthday money 来支付,因此符合答案的只有 A。

12. What is the man going to do for the woman to help her manage her money?

【答案】B

【详解】在对话的最后,the man 准备和 the woman 一起坐下来商讨怎样作出预算以解决问题,因此选 B。budget 预算。

**Questions 13 to 15 are based on the conversation you have just heard.**

13. About how old was the man when he returned to the United States?

【答案】C

【详解】The man 从出生开始一直在自己的家乡 Fresno, California 呆到 7 岁,之后随父亲工作在国外生活了 10 年,所以当他回国的时候应该是 17 岁。

14. What is the man studying?

【答案】C

【详解】这是一个细节题,The man 在对话中说他的专业是心理学,因此选 C。

15. What is the woman's job?

【答案】A

【详解】在对话的最后 Jenny 说她的工作是商品经销代理,因此选 A。

**Section B**

### Passage One

Ants have nurses for their babies just as some people have nurses for their children.

After the eggs are laid by the ant queen, they are picked up by the nurses and transported to another room. There they are watched until they hatch. At that time they are removed to still another special room, which is somewhat like a nursery.

When the ants are first hatched, they are very small and helpless. They can do nothing but wiggle about on the ground. The nurses must wash and feed them and keep them warm and safe. The nurses lick the babies to keep them clean, feed them several times a day, and at night carry them down to the lower chambers where it is warm. In the morning the nurses return the babies to the upper part of the nest. If the day is warm, the attendants may take them outside into the sunshine.

Baby ants need a great deal of care and attention as do some other baby insects and animals.

### Passage Two

An artist went to a beautiful part of the country for a holiday and stayed with a farmer. Every day he went out with his paints and his brushes and painted from morning to evening. Then when it got dark, he went back to the farm and had a good dinner before he went to bed.

At the end of his holiday he wanted to pay the farmer, but the farmer said, "No, I do not want money—but give me one of your pictures. What is money? In a week it will be finished, but your painting will still be here."

The artist was very pleased and thanked the farmer for saying such kind things about his paintings.

The farmer smiled and answered. "It is not that. I have a son in London. He wanted to become an artist. When he comes here next month, I will show him your picture, and then he will not want to be an artist any more, I think."

### Passage Three

Paper is one of the most important products ever invented by man. While in its spread use a written language would not have been possible without some cheap and practical material to write on. The invention of paper meant more people could be educated because more books could be printed and distributed. Together with the printing press paper provided an extremely important way to communicate knowledge.

How much paper do you use every year? Probably, you can not answer that question quickly. In 1900, the world's use of paper was about one kilogram for each person a year. Now some countries use as much as fifty kilograms for each person a year. Countries like U. S., England and Sweden use more paper than other countries.

Paper, like many other things that we use today, was first made in China. In Egypt and the West, paper was not very commonly used before the year 1400. The Egyptians worked on a kind of material made of a water bran. Europeans used parchment for hundreds of years. Parchment was very strong. It was made from the skin of certain young animals. We have learned of the most important facts of European history from records that were kept on parchment.

**Questions 16 to 18 are based on the passage you have just heard.**

16. What is the passage about?

【答案】B

【详解】由文章的第一句话可知本文主旨是 B。

17. What can you tell while not stated in the passage?

【答案】B

【详解】由文段"After the eggs are laid by the ant queen, they are picked up by the nurses..."可知。

18. Which of the following statements is not true?

【答案】C

【详解】由文段"They can do nothing but wiggle about on the grand."可知幼蚁并非一直睡觉。

**Questions 19 to 21 are based on the passage you have just heard.**

19. Why did the farmer want one of the artist's picture?

【答案】C

【详解】由末段"I have a son in London...I will show him your picture..."可知答案为 C。

20. How did the artist feel at first?

【答案】B

【详解】由"The artist was veyr pleased and thanked the farmer for saying such..."可知画家起初很高兴。

21. What did the farmer's last words mean?

【答案】B

【详解】最后一句话意为:他很想成为画家,我把你的画拿给他看,之后他就再也不会想成为画家了。言外之意就是说画家的画不好。

**Questions 22 to 25 are based on the passage you have just heard.**

22. What does the passage mainly talk about?

【答案】C

【详解】由首句"Paper is one of the most important products..."可知。

23. Which of the following is not mentioned in the passage?

【答案】D

【详解】由文段"The invention of paper meant..."可知 A、B、C 三项在文中都提及,只有 D 项文中不曾体现。

24. When did Egyptians begin to use paper widely?

【答案】B

【详解】由"In Egypt and the West, paper was not very commonly used before the year 1400."可知 B 对。

25. According to the passage, now how much paper does each person use a year in some countries?

【答案】A

【详解】由文句"Now some countries use as much as fifty kilograms for each person a year."可知。

## Section C

Visual aids offer several advantages. The (26) primary advantage is clarity. If you are discussing an object, you can make your (27) message clearer by showing the object or some (28) representation of it. If you are citing statistics, showing how something works, or demonstrating a (29) technique, a visual aid will make your (30) information more vivid to your (31) audience. After all, we live in a visual age. Television and movies have conditioned us to (32) expect a visual image. By using visual aids in our speeches, you often will make it easier for listeners to understand exactly what you are trying to (33) communicate.

Another advantage of visual aids is interest. (34) The interest generated by visual images is so strong that visual aids are now used routinely in many areas, not just speechmaking. Still another advantage of visual aids is retention. (35) Visual images often stay with us longer than verbal ones. We've all heard that works can "go in one ear and out the other." Visual images tend to last.

(36) For all these reasons, you will find visual aids of great value in your speeches. Let us look first at the kinds of usual aids you are likely to use, then at guidelines for preparing visual aids, and finally at some tips for using visual aids effectively.

## Section C

26. primary
27. message
28. representation
29. technique
30. information
31. audience
32. expect
33. communicate
34. The interest generated by visual images is so strong that visual aids are now used routinely in many areas
35. Visual images often stay with us longer than verbal ones
36. For all these reasons, you will find visual aids of great value in your speeches

# Unit 9

### 录音文字材料

## Section A

1. W: You don't feel very well, do you? You look pale. Have you got a high fever?
   M: Oh, no. But my stomach aches. Maybe the seafood doesn't agree with me.
   Q: What probably caused the man's stomachache?
2. M: I wish I could see Hans here.
   W: He was planning to come. But a few minutes ago his neighbor called to say that he had to take his father to the hospital.
   Q: Who was ill?
3. W: I need a car this Sunday, but mine has broken down.
   M: I'm sorry to hear it. But you can always rent one if you have a license.
   Q: What does the man mean?
4. W: Did you go to the concert last night?
   M: Oh, yes. It was supposed to start at 7:30. But it was delayed a quarter.
   Q: When did the concert start?
5. M: Is there anything you want me to get? I'm leaving now.
   W: Pick up a bottle of milk and a loaf of bread, please.
   Q: Where is the man probably going?
6. M: If I were you, I'd live in the city instead of going to work by train.
   W: But, you know, the country is so beautiful in spring and fall.

### 答案与详解

1.【答案】B
【详解】由男士的话"Maybe the seafood doesn't agree with me."可知男士的胃疼是由海鲜引起的。

2.【答案】B
【详解】由"he had to take his father to the hospital."可知 Han's father was ill.

3.【答案】D
【详解】由女士的话"But you can always rent one if you have a license."可知她并没有借车给那位男士的意思,而是建议他租用。

4.【答案】B
【详解】这是一道计算题,由"supposed to start at 7:30, was delayed a quarter"可知 7:30 加上 15 分钟即可。

5.【答案】D
【详解】由"pick up a bottle of milk and a loaf of bread"可知他将去买吃的和喝的,运用排除法,只能选 D 去杂货店。

6.【答案】B
【详解】由首句"If I were you, I'd live in the city instead of going to work by train"可知男士愿呆在城市,而女士的话中

Q: Where does the woman prefer to live?

7. M: Please buy a pack of cigarettes for me while you are at the store.

W: I'm not going to any store. I'm going to see Uncle Tom. But I will get them for you at the gas station.

Q: Where will the woman stop on her way?

8. M: Why didn't you stop when I first signaled?

W: Sorry. Would I have to pay a fine?

Q: What's the probable relationship between the man and woman?

**Conversation One**

M: Mom. Can I go outside to play?

W: Well, did you get you Saturday's work done?

M: Ah, Mom. Do I have to?

W: Well, you know the rules. No playing until the work is done.

M: So, what is my work?

W: Well, first you have to clean the bathroom including the toilet. And don't forget to scrub the bathtub.

M: No, I want to do the family room.

W: Well, okay, but you have to vacuum the family room and the hall, and be sure to dust everything. Oh, and don't forget to wipe the walls and clean the baseboards. [Okay.] And after that. [Oh, no.] Next, sweep and mop the kitchen floor and be sure to polish the table in the living room.

M: Okay. Okay.

W: And make your bed and pick up all your toys and put them away. And...

M: More?

W: Yeah. And then, how about going out for lunch and getting a big milk shake, but you probably don't want to do that.

M: No, No. I want to.

W: Okay. While you're doing your work, I'll wash clothes.

**Conversation Two**

W: Hello, Ultimate Computers. May I help you?

M: Yes, this is Jack Kordell from Hunter's Office Supplies. May I speak to Elaine Strong, please?

W: I'm sorry, but she's not in right now.

M: Okay, do you know when she'll be back?

W: Uh, yes, she should be here later on this afternoon maybe about 4:30. May I take a message?

M: Yes. Ms. Strong sent me a brochure detailing your newest line of laptop computers with a description of other software products, but there wasn't any information about after-sales service.

W: Oh, I'm sorry. Would you like me to fax that to you?

M: Yes, but our fax is being repaired at the moment, and it won't be working until around 2:30. Hum... could you try sending that information around 3:30? That should give me time to look over the material before I call Ms. Strong, say, around 5:00.

W: Sure. Could I have your name, telephone number, and fax number, please?

M: Yes. Jack Kordell and the phone number is 560-1287. And the

有转折词"But",可知她更愿在乡村,答案选择 B。

7.【答案】C

【详解】由女士的话"But I will get them for you at the gas station."可知她在加油站会停下来。

8.【答案】A

【详解】由女士的问话"Would I have to pay a fine?"可知她问男士是否要付罚款,再由排除法可知只有 A 对。

**Questions 9 to 12 are based on the conversation you have just heard.**

9. What does the boy want to do at the beginning of the conversation?

【答案】A

【详解】这样的题目要求注意听开始部分,"Mom? Can I play outside?"答案是 A。

10. What is one thing the boy is NOT assigned to do around the house?

【答案】A

【详解】对话中出现的 assignment 有 clean the bathroom including the toilet, vacuum the family room and the hall, dust everything, wipe the walls and clean the baseboards. 等,但 clean the bathroom including the toilet 是被 the boy 拒绝了。I want to do the family room. 所以是 A 答案。

11. What does the boy have to do in his bedroom?

【答案】B

【详解】关于 the boy's bedroom,他妈妈说要 make your bed and pick up all your toys and put them away,也就是 B。

12. What is the mother going to do while the boy is doing his household chores?

【答案】C

【详解】他妈妈说"While you're doing your work, I'll wash clothes."可见她要洗衣服,所以答案是 C。

**Questions 13 to 15 are based on the conversation you have just heard.**

13. Why can't Elaine Strong answer the phone?

【答案】B

【详解】男士想找到 Elaine Strong,但所得到的回答是:I'm sorry, but she's not in right now. 但并没说去干什么了,答案是 B。

14. What does the man want the woman to send?

【答案】A

【详解】男士说"there wasn't any information about after-sales service."即男士缺少售后服务的信息。所以需要女士发给他关于售后服务的信息。答案是 A。

15. What is the man's name?

【答案】C

【详解】文中已明确讲到:"It's Kordell with a 'K' and two 'l's. K-o-r-d-e-l-l.'"报名字时要注意别把字母顺序弄错了。所以答案是 C。

fax number is 560 – 1288.

W: Okay. Jack Kordell. Is your name spelled C – o – r – d – e – l?

M: No. It's Kordell with a "K" and two "l"s. "K – o – r – d – e – l – l."

W: All right, Mr. Kordell. And your phone number is 560 – 1287, and the fax number is 560 – 1288. Is that correct?

M: Yes, it is.

W: All right. I'll be sure to send you the fax this afternoon.

M: Okay, bye.

**Section B**

Passage One

A newlywed farmer and his wife were visited by her mother, who immediately demanded an inspection of the place. The farmer had genuinely tried to be friendly to his new mother-in-law, hoping that it could be a friendly relationship. However, as she kept nagging them at every opportunity, demanding changes, offering unwanted advice, and generally making life unbearable to the farmer and his new bride.

While they were walking through the barn, during the forced inspection, the farmer's mule suddenly reared up and kicked the mother-in-law in the head, killing her instantly. It was a shock to all no matter their feelings toward her demanding ways.

At the funeral service a few days later, the farmer stood near the casket and greeted folks as they walked by. The pastor noticed that whenever a woman would whisper something to the farmer, he would nod his head "yes" and say something. Whenever a man walked by and whispered to the farmer, however, he would shake his head "no", and murmur a reply.

Very curious as to this strange behavior, the pastor later asked the farmer what that was all about. The farmer replied, "the women would say, 'What a terrible tragedy' and I would nod my head and say 'Yes, it was.' The men would then ask, 'Can I borrow that mule?' and I would shake my head and say, 'Can't, it's all booked up for a year.'"

Passage Two

Psychology and sociology are both categorized as social sciences, and both study human behavior. However, psychology is the study of individual behavior. Psychology deals with the possible problems an individual might have in social interaction with other individuals, but the main concern of sociology is the ways that different societies with different cultures deal with each other.

Sociology asks and tries to answer questions like these: why does one society progress rapidly and another one remain primitive for centuries? What is the main reason for revolution in a society? What is the role of religion or art in a society?

Psychology asks and tries to answer questions like these: why does an individual adapt easily to a changing environment and another individual become mentally disturbed? What are the causes of antisocial behavior? What role does religion or art play in an individual's mental and emotional life?

Psychology and sociology often work together in their study of human behavior. It is assumed that by better understanding individual motivation and behavior, more will be learned about group motivation

**Questions 16 to 19 are based on the passage you have just heard.**

16. What was the mother's attitude to the new couple?

【答案】A

【详解】由文句"However,... demanding changes, offering unwanted advice,..."可知这位母亲对新婚夫妇要求很高。

17. Who killed the farmer's mother-in-law?

【答案】D

【详解】由第二段"while they were walking through the barn（棚舍）... killing her instantly."可知是D. A mule(一头骡子)杀死了岳母。

18. Who was curious about the farmer's strange behavior at the funeral service?

【答案】B

【详解】由"very curious as to this strange behavior, the pastor later asked..."可知在葬礼上对农民的举动觉得奇怪的是牧师。

19. Why couldn't the farmer lend his mule to the men at the funeral service?

【答案】C

【详解】由最后一句话"Can't, it's all booked up for a year."可知C对。

**Questions 20 to 22 are based on the passage you have just heard.**

20. What is the passage about?

【答案】A

【详解】游览全文可知全篇讲的都是关于 psychology 和 sociology 由此可选 A。

21. Which of the following subjects is the concern of psychology?

【答案】D

【详解】由第三段可知心理学主要研究引起社会现象的原因以及动机，由排除法可知 D(社会现象引起的原因)属于其范畴。

22. What is the relationship between psychology and sociology?

【答案】D

【详解】由末段首句"Psychology and sociology often work together in their study of human behavior."可知 D 对。

and behavior. The reverse is also assumed: if scientists can learn more about social groups, they will learn more about individuals.

### Passage Three

The first navigational lights in the New World were probably lanterns hung at harbor entrances. The first lighthouse was put up by the Massachusetts Bay Colony in 1716 on Little Brewster Island at the entrance to Boston Harbor. By 1776 there were only a dozen or so true lighthouses in the colonies. Little over a century later there were 700 lighthouses. The first eight were on the West Coast in the 1850's featured the same basic New England design: a Cape Cod dwelling with the tower rising from the center or standing close by. Most American lighthouses shared several features: a light, living quarters, and sometimes a bell. They also had something else in common: a keeper and usually the keeper's family. The keeper's essential task was maintaining a steady and bright flame of the lantern. The earliest keepers came from every walk of life and appointments were often handed out by local customs commissioners.

### Section C

Every person writes a different handwriting. A (26) graphologist can determine a person's (27) personality from his or her handwriting. (28) Hard-to-read handwriting is often the sign of a person who really doesn't want to (29) communicate with others. Such a person may have a hard time getting along with people. Sometimes, though, handwriting is not easy to read because the person writes too fast. To a graphologist, this means that the writer has a lot of (30) energy.

People who are feeling happy are likely to write (31) uphill. With generally happy people, the dots of their i's are likely to be the right of the letter (32) stems. If one's t-bars are wavy and i-dots are more like (33) curves than like dots, or the endings of words are upcurved, this person often has a sense of humor. (34) Also, t-bars that are heavy on their right-hand ends are connected with bad temper. (35) If one likes to make a splash, do things in a big way, his writing is often large. (36) Small writing may be a sign that the writer's power of concentration is quite strong.

**Questions 23 to 25 are based on the passage you have just heard.**

23. When was the first lighthouse put up?

【答案】C

【详解】由"The first lighthouse was put up by the Massachusetts... Boston Harbor"可知选 C。

24. How many lighthouses were there in the United States by 1776?

【答案】B

【详解】由"By 1776 there were only a dozen or so true lighthouses in the colonies."可知仅有 12 个左右，由排除法选 B。

25. Which is not the common feature shared by most American lighthouses?

【答案】C

【详解】由文段"Most American lighthouses shared several features:... There also had something else in common..."可知只有 C 不属于 Common feature shared 范畴。

### Section C

26. graphologist
27. personality
28. Hard-to-read
29. communicate
30. energy
31. uphill
32. stems
33. curves
34. Also, t-bars that are heavy on their right-hand ends are connected with bad temper
35. If one likes to make a splash, do things in a big way, his writing is often large
36. Small writing may be a sign that the writer's powers of concentration is quite strong

# Unit 10

## 录音文字材料

### Section A

1. M: Six airmail stamps and two regular stamps, please.
   W: Here you are. That will be one dollar and eighteen cents.
   Q: Where did this conversation most probably take place?

2. M: I suppose one reason so many tourists come here is because everything is so cheap.
   W: Cheap? Nothing is really cheap in Tokyo.

## 答案与详解

1.【答案】C

【详解】本题主要考查对对话地点的判断。从男士说"Six airmail stamps and two regular stamps, please."可知对话明显发生在邮局，所以 C 为正确答案。

2.【答案】B

【详解】本题主要考查对对话含义的理解。男士说很多游客到这里的原因是这里的东西很便宜，而女士回答"Noth-

Q: How did the woman react?

3. W: You have traveled very much, right?

M: Yes, I've just been to Fiji. And I've visited almost every country in Europe.

Q: Where hasn't the man been?

4. W: It's a shame I got such a bad start in the last race. It was so hard to catch up. Though I tried very hard, all I could see was the backs of the others' heads.

M: Let's work on your start. The most important thing is concentration.

Q: What is the probable relationship between the two people?

5. W: I like Joan, she's not one of the quiet, submissive girls, is she? Go and say sorry. Make it up that way.

M: Here's the waitress. Get me some coffee, would you?

Q: What do you think the relationship between the woman and the man probably is?

6. M: Good morning, I want some information about the flights to Paris next weekend.

W: Well, there are two flights in the morning and three in the afternoon. Here's the timetable.

Q: How many flights to Paris on the weekend?

7. W: Our dean has been busy since last month. He never stops working until twelve o'clock p. m. as I know.

M: I should never have troubled him so much, had I known he was so busy.

Q: What do we learn from this conversation?

8. M: Excuse me, I'd like to send some flowers to my friend.

W: Let's see. These red roses are very nice.

Q: Where does this conversation most likely take place?

**Conversation One**

W: Dad. You love me, don't you?

M: Of course, I do. Why do you ask... Ah, what's on your mind?

W: Well, I saw this great offer for a free cell phone here in the newspaper, and...

M: Free? Nothing's ever free.

W: Well, the phone is free... after a $50 mail-in rebate.

M: Ah, so that's the catch. And why do you need a cell phone anyway?

W: Dad. All my friends have one, and I can use it to call you in case the car breaks down.

M: Ah, I don't know. There are always so many fees.

W: But the monthly charge for this service is only $29.99, with 1,000 free weekday minutes nationwide, and unlimited weekend minutes. Plus, unlimited, anytime minutes for anyone using the same service.

M: I don't know.

W: And you can roll over the extra minutes to the next month instead of just losing them. What do you think of that?

M: Yeah, but what is the term of the service agreement?

ing is really cheap in Tokyo."所以女士是不同意男士说法的,B 为正确答案。

3.【答案】C

【详解】本题主要考查对对话内容的判断。从男士回答"I've been to Fiji, and I have visited almost every country in Europe."可知游览过大多数欧洲国家,选项 C 不是欧洲国家,所以 C 为正确选项。

4.【答案】D

【详解】本题主要考查对人物关系的判断。从两人的谈话中可知他们是在谈论比赛,其中的关键词是"race"。男士回答"Let's work on your start."可知他们在训练,所以他们之间应该是教练和运动员的关系。D 为正确答案。

5.【答案】C

【详解】本题主要考查对人物关系的判断。主要是理解女士所说话"Go and say sorry. Make it up that way."女士要他给 Joan 道歉弥补过错,可知他们之间是朋友关系,C 为正确答案。

6.【答案】D

【详解】本题主要考查对数字的理解。从女士的回答"there are two flights in the morning and three in the afternoon."可知 3+2=5,所以在周末有 5 次航班到达巴黎,D 为正确答案。

7.【答案】B

【详解】本题主要考查对对话含义的理解。男士回答"I should never have troubled him so much, had I known he was so busy."这是虚拟语气句,意思是"如果我知道主任有这么忙,我绝不会这么麻烦他的"。那么 B 项正好说明了他麻烦于主任很多。其他选项都不正确,所以 B 为正确答案。

8.【答案】C

【详解】本题主要考查对对话地点的判断。主要抓住关键词"send flower to my friend","red roses"。可明显看出对话是发生在花店,C 为正确答案。

**Questions 9 to 12 are based on the conversation you have just heard.**

9. What reason does the girl give for needing a cellphone?

【答案】B

【详解】女孩要她爸爸给买一部手机,因为她的朋友们都有手机了,而且万一车子出故障的时候她还可以拿来联系,因此答案是 B。

10. What is the term of service for this plan?

【答案】B

【详解】这是一道关于时间的细节题,答案是 B。

11. Why does the girl suggest that her father buy a new car too?

【答案】A

【详解】女孩让她爸爸买辆新车因为手机和他们以前开的旧车不相配。"the new car you'll need to buy so I can use the cell phone. I mean, what's gonna look like if I'm using a cell phone in our old lemon."因此选 A。

12. What will probably happen next?

【答案】D

W：It's only for six months.

M：But what if you cancel early?

W：Um...Ah, there's a cancellation fee of $200, but with...

M：Two hundred bucks!

W：Yeah, but you won't have to worry about me while I'm driving the new car.

M：New car? What new car?

W：The new car you'll need to buy so I can use the cell phone. I mean, what's gonna look like if I'm using a cell phone in our old lemon.

M：Teenagers. What'll they think of next?

**Conversation Two**

M：Honey, the basketball game is about to start. And could you bring some chips and a bowl of ice cream? And...uh...a slice of pizza from the fridge.

W：Anything else?

M：Nope, that's all for now. Hey, hon, you know, they're organizing a company basketball team, and I'm thinking about joining. What do you think?

W：Humph!

M："Humph", What do you mean "Humph"? I was the star player in high school.

W：Yeah, twenty - five years ago. Look, I just don't want you having a heart attack running up and down the court.

M：So, what are you suggesting? Should I just abandon the idea? I'm not that out of shape.

W：Well...you ought to at least have a physical before you begin. I mean, it has been at least five years since you played at all.

M：Well, okay, but...

W：And you need to watch your diet and cut back on the fatty foods, like ice cream. And you should try eating more fresh fruits and vegetables.

M：Yeah, you're probably right.

W：And you should take up a little weight training to strengthen your muscles or perhaps try cycling to build up your cardiovascular system. Oh, and you need to go to bed early instead of watching TV half the night.

M：Hey, you're starting to sound like my personal fitness instructor!

W：No, I just love you, and I want you to be around for a long, long time.

**Section B**

Passage One

Washington is a state of the north-west USA on the Pacific coast. Its area is 68,192 square miles and its population more than 3,409,000. Its capital is Olympia and its largest city is Seattle. The state is mountainous with the Cascade range running from north to south. The main agricultural products are apples and other fruits, and grain crops. Various minerals are mined, including zinc, magnesite, uranium, gold and silver. Industries include aircraft, wood and paper products, cement and chemicals. The area was explored by the Lewis and Clarke Expedition in 1805, and was jointly controlled by Britain and the USA from 1818 until the border was fixed in 1846. It was made a state of the Union in 1889. The population grew as lumbering and fishing were developed, with the discovery of

【详解】在对话的最后爸爸说的一句话是"Teenagers. What'll they think of next?"说明他对女儿的这种奢侈的想法不满，接着可能会对女儿进行一番说服教育，因此选D。

**Questions 13 to 15 are based on the conversation you have just heard.**

13. What does the man want to do?

【答案】A

【详解】他和妻子在商量着:you know, they're organizing a company basketball team, and I'm thinking about joining，他想要加入公司篮球队。

14. What is the woman's main concern?

【答案】C

【详解】他妻子说:I just don't want you having a heart attack running up and down the court. 可见他妻子最担心是他的身体能否承受得了运动强度，担心他的健康。答案是C。

15. Why does the man's wife recommend cycling?

【答案】B

【详解】因为 try cycling to build up your cardiovascular system.，能够增强心血管系统功能，所以选择B。

**Questions 16 to 18 are based on the passage you have just heard.**

16. According to the passage, what's Washington?

【答案】D

【详解】此题主要考查对文章信息细节的理解。文中第一句"Washington is a state of north west USA on the Pacific coast."可判断D项符合原文，A项明显错误。文中所指的Washington是指华盛顿州，它并不是美国的首都，C项"of the continent"明显与"...on the Pacific coast"不符，所以C项也错误。D为正确选项。

17. How old is Washington as a state?

【答案】C

【详解】此题主要考查对文章信息细节的掌握，文中的

gold in 1852, and with the arrival of the railways.

## Passage Two

US law is derived from English law and is based on common law, statute law and the Constitution. The USA has two separate sets of courts, state and federal. So as well as the central federal system, each state has its own laws, courts, police and prisons. Cases involving federal laws are first heard before a federal district judge in a district court; appeals may be made to the Courts of Appeals and possibly to Supreme Court. The federal legal system has its own police force, the F. B. I. (Federal Bureau of Investigation). Ordinary criminal and civil matters are dealt with by the state systems in local, district and country courts. Some states also have a state Supreme Court.

## Passage Three

Both radio and TV are privately controlled in the US, and are run by commercial companies. The three main national TV networks are ABC, CBS and NBC, and there are also many local TV stations. There are as many TV sets in the US as in all the rest of the world: over 60 million. 95 per cent of all households have at least one TV receiver, which is on for an average of four hours daily. Americans also continue to buy more radios: there are over 300 million in the USA, many of them in motor cars. Most American TV and radio programs are light, popular entertainment, but broadcasting is also an important medium of education. TV developed rapidly after the Second World War; the first color TV transmissions were in 1954. TV has now replaced newspapers as the most important immediate source of news; the big daily news programs are watched every evening by 30 million people.

关键句"It was made a state of the Union in 1889."所以至今华盛顿成为州已有115年,答案C最贴近原文,所以C为正确选项。

18. Which of the following respects has nothing to do with the growth of its population?

【答案】A

【详解】此题主要考查对文章信息细节的掌握,从文中最后一句"The population grew was lumbering and fishing were developed, with the discovery of gold in 1852, and with the arrival of the railways."可找到B、C、D三项,所以A项不是促进其人口发展的因素,A为正确选项。

**Questions 19 to 21 are based on the passage you have just heard.**

19. According to the passage, which of the following statements is NOT true?

【答案】D

【详解】此题主要考查对文章信息细节的判断。从文中"US law is derived from English law and is based on common law,.., state and federal. 可以分别判断A、B、C三项都与原文相符,D项明显错误,所以D为正确选项。"

20. According to the passage, what are ordinary civil matters dealt with by?

【答案】D

【详解】此题主要考查对文章信息细节的掌握。解决此类题型主要是找到文中与之相关的关键句,无论文章的长短,抓住主要问题所在。此题在文中相关的句子"Ordinary criminal and civil matters are dealt with by the state system in local, district and country courts."可很明显的找到D为正确答案。

21. What's the topic of the passage?

【答案】B

【详解】此题主要考查对文章主旨大意的理解。A、C、D三项都只是文中讲美国法律的一个方面,只有B项能对全文进行概括。B为正确选项。

**Questions 22 to 25 are based on the passage you have just heard.**

22. According to the passage, which of the following statements is NOT true?

【答案】C

【详解】此题主要考查对文章信息细节的判断。从文章第一句"Both radio and TV are privately controlled in the US, and are run by commercial companies."可判A、B两项符合原文。从文章"The three main... many local TV stations."可判断C项错误,D项符合原文。本题是要找出表述错误的,所以C为正确选项。

23. According to the passage, what's the average time for most Americans to watch TV every day?

【答案】C

【详解】此题主要考查对文章关键句的掌握,从文中"95 percent of all households... of four hours daily."可找到答案。$4 \times 60 = 240$ minutes,所以C为正确选项。

24. What is implied but not stated in the passage?

**Section C**

In a (26) competitive economy, the consumer usually has the choice of several different (27) brands of the same product. Yet underneath their labels, the products are often nearly (28) identical. One manufacturer's toothpaste (29) tends to differ very little from another manufacturer's. Thus, manufacturers are (30) confronted with a problem—how to keep sales high enough to stay in business. Manufacturers solve this problem by advertising. They try to appeal to consumers in various ways. In fact, advertisements may be (31) classified into three types according to the kind of appeals they use.

One type of advertisement tries to (32) appeal to the consumer's reasoning mind. It may offer a claim that seems scientific. For example it may say the dentists (33) recommend flash toothpaste. (34) In selling a product, the truth of the advertising may be less important than the appearance of truth. A scientific approach gives the appearance of truth.

Another type of advertisement tries to amuse the potential buyer. (35) Products that are essential boring, such as insecticide, are often advertised in an amusing way.

One way of doing this is to make the products appear alive. For example, the advertisers may personify cans of insecticide, and show them attacking mean-faced bugs. Ads of this sort are silly, but they also tend to be amusing. (36) Advertisers believe that consumers are likely to remember and buy products that the consumers associate with fun.

【答案】B

【详解】此题主要考查对文章信息的理解。从文中"Americans also continue to buy more radios: there are over 300 million in the USA, many of them in motor cars."可看出其中暗指许多美国人开车时喜欢听广播,可判断B项正确,其它几项文中均没有描述。

25. Which of the following titles best sums up the passage?

【答案】D

【详解】此题主要考查对文章主旨的理解。从全文可以看出主要是讲美国电视和广播,那么综上所述就是美国的电视广播情况,所以D为正确选项。

**Section C**

26. competitive 27. brands 28. identical
29. tends 30. confronted 31. classified
32. appeal 33. recommend
34. In selling a product, the truth of the advertising may be less important than the appearance of truth
35. Products that are essential boring, such as insecticide, are often advertised in an amusing way
36. Advertisers believe that consumers are likely to remember and buy products that the consumers associate with fun

# Unit 11

## 录音文字材料

**Section A**

1. M: To teach those students English, do you have to speak their language quite well?
   W: Quite the contrary. They benefit most when the class is conducted entirely in the foreign language.
   Q: What language is used in the woman's classes?
2. M: Hello! Can you give me some information about the apartments you're going to rent?
   W: All right. The rent is due on the first day of the month. You have to sign a one-year lease with us and pay the security deposit and cleaning fee. One more thing, utilities are not in-

## 答案与详解

1.【答案】A

【详解】关键句是"Quite the contrary"。女士说全外文课堂让学生们获益最多。

2.【答案】C

【详解】女士说要签一个一年的租约并付保证金和清洁费,另外公共设施不含在租金里。符合的选项只有C。

cluded in the rent.

Q: If the man rents the apartment, what must he do?

3. M: Open wide. Now show me where it hurts.
   W: Here, on the bottom. Especially when I bite into something hot or cold.
   Q: Who is the man?

4. M: When does the next train leave?
   W: You have just missed one by 5 minutes. Trains leave every fifty minutes, so you have to wait for a while.
   Q: How long does the man have to wait for the next train?

5. M: Well, Mary, what can I do for you today? How is your mother feeling?
   M: She's well, thanks. She wants to know whether you can send an order over to our house right away.
   Q: Where does the above conversation most probably take place?

6. W: Can you tell what time flight 318 arrives?
   M: Yes, it was scheduled to arrive at 6:00 p. m., but it has been delayed for two hours.
   Q: When is the plane expected to arrive?

7. W: I'm so tired today that I could hardly stay up late tonight.
   M: Just leave the work to me.
   Q: What does the man offer to do?

8. W: I don't understand how you got a ticket. I always thought you were a careful driver.
   M: I usually am, but I thought I could make it before the light turned.
   Q: Why did the man get a ticket?

**Conversation One**

W: Hello. 24th Precinct. Officer Jones speaking.

M: Help. Yeah, uh, it was wild, I mean really bizarre.

W: Calm down sir! Now, what do you want to report?

M: Well, I'd like to report a UFO sighting.

W: A what?

M: What do you mean "what?" An unidentified flying object!

W: Wait, tell me exactly what you saw.

M: Well, I was driving home from a party about three hours ago, so it was about 2:00 a. m, when I saw this bright light overhead.

W: Okay. And then what happened?

M: Oh, man. Well, it was out of this world. I stopped to watch the light when it disappeared behind a hill about a kilometer ahead of me.

W: Alright. Then what?

M: Well, I got back in my car and I started driving toward where the UFO landed.

W: Now, how do you know it was a UFO? Perhaps you only saw the lights of an airplane [No], or the headlights of an approaching car [No]. Things like that happen, you know.

M: Well if it was that, how do you explain "the BEAST"?

W: What do you mean, "the BEAST"?

M: Okay. I kept driving for about five minutes when all of a sudden, this giant, hairy creature jumped out in front of my car.

W: Oh, yeah. Then what?

3.【答案】C
【详解】关键词是 bite。女士回答"Especially when I bite..."和 bite 有关的是牙齿,可推知男士为牙医。

4.【答案】A
【详解】计算题。火车每 50 分钟走一趟,男士 5 分钟前错过一趟,还需要等 50 - 5 = 45 分钟。

5.【答案】A
【详解】由"what can I do for you today?"和 order 可推知对话发生在一个商店里。符合的只有选项 A。

6.【答案】B
【详解】计算题。原本晚上 6 点到,晚点 2 小时,预计晚上 8 点到。

7.【答案】C
【详解】"Just leave the work to me."意思为把工作交给我做吧。

8.【答案】A
【详解】由"I thought I could make it before the light turned"知是闯了红灯。

**Questions 9 to 12 are based on the conversation you have just heard.**

9. Where was the man coming from when he first saw the UFO?
【答案】A
【详解】女士问男士怎么回事,他就开始叙述事件的经过,当时"I was driving home from a party about three hours ago",然后看见了 UFO,所以应该选择 A。

10. What jumped out in front of the man's car?
【答案】C
【详解】当他解释这个 beast 时,他说了这样一句话,"this giant, hairy creature jumped out in front of my car,"说出了这个外星生物的 hairy 特征。

11. What happened next to the man?
【答案】C
【详解】根据当事人的描述:"the beast opened the car door, carried me on his shoulders to this round - shaped flying saucer,"所以说:他是被强制带到了飞碟内了。答案为 C。

12. What does the police officer suggest at the end of the story?
【答案】B
【详解】最后警察建议:"We have a great therapist that deals with these kinds of cases."这里 therapist 是指咨询社会心理的医师。从这里可以推断出,该男子需要咨询,即是"seek counseling"。

M：Well, then, the beast picked up the front of my car and said, "Get out of the car. I'm taking you to my master!" Something like that.

W：Wow? A hairy alien who can speak English! Come on!

M：I'm not making this up, if that's what you're suggesting. Then, when I didn't get out of the car, the beast opened the car door, carried me on his shoulders to this round – shaped flying saucer, and well, that's when I woke up alongside the road. The beast must have knocked me out and left me there.

W：Well, that's the best story I've heard all night, sir. Now, have you been taking any medicine, drugs, or alcohol in the last 24 hours? You mentioned you went to a party.

M：What? Well, I did have a few beers, but I'm telling the truth.

W：Okay, okay. We have a great therapist that deals with these kinds of cases.

M：I'm not crazy.

W：Well, we'll look into your story. Thank you.

**Conversation Two**

W：Today's guest on "Science Update" is David Brown. Dr. Brown, you and your team have found bacteria far below the Earth's surface. You must be thrilled about your discovery.

M：Well, yes, it's very exciting. For a long time we'd suspected the presence of such organisms, but we lacked substantial evidence.

W：How did you confirm the existence of the bacteria?

M：Well, technology helped. Our drilling techniques have improved significantly, and so the risk that surface bacteria could be mistaken for those found at much greater depth was reduced. With the new techniques, we could get much deeper into the Earth.

W：How far down did you actually get?

M：In one case, about three kilometers. We were surprised, I must tell you, that there were organisms that far down.

W：You know, it sounds like fiction, something like a lost world.

M：Let's call it a hidden biosphere, and it's probably a very extensive one. The mass of the living organisms below the surface may be equal in size to the mass of the surface bacteria.

W：Have you found any unique life-forms?

M：Yes. One of the organisms is the first anaerobic bacillus ever discovered. That means it can live and grow only where there is no oxygen.

W：Is there any danger of these bacteria infecting people when you bring them to the surface?

M：The bacteria in question were adapted to an environment that's hostile and alien to humans. Conversely, these anaerobic bacteria could not survive in our environment. So we really don't need to worry about these bacteria causing illness in people.

**Section B**

**Passage One**

Poisonings that cause death happen most of the time to children between the ages of one and three. Some doctors call this stage the "Age of Accidents". Children want to look at things and taste them. They will eat or drink anything they can find, even if it tastes bad. You must make your home safe for children and protect them form poison-

**Questions 13 to 15 are based on the conversation you have just heard.**

13. What is the main topic of the interview?

【答案】A

【详解】这段访谈谈论的主题是地表以下的细菌，与之意思相符的只有选项A。通常听力材料都会开宗明义，所以考生一定要注意听力材料的开头。

14. What aspect of the hidden biosphere does the man discuss?

【答案】B

【详解】在答听力题时，考生最好能够先浏览一遍问题，这样可以做到有的放矢。例如，在这道题里，考生如果能够先浏览问题，在听时格外留意"hidden biosphere"（隐藏的生物圈）的出现，就不难发现男士讨论的是这个生物圈的广阔性。

15. According to the man, why is there no danger of infection by the bacteria?

【答案】B

【详解】"anaerobic bacillus"是指"厌氧细菌"，这是一个生僻词组，考生不知道的可能性很大。但男士随后又对它进行了解释，指出这种细菌只能生存在没有氧气的地方，可见，它不可能给人类造成任何感染，因为它不能与人类共存在一个环境里，与之意思相近的只有选项B。

**Questions 16 to 18 are based on the passage you have just heard.**

16. What is the "Age of Accidents"?

【答案】A

【详解】开始部分介绍"the ages of one and three"为"Age of Accidents"。

17. Why do children often get into poisons?

ing.

Here are three things you can do. One, know which things around the home are poisons. Two, keep poisons out of your child's reach at all time. Three, be aware of how clever children are when it comes to finding poisons.

Nearly all chemicals and drugs in the home contain things that can poison someone, be sure to read the labels on products you bring into the home. Look for the words which are meant as warnings. These warnings will read Poison, For External Use Only, and Keep Out of Reach of Children. Look around your home for bottles and jars which bear these warnings. Then, put them away in a place where a child cannot reach or find them.

### Passage Two

At 5:13 on the morning of April 18th, 1906, the city of San Francisco was shaken by a terrible earthquake. A great part of the city was destroyed and a large number of buildings were burnt. The number of people who lost homes reached as many as 250,000, about 700 people died in the earthquake and the fires.

Another earthquake shook San Francisco on October 17th, 1989. It was America's second strongest earthquake about 100 people were killed. It happened in the evening as people were traveling home. A wide and busy road which was built like a bridge over another road fell onto the one below. Many people were killed in their cars, but a few lucky ones were not hurt.

Why do earthquakes happen? Scientists explain that the outside of the earth is made of a number of different plates. At San Francisco the Pacific plate is moving very slowly—at 5.3 centimetres a year. Sometimes these two plates stop and do not move for years. Then suddenly, they jump and an earthquake is felt. As a result of the movement of these plates, west America near the sea has always been a bad place for earthquakes. When the 1906 earthquake happened, the Pacific plate jumped 5—6 metres to the north.

### Passage Three

When Christopher Columbus and his friends reached America in 1492, they discovered the plant "corn" there. At least 450 years ago corn was brought to China. By 1555 it had already become important here. Corn is a useful plant that can be eaten by both people and animals. In the 17th century corn was grown a lot in Tibet and Sichuan. At that time the land along the Changjiang River was becoming very crowded; there was not enough room for the population. Farmers had to move onto the hills, but they couldn't grow rice there. They needed a plant which didn't need as much water as rice. Luckily they were able to grow this new corn.

Today, corn is found all over the world. It is a very useful plant that can be prepared in many different ways. People in the West often boil it and eat it with salt and butter. Sometimes they cook it whole over an open fire. In many parts of the world corn is made into powder. The powder is then mixed with water and other things, and made into different kinds of food.

Corn was not the only food that was taken to Europe. A number of other plants were found in America, for example, beans, potatoes and different fruits. The potato is another plant that was taken back by ear-

【答案】A

【详解】由"Children want to look at things and taste them. They will eat or drink anything they can find, even if it tastes bad."得知因为儿童很好奇。

18. Which warning is not mentioned in the passage?

【答案】D

【详解】材料中只提到 poison, for external use only 和 keep out reach of children.

**Questions 19 to 21 are based on the passage you have just heard.**

19. When did the America's second strongest earthquake take place?

【答案】B

【详解】材料中提到"Another...on October 17th, 1989. ...It happened in the evening...",可知答案选 B。

20. How many people died in the 1906's earthquake in San Francisco?

【答案】C

【详解】1906 年的地震造成 250,000 人失去住所,大约 700 人丧生。1989 年的地震造成大约 100 人丧生。注意不同数据所对应的时间及事件。

21. Why do earthquakes happen?

【答案】A

【详解】材料说"As a result of the movement of these plates, west America near the sea has always been a bad place for earthquakes."排除选项 B、D,而选项 C 未在材料中提及。答案为 A。

**Questions 22 to 25 are based on the passage you have just heard.**

22. Why did many farmers along the Changjiang River in the 17th century move into the hills?

【答案】C

【详解】因为"there was not enough room for the population",所以答案为 C。

23. Why was corn grown a lot in Tibet and Sichuan?

【答案】A

【详解】农民大量种玉米因为"They needed a plant which didn't need as much water as rice."可知答案为 A。

24. How is corn cooked in the West according to the passage?

【答案】A

【详解】由"People in the west often boil it and eat it with salt and butter."可知答案为 A。

25. Which plant was found in America but was not mentioned in the passage?

【答案】B

【详解】Potato, corn, bean 都有提到,只有 tomato 未谈及。

ly traveler. It is a very useful plant; it can be grown in places where it is too cold to grow rice.

**Section C**

Editing and (26) proofreading benefit richly from (27) word-processing. Instead of crossing out mistakes, or (28) rewriting an entire paper to correct (29) numerous errors, you can make all (30) necessary changes within the most recent draft. If you find (31) editing or proofreading on the screen hard on your eyes, print out a copy. Mark any (32) corrections on that copy, and then transfer them to the final draft.

If the word-processing (33) package you're using includes spelling and grammar checks, by all means use them. (34) The spell-check function tells you when a work is not in the computer's dictionary. Keep in mind, however, that the spell-check can't tell you how to spell a name correctly or when you have mistakenly used. (35) Also, use the grammar check with caution. Any errors it doesn't uncover are still your responsibility.

A word-processed paper, with its clean appearance and handsome formatting, looks so good that you may feel it is in better shape than it really is. Do not be fooled. (36) Take sufficient time to review your grammar, punctuation, and spelling carefully.

**Section C**

26. proofreading 27. word – processing
28. rewriting 29. numerous
30. necessary 31. editing
32. corrections 33. package
34. The spell-check function tells you when a work is not in the computer's dictionary
35. Also, use the grammar check with caution
36. Take sufficient time to review your grammar, punctuation, and spelling carefully

# Unit 12

## 录音文字材料

**Section A**

1. M: Well, what did you think of the movie?
   W: I don't know why I let you talk me into going. I just don't like violence. Next time you'd better choose a comedy.
   Q: What kind of movie does the woman prefer?
2. W: So you haven't had time to see much outside London.
   M: Oh, I've spent two or three holidays in England with my mother's relatives who live in the Midlands, so I've gotten to know Oxford and Cambridge, Stratford and various other places fairly well.
   Q: What does the man mean?
3. M: We've been waiting for him to take our order for over twenty minutes.
   W: Yes, and I have an appointment at 2:00.
   Q: Where is this conversation taking place?
4. W: I'm going away to the seaside for a few days and I'd like you to keep an eye on my home while I'm away.
   M: Certainly, no problem.
   Q: What is the man asked to do?
5. M: I'm trying to figure out how to work the new machine. But look at this instruction sheet!
   W: What a useless piece of paper.
   Q: What does the woman mean?
6. W: Fasten your seat belt, Sir.

## 答案与详解

1.【答案】B
【详解】关键词为 comedy。女士希望男士下一次最好选喜剧片，可知答案为 B。

2.【答案】C
【详解】男士说"I've gotten to know Oxford and Canbridge, stratford and various other places fairly well."可知他去过很多地方。做此题不要被女士问话迷惑。

3.【答案】C
【详解】符合"waiting for him to take our order"的只有选项 C。

4.【答案】B
【详解】keep an eye on 的意思是留意，照看。

5.【答案】D
【详解】男士显然还不会操作新机器，可见他没看懂说明，由 useless 看出女士也没看懂。

6.【答案】A

M: Of course, I didn't realize that we were going to land so soon.

Q: Where does this conversation most probably take place?

7. W: I hope you have a good flight.

M: The weather's supposed to be clear all down the coast, so it should be pretty smooth.

Q: Where is this conversation probably taking place?

8. M: Did you buy a birthday present for your brother?

W: Not yet, but I've been thinking about getting him a record. He likes classical music.

Q: Which record would the woman's brother like best?

**Conversation One**

M: Well, hi. What are you looking for today?

W: Uh, I'm just looking.

M: Well, how about a ring from someone special?

W: There is no one special.

M: Well, take a look at this CD player. A great bargain today only.

W: Nah. I already have one, plus the handle is cracked.

M: Okay. Well, what about this genuine leather jacket? It would look great on you.

W: Hum. Let me take a look at it.

M: Sure.

W: Umm. There are stains on the sleeves. I'll pass.

M: Well okay. Well, wouldn't you like to walk home with some of these great records? Some of the best hits from the 1960's.

W: Yeah, let's see. Now here's something I'd... Ah, these records are scratched.

M: Just in a couple places. Listen. I'll sell you these ten records for fifty dollars. A steal!

W: Whoa! They're too expensive. I'll give you twenty-five bucks for them.

M: Ah, come on. I can't charge you less than thirty dollars and break even.

W: Well, that guy over there is selling similar records for a much better price, so thanks anyway.

M: Wait, wait, wait, wait. You drive a hard bargain. Twenty-eight dollars, and that's my final offer.

W: Huh... I'll think about it.

M: Wait, wait, wait, wait. Listen. I'll even throw in this vase.

W: Now what am I going to do with a vase?

M: Well, you can give it to that someone special when you find her... and this ring would look great with it.

W: Oh, I'll stick with the records.

**Conversation Two**

M: Hello.

W: Hello Roger? This is Ann.

M: Oh hi, Ann. How have you been? And how's your new apartment working out?

W: Well, that's what I'm calling about. You see, I've decided to look for a new place.

M: Oh, what's the problem with your place now? I thought you liked the apartment.

W: Oh, I do, but it's a little far from campus, and the commute is

【详解】关键词为 land,提到降落可推知对话发生在飞机里。

7.【答案】C

【详解】对话的主题是 flight,所以最符合的选项是 C。

8.【答案】D

【详解】对话提到"He likes classical music"。

**Questions 9 to 12 are based on the conversation you have just heard.**

9. Why did the man not buy the ring for sale?

【答案】C

【详解】顾客的回答:Uh, I'm just looking. There is no one special. 先说看看,看后说没有特别的。说明没有买戒指的需要。选 C。

10. What is the problem with the CD player?

【答案】C

【详解】顾客回答是 Nah. I already have one, plus the handle is cracked. cracked 是裂开的意思,所以答案为 C。

11. Why is the man not interested in the leather jacket?

【答案】A

【详解】There are stains on the sleeves,所以顾客对那件衣服没有丝毫的兴趣。答案为 A。

12. From the conversation, what does the customer probably purchase from the merchant in the end?

【答案】A

【详解】这里许多人会选择 C,要注意商人是说 throw in the vase,意思是做添头的,买唱片免费赠送的,所以最后应该只买唱片,送的 vase。

**Questions 13 to 15 are based on the conversation you have just heard.**

13. What is the main topic of the conversation?

【答案】B

【详解】综合对话的内容,可以知道,都是提出或解决关于寻找房子的问题,只有 B 项符合题目意思。选 B。

14. What kind of place is the girl looking for?

【答案】A

【详解】I'd like to share an apartment with one or two roommates within walking distance to school. 这句话表达了她对

just killing me. Do you think you could help? I thought you might know more about the housing situation near the university.

M: Alright. So, what kind of place are you looking for?

W: Well, I'd like to share an apartment with one or two roommates within walking distance to school.

M: Okay, what's your budget like? I mean how much do you want to spend on rent?

W: Uh, somewhere under \$200 a month, including utilities, if I could. Oh, and I'd prefer to rent a furnished apartment.

M: Hmm. And anything else?

W: Yeah, I need a parking space.

M: Well, I know there's an apartment complex around the corner that seems to have a few vacancies. I'll drop by there on my way to class today.

W: Hey, thanks a lot.

M: No problem.

**Section B**

Passage One

There are several things that every defensive driver should learn. To begin with, the defensive driver should learn to drive in good manners. That is, he should always let the other person have the right-of-way if there is any doubt. Also, the defensive driver should learn to anticipate, or guess, what the other driver is going to do next. This gives him time to get his car into a safer position, if necessary. Furthermore, every defensive driver should learn to give the proper signal before changing directions, allowing enough time for other drivers to react to it. Lastly, every defensive driver should learn to keep a safe distance between his car and the car ahead.

Passage Two

Once a neighbor stole one of Washington's horses. Washington went with a police officer to the neighbor's farm to get the horse but the neighbor refused to give the horse up, he claimed that it was his horse. Washington placed both of his hands over the eyes of the horse and said to the neighbor. "If this is your horse, then you must tell us in which eye he is blind."

"In the right eye!" the neighbor said.

Washington took his hand from the right eye of the horse and showed the police officer that the horse was not blind in the right eye.

"Oh, I have made a mistake," said the neighbor. "He is blind in the left eye."

Washington then showed that the horse was not blind in the left eye, either.

"I have made another mistake," said the neighbor.

"Yes," said the police officer, "and you have also proved that the horse does not belong to you. You must return it to Mr. Washington."

Passage Three

For good or bad, computers are now part of our daily lives. With the price of a small home computer now being lower, experts predict

理想房子的要求,想离学校近一些的房子,选 A。

15. How is the boy going to help her?

【答案】C

【详解】男孩向她提供帮助 I know there's an apartment complex around the corner that seems to have a few vacancies, vacancies 是空白,空缺的意思,符合意思的是 C。

**Questions 16 to 18 are based on the passage you have just heard.**

16. How many points are mentioned for a defensive driver to learn?

【答案】B

【详解】听力材料中共讲到 4 点。注意抓住"To begin with","Also","Furthermore","Lastly"这几个关键词。

17. What should a driver do before he changes directions?

【答案】A

【详解】改变行车方向要发出适当的信号。

18. What is the purpose of this article?

【答案】B

【详解】文章主旨在第一句话"There are several things that every defensive driver should learn." defensive 在这里指避免交通事故的。所以选 B。

**Questions 19 to 21 are based on the passage you have just heard.**

19. What happened to Washington's horse?

【答案】A

【详解】材料开始便给出"Once a neighbor stole one of Washington's horses"。

20. How many people are mentioned in the story?

【答案】C

【详解】三个人分别为:Washington, the neighbor, a police officer。

21. How did the police officer find the horse belonged to Washington?

【答案】B

【详解】依据他的所见所闻。

**Questions 22 to 25 are based on the passage you have just heard.**

22. What is this passage mainly about?

that before long all schools and businesses and most families in the richer parts of the world will own a computer of some kind. Among the general public, computers arouse strong feeling—people either love them or hate them.

The computer lovers talk about how useful computers can be in business, in education and in the home—apart from all the games, you can do your accounts on them, use them to control your central heating, and in some places even do your shopping with them. Computers, they say, will also bring more leisure, as more and more unpleasant jobs are taken over by computerized robots.

The haters, on the other hand, argue that computers bring not leisure but unemployment. They worry, too, that people who spend all their time talking to computers will forget how to talk to each other. And anyway, they ask, what's wrong with going shopping and learning languages in classroom with real teachers? But their biggest fear is that computers may eventually take over from human beings altogether.

**Section C**

Let us (26) suppose that you are in the position of a parent. Would you allow your children to read any book they wanted without first checking its (27) contents? Would you take your children to see any film without first finding out whether it is (28) suitable for them? If your answer to these questions is "yes", then you are just plain (29) irresponsible. If your answer is "no", then you are exercising your right as a parent to (30) protect your children from what you consider to be (31) undesirable influences. In other words, by acting as an (32) examiner yourself, you are (33) admitting that there is a strong case for censorship.

Now, of course, you will say that it is one thing to exercise censorship where children are concerned and quite another to do the same for adults. Children need protection and it is the parents' responsibility to provide it. But what about adults? Aren't they old enough to decide what is good for them? (34) The answer is that many adults are, but don't make the mistake of thinking that all adults are like yourself. Censorship is for the good of society as a whole. Like the law, it contributes to the common good.

Some people think that it is a shame that a censor should interfere with works of art. (35) But we must bear in mind that the great proportion of books, plays and films which come before the censor are very far from being "works of art".

When censorship laws are relaxed, dishonest people are given a chance to produce virtually anything in the name of "art". One of the great things that censorship does is (36) to prevent certain people from making fat profits by corrupting the minds of others. To argue in favour of absolute freedom is to argue in favour of anarchy. Society would really be the better if it were protected by correct censorship.

【答案】C

【详解】材料主要讲的是人们对电脑不同的态度。

23. According to the passage, what is not mentioned about computers?

【答案】A

【详解】材料谈到"computers can be useful in business",但没有谈到电脑会带来金融问题。

24. What is the biggest fear of the computer-haters?

【答案】D

【详解】材料最后说到"But their biggest fear is that computers may eventually take over from human beings altogether"。

25. What's the speaker's attitude to computers?

【答案】D

【详解】说话者主要陈述了人们对电脑不同的态度,并不带感情色彩,因此是中立的。

**Section C**

26. suppose　27. contents　28. suitable
29. irresponsible　30. protect　31. undesirable
32. examiner　33. admitting

34. The answer is that many adults are, but don't make the mistake of thinking that all adults are like yourself
35. But we must bear in mind that the great proportion of books, plays and films which come before the censor are very far from being "works of art"
36. to prevent certain people from making fat profits by corrupting the minds of others

# Unit 13

## 录音文字材料

**Section A**

1. W: I was hoping to get some books from the library before it closes.
   M: My watch says 6:50, so we have around 40 minutes left to get there.
   Q: What time does the library close?
2. W: Mary would have been sick if she had eaten all those sweets.
   M: She told me she had a toothache last night.
   Q: What do we know about Mary?
3. M: Come and watch TV, dear.
   W: No, I've written to Aunt Mary and I've said bad things about Aunt Agate. Now I'm writing to Aunt Agate and I'm saying bad things about Aunt Mary! I wrote two letters but put the same ideas in both of them. It's easy!
   Q: Whom is the woman writing to?
4. W: Did Bill buy the car which we saw last week?
   M: Yes, the price of the car was so low that it made Bill suspicious of its value at first.
   Q: What did Bill think of the car at first?
5. W: What's the matter with Mary? Why did she leave the examination half-way through?
   M: She said she had not been well prepared.
   Q: What would probably be the result if she finished the examination.
6. M: Do you want to send it by air mail or by ordinary mail?
   W: By air mail, please. It's much quicker than that.
   Q: Where does this conversation most probably take place?
7. W: You're going to Chicago tomorrow, aren't you?
   M: Yes, I thought I'd fly, but then I decided that taking a bus would be cheaper than driving or flying.
   Q: How would the man get to Chicago?
8. W: I thought the boat only held four. How did you reserve five seats?
   M: They've enlarged it to carry six. I'll reserve the last seat for you.
   Q: How many people are going on the boat if the woman is coming?

**Conversation One**

M: Thanks for stopping by, Ann. I'd like to talk to you about a research project I thought you might be interested in. A friend of mine is working at Yellowstone National Park this summer.

W: Yellowstone! I've always wanted to spend some time out in Wyoming.

M: Wait till you hear what the project is. She's working with the buffalo population. The herds have been increasing in size lately, which is good in theory.

W: Yeah, but I thought they were in danger of becoming extinct.

M: Well, apparently, because of all the winter tourists, paths are created in the snow. More buffalo are surviving the harsh winters be-

## 答案与详解

1.【答案】C
【详解】现在时间是6:50,还有大约40分钟留给他们,所以应是7:30关门,选C。

2.【答案】C
【详解】女士说Mary如果吃了糖她会出问题的,男士接着说Mary牙疼了一晚,显然此句是说因为吃了糖所以牙疼。

3.【答案】B
【详解】此题要重点抓住信是给谁写的,女士说的第一句话就已告诉我们是答案B中的Aunt Agate。

4.【答案】A
【详解】根据对话中男士所说的,Bill一开始对车子价格如此低廉而感到吃惊。

5.【答案】A
【详解】由对话知Mary因为没有准备好考试中途就离开了,所以不及格可能性比较大,选A。

6.【答案】D
【详解】对话主要是在讲寄信,故选D,应该在邮局。

7.【答案】B
【详解】男士先说他想坐飞机,但后来决定搭长途汽车,因为比较便宜,所以选B。

8.【答案】C
【详解】由女士话中知已订了五个位置,男士又说船能载六人,他会留最后一个位置给女士,故选C,六人。

**Questions 9 to 12 are based on the conversation you have just heard.**

9. What did the professor want to talk to Ann about?
   【答案】D
   【详解】这段对话一开头,教授就向Ann提及了找她谈话的原因是为了一个研究项目,因此选D。
10. According to the professor, why is the buffalo population increasing?
   【答案】A
   【详解】据教授说,由于冬天游客的到访,雪地里有了路,这些路方便了野牛行动和寻找食物,也增加了它们的生存机会,其结果就是野牛数目增多。符合教授这一说法的只

cause the paths make it easier for the buffalo to move around and find food. But it turns out that some of the herds are infected with a bacteria.

W: Oh yeah, I heard about that. Bru –

M: Brucella abortus.

W: Right. It's been around for quite a while.

M: Yes it has. And because the buffalo population is increasing, they've been roaming more than usual, and the disease has begun to spread to the cattle ranches that border the park.

W: That's bad news! Isn't that the disease that causes animals to abort their young?

M: Yes, and it's caused a lot of controversy. Some of the ranchers even want to destroy the buffalo herds.

W: That's awful! Have they made much progress with the research?

M: So far, they've been collecting tissue samples from dead buffalo to see if the bacteria's present.

W: I'd really be interested in working on this. You know I've been researching diseased animal populations.

M: That's why I thought of you I took the liberty of mentioning your name to my friend. She's hoping you'll be able to spend the whole summer out there.

W: Well, I was going to work on my thesis a lot in July, but I'm sure my adviser wouldn't want me to pass up this opportunity.

**Conversation Two**

W: Hi. How can I help you?

M: Yes, I'd like to return this sweater for a refund. I bought a week ago.

W: Well, first of all, what seems to be the problem?

M: Well, isn't it obvious by just looking at it? The first time I washed and dried it, the thing shrank at least five sizes. It wouldn't even fit an emaciated snake.

W: Uh, I see what you mean, but did you follow the washing instructions? I think it says here... yeah... right here on the label to hand wash it and then to dry it on low heat.

M: How was I supposed to know that? The label is written in Chinese! And something else: The stitching is coming undone and the color faded from a nice dark blue to a seaweed green. What kind of merchandise are you trying to sell here anyway?

W: Listen, sir. We take a lot of pride in our clothing. What I can do is allow you to exchange the sweater for another one.

M: I don't want to exchange it for anything! I just want my money back!

W: Well, I can give you credit on your next purchase, and since the item you purchased was on clearance [No wonder!], we can't give you a refund.

M: A clearance item! There wasn't anything on the price tag or on the clothing rack that said anything about that.

W: I guess you didn't read the fine print in our ad. Look. Here's the ad, and the information about the clearance sale is right here at the bottom on the back page.

M: Where? [Here] There? What? That small print. You'd need an electronic microscope to see those words. I want to talk to the

有选项 A。

11. Why does the professor think Ann would be interested in going to Yellowstone?

【答案】C

【详解】对于教授的提议,Ann 表示很感兴趣,因为她一直在研究患病动物这一课题,而教授也正是因为这一点才想到 Ann 的,可见正确答案是 C。

12. How will Ann probably spend the summer?

【答案】A

【详解】"pass up"是"拒绝、放弃"的意思。Ann 认为自己的导师不会希望自己放弃这一机会,从而可以推测出 Ann 很可能会接受邀请,前往黄石国家公园进行研究工作,与之意思最接近的是选项 A。

**Questions 13 to 15 are based on the conversation you have just heard.**

13. What reason is NOT mentioned why the man wants to return the item?

【答案】D

【详解】A,B,C 三项男士都提到过,D 项没有涉及。

14. What reason is NOT mentioned why the sales clerk can't help the customer with his request?

【答案】C

【详解】题干中 A,B,D 三项都出现过,唯有 C 是不符合事实的。所以应该选 C。

15. How does the conversation end?

【答案】C

【详解】Take your sweater. You should open up a pet store and sell it as a dog sweater. 可见他是什么也没有换成,所以应该选 C。

manager.

W: Uh, he's not here at the moment.

M: Look. This is ridiculous.

W: And anyway, you can only return items with a receipt within six days, and unfortunately, that was yesterday in your case.

M: But, your store was closed yesterday because of the national holiday. [Sorry] What a rip off. Listen. I give up. Your store policies are completely unreasonable, the quality of your merchandise is shoddy at best, and your service, well, is non-existent. And how do you expect people to shop here?

W: You did... Ha, ha...

M: Here. Take your sweater. You should open up a pet store and sell it as a dog sweater.

**Section B**

Passage One

Mr. Green left his car outside his apartment one night, as usual, but when he came down the next morning to go to his office, he discovered that the car wasn't there. He called the police and told them what had happened, and they said that they would try to find the car.

When Mr. Green came home from his office that evening, the car was back again in its usual place in front of his house. He examined it carefully to see whether it had been damaged, and found two theater tickets on one of the seats and letter which said, "We are very sorry. We took your car because of an emergency."

Mr. Green and Mrs. Green went to the theater with the two tickets the next night and enjoyed themselves very much.

When they got home, they found that thieves had taken almost everything they had in their apartment.

Passage Two

These days we are so accustomed to telegraph messages that it is hard for us to imagine the excitement that was felt in the nineteenth century when the first cables were laid.

Cable laying proved to be immensely difficult. The cable which in the autumn of 1850 carried the first telegraph messages between England and France had a very short life. The day after, a fisherman "caught" the cable by mistake. Thinking that the copper wire at the center of the thick cable was gold, he cut a piece off to show his friends. However, a new cable was put down and soon news could travel quickly across Europe. But there was still no way of sending messages between Europe and America.

When the Atlantic Telegraph Company was formed in 1856, a serious attempt was made to "join" Europe to America with no less than 2300 miles of cable. As no single ship could carry such a weight, the job was shared by two sailing ships, the Agamemnon and the Niagara. The intention was that after setting out in opposite directions, they should meet in the middle of the Atlantic Ocean where the two cables would be connected together. But the ships had hardly covered 300 miles when the cable broke. In 1858, a second attempt was made. This time, greatly hindered by storms, the ships were again unsuccessful. There was great rejoicing a few months later, when after the combined efforts of both ships, Britain and America were at last connected by cable and the Queen of England was able to speak to the President of the

**Questions 16 to 18 are based on the passage you have just heard.**

16. What happened to Mr. Green's car one morning?

【答案】C

【详解】由文中"he discovered that the car wasn't there"一句知他的车不见了,选 C。

17. What did Mr. Green find when he examined the car?

【答案】B

【详解】由"and found two theater tickets on one of the seats"知有两张电影票在车里,选 B。

18. Why did the person who took the car give them tickets?

【答案】C

【详解】由最后一句知,偷车人给他们票是为了趁他们去剧院时到他们家行窃。选 C。

**Questions 19 to 21 are based on the passage you have just heard.**

19. When were the first cables in the world laid?

【答案】A

【详解】由文章第一句"...in the nineteenth century when the first cables were laid."知世界上第一根电缆是在 19 世纪安装的,选 A。

20. What happened to the first cable between England and France?

【答案】D

【详解】由文中"The day after, a fisherman"caught" the cable by mistake. Thinking that the copper wire at the centre of the thick cable was gold, he cut a piece off to show his friends."两句可知英法两国间的电缆被渔民砍了,选 D。

21. How many principal attempts were made before people could send messages between England and America?

【答案】B

【详解】此题需综合分析,由"...in 1856, a serious attempt was made to ..."可知第一次尝试是在 1856 年。"In 1858, a second attemp was made." 第二次尝试是在 1858 年。"There was great rejoicing a few months later, ...the Queen of England was able to speak to the President of the United States."一句知第三次尝试成功了。故应有 3 次尝试,选 B。

United States.

Passage Three

We are all familiar with magnets. The Chinese knew about them in the eleventh century. But the earth itself is also a magnet, with a magnetic North and a magnetic South.

No one really understands very much about the earth's magnetism, although many people have written about it. Scientists believe the center of the earth is like an enormous magnet, giving out a steady magnetic force. This slowly moves away from the center and up to the surface. By the time it reaches the surface the force is really very weak.

Many scientists now believe the magnetism of the earth may have a strong influence on life itself. Animals, birds and even people are to some extent controlled by magnetic forces. When there is a sudden increase in magnetic strength, many animals have difficulty finding their way. Mice, placed near a strong magnet, lose their hair and die early. And when the earth's magnetism suddenly decreases, the number of men or women who kill themselves increases.

How much does the earth's magnetism control our lives? We cannot know for certain. We do know this magnetism is getting weaker and we know that in 2500 years the magnetic poles may change position again. And we know when this happened in the past there were enormous changes in life on Earth.

**Questions 22 to 25 are based on the passage you have just heard.**

22. What do people know about the earth's magnet?

【答案】A

【详解】由文中"But the earth itself is also a magnet, with a magnetic North and a magnetic South."一句可选出答案 A。人对地球磁性的了解是有南北两极。

23. What do we know about the center of the earth?

【答案】C

【详解】"Scientists believe the centre of the earth is like an enormous magnet, giving out a steady magnetic force. By the time it reaches the surface the force is really very weak."说明地球核心磁力最大,越到表面越弱,应选 C。

24. According to the passage, what will happen if the magnetic force of the earth increases suddenly?

【答案】A

【详解】"When there is a sudden increase in magnetic strenth, many animals have difficulty firding their way."一句可从 A、B、C、D 四项中分析出 A 为正确答案。

25. According to the passage, what happens to the magnetic force of the earth?

【答案】C

【详解】由"We do know this magnetism is getting weaker..."一句可知地球的磁力变得越来越弱。故选 C。

**Section C**

Most people think that the older you get, the (26) harder it is to learn a new language. That is, they (27) believe that children learn more easily than (28) adults. Thus, at some (29) point in our lives, maybe around age twelve or thirteen, we lose the (30) ability to learn language well. Is it true that children learn a foreign language more (31) easily than adults? One report showed that the (32) teenagers learned more, in less time, than the younger children. Another report showed that the ability to learn a language increases as the age increases, from (33) childhood to adulthood. (34) There are several possible explanations for these results. For one thing, adults know more about the world and therefore are able to understand meanings more easily than children. (35) Moreover, adults can use logical thinking to help themselves in learning a new language. Finally, adults are more self-controlled than children. (36) Therefore, it seems that the common belief that children are better learners than adults may not be true.

**Section C**

26. harder 27. believe 28. adults
29. point 30. ability 31. easily
32. teenagers 33. childhood
34. There are several possible explanations for these adults
35. Moreover, adults can use logical thinking to help themselves in leaning a new language
36. Therefore, it seems that the common belief that children are better learners than adults may not be true

# Unit 14

## 录音文字材料

**Section A**

1. W: Has the *Time Magazine* come yet? It is already the 15th of the month.

## 答案与详解

1.【答案】B

【详解】女士说已经 15 号了,但男士说要等到后来杂志才

M: Sorry, it's late. Probably not till the day after tomorrow.
Q: When will the magazine probably come?

2. W: I've had four colds this winter and I think I'm catching another one.
M: I've only had half that many, but my wife has had six.
Q: What happened to the woman?

3. W: I think it will rain today.
M: Really? I don't, at least this morning.
Q: When does the man think it will rain?

4. M: What did you think of the final exam?
W: I was expecting it to be easy, but at the end of the first hour, I was still on the first page. I barely had time to get to the last question.
Q: What can we conclude from the above conversation?

5. M: How do Jane and Bill like their new home?
W: It's really comfortable, but they're tired of having to hear the jets go over their house at all hours.
Q: What is located close to Jane and Bill's new home?

6. W: Do you think that Bob is serious about Sally?
M: Well, I know this. I've never seen him go out so often with the same person.
Q: What conclusion does the man want us to draw from his statement?

7. W: How often should I take these pills and how many should I take?
M: Take two pills every six hours.
Q: How many pills should the woman take in twenty-four hours?

8. W: Can you tell me how long a formal visit lasts in the USA?
M: It depends. Usually half an hour. It can last a bit longer if tea or coffee is brought in. But never longer than an hour. Remember in no circumstances should you get your host tired of seeing you again.
Q: In most circumstances, how long does a visit last?

**Conversation One**

M: Hi. Uh, haven't we met before? You look so familiar.
W: Yeah. We met on campus last week, and you asked me the same question.
M: Oh, oh really? I'm sorry, but I'm terrible with names. But, but, but... Let me guess. It's Sherry, right?
W: No, but you got the first letter right.
M: I know, I know. It's on the tip of my tongue. Wait. Uh, Sandy, Susan. Wait, wait. It's Sharon.
W: You got it... and only on the fourth try.
M: So, well, Sh..., I mean Sharon. How are you?
W: Not bad. And what was your name?
M: It's Ben, but everyone calls me B. J. And, uh, what do you do, Sh... Sharon?
W: I'm a graduate student majoring in TESL.
M: Uh, TESL... What's that?
W: It stands for teaching English as a second language. I want to teach English to non – native speakers overseas.
M: Oh, yeah. I'm pretty good at that English grammar. You know,

到,故应为 17 号,选 B。

2.【答案】D
【详解】女士说自己这个冬天得了 4 次感冒,估计现在又得了另一场。所以选答案 D。

3.【答案】C
【详解】男士认为至少今天早上不会下雨,言下之意则是可能晚一点会下雨。

4.【答案】B
【详解】女士说她原本以为考试很简单,可是一个小时她第一张卷子还没做完,最后也没做完整套卷子,可知考试应该是很难的。选 B。

5.【答案】C
【详解】总是有飞机经过,而推测出附近是飞机场。

6.【答案】B
【详解】男士说他从未见过 Bob 与同一个女孩如此频繁的约会,说明 Bob 对 Sally 是认真的,选 B。

7.【答案】A
【详解】每 6 小时 2 颗药,即 24 小时 8 颗药,属于简单的数学运算题。选 A。

8.【答案】A
【详解】Usually half an hour 可知答案选 A。

**Questions 9 to 12 are based on the conversation you have just heard.**

9. Where did the man and woman first meet?
【答案】B
【详解】女士说 We met on campus last week, campus 是校园,意指在学校认识的,选 B。

10. What is the woman's name?
【答案】A
【详解】此题考查辨音能力。答案为 A。

11. In what field is the woman majoring?
【答案】C
【详解】TESL 就是她的专业,根据解释我们知道 It stands for teaching English as a second language. I want to teach English to non – native speakers overseas. 属于教育而不是单纯的英语。

12. What major is the man considering most at this time?
【答案】C
【详解】在 but 之后的 I guess I'm now leaning towards a degree in marketing 才是重点所在。所以选 C。

verbs and adjectives, and uh... Hey, that's sound really exciting. And do you need some type of specific degree or experience to do that? I mean, could I do something like that?

W: Well, most employers overseas are looking for someone who has at least a Bachelor's degree and one or two years of experience. [Oh!] And what do you do? Are you a student on campus?

M: Yeah, but, uh... I guess I'm mulling over the idea of going into accounting or international business, but I guess I'm now leaning towards a degree in marketing.

W: Oh, uh,... Well, I have to run. I have a class in ten minutes.

M: Oh, okay. And, uh, by the way, there's this, uh, dance on campus at the student center tonight, and I was wondering if you'd... you know... like to come along.

W: Oh really? Well, perhaps...

M: Okay, well, bye.

**Conversation Two**

M: Next. Uh, your passport please.

W: Okay.

M: Uh, what is the purpose of your visit?

W: I'm here to attend a teaching convention for the first part of my trip, and then I plan on touring the capital for a few days.

M: And where will you be staying?

W: I'll be staying in a room at a hotel downtown for the entire week.

M: And uh, what do you have in your luggage?

W: Uh, well, just, just my personal belongings um,... clothes, a few books, and a CD player.

M: Okay. Uh, please open your bag.

W: Sure.

M: Okay... Everything's fine. [Great]. Uh, by the way, is this your first visit to the country?

W: Well, yes and no. Actually, I was born here when my parents were working in the capital many years ago, but this is my first trip back since then.

M: Well, enjoy your trip.

W: Thanks.

**Section B**

Passage One

The World Expo is a large-scale, global non-commercial Expo. It aims to promote the exchange of ideas and development of the world's economy, culture, science and technology, to allow exhibitors to publicize and display their achievements and improve international relationships. Accordingly, the World Expo with its 150-year history is regarded as the Olympic Games of economy, science and technology.

The first World Expo was held in London on 1 May 1851. At that time, Britain was the greatest power in the world. To display the power and pride of the country, the British government built a 1,700 feet long and 100 feet high "Crystal Palace" in Hyde Park with 4,500 tons of steel and 300,000 blocks of glass. Queen Victoria invited ten countries to exhibit products inside this "Palace". A total of 6.3 million people attended the Expo throughout the 160 days of its duration.

There are two types of World Expo, the Comprehensive World Expo and the Special World Expo. Over the past 50 World Expo, most

**Questions 13 to 15 are based on the conversation you have just heard.**

13. What is the purpose of the woman's visit?

【答案】C

【详解】当男士问女士来这里的目的是什么时,女士回答道她首先要参加一个教学会议,然后计划在首都游玩几天,因此她的目的有两方面,答案选 C。

14. What things are in the woman's luggage?

【答案】B

【详解】男士问女士的行李箱里有什么东西,女士回答说都是私人物品,包括一些衣物,几本书和一个 CD 机,并且通过了检查,所以答案选 B。

15. What other piece of information do we learn about the woman?

【答案】C

【详解】根据对话内容,这位女士多年前是在这里出生的,但这是她第一次回来这里,因此只有 C 符合答案。

**Questions 16 to 18 are based on the passage you have just heard.**

16. What is the World Expo?

【答案】A

【详解】由文章第一句 The World Expo is a large-scale, global non-commercial Expo. 可知 the World Expo 是一个国际性的非商业展览。选 A。

17. According to the passage, why did the British Government build "Crystal Palace"?

【答案】B

【详解】由文中"To display the power and pride of the country the British government built a 1,700 feet long and 100 feet high 'Crystal Palace' in ..."一句知是为了显示国家的强大与骄傲。

18. Which of the following countries has not yet hosted the World Expo?

【答案】D

have been comprehensive. The products exhibited at Comprehensive World Expos covered items of every nature. Comprehensive World Expos are held once every five years. The products exhibited at Special World Expos are specialized. The 1999 Kunming World Expo was a Special World Expo. To date, a total of 24 cities in 13 countries have hosted the World Expo, including Britain, France, America, Germany, Belgium, Canada, Japan, Australia, Spain, Italy, Korea, Portugal and China.

### Passage Two

Several years ago a college professor took a well-organized speech and changed the order of its sentences at random. He then had a speaker deliver the original version to one group of listeners and the changed version to another group. After the speeches, he gave a test to see how well each group understood what they had heard. Not surprisingly, the group that heard the original, unchanged speech scored much higher than the other group.

A few years later, two professors repeated the same experiment at another school. But instead of testing how well the listeners comprehended each speech, they tested to see what effects the speeches had on the listeners' attitudes toward the speakers. They found that people who heard the well-organized speech believed the speaker to be much more competent and trustworthy than did those who heard the changed speech.

These are just two of many studies that show the importance of organization in speechmaking. You realize how difficult it is to pay attention to the speaker, much less to understand the message. In fact, when students explain what they hope to learn from their speech class, they almost always put the ability to organize their ideas more effectively at the top of the list. This ability is especially vital for speechmaking. A speaker must be sure listeners can follow the progression of ideas from beginning to the end. This requires that speeches be organized strategically.

The first step in developing a strong sense of speech organization is to gain command of the three basic parts of a speech—introduction, body, and conclusion—and the strategic role of each. The body is the longest and most important part. Also, you will usually prepare the body first. The process of organizing the body of a speech begins when you determine the main points.

### Passage Three

There are four types of college degrees, starting with the associate degree. The associate degree takes about two years to complete when one is enrolled full time. The bachelor's degree takes four years when one is enrolled full time, with the master's taking one to two years, and the doctor's three to four years. The associate degree may be substituted for the first two years of a bachelor's degree if it is a transfer degree. Not all associate degrees are designed for transfer. Some are technical degrees that are called terminal degrees, which means they do not count toward a bachelor's. The bachelor's is normally required before one can work at the master's level. Likewise, the master's is normally required before one can work at the doctor's level.

【详解】"To date, a total of 24 cities in 13 countries have hosted the World Expo, including Britain, France, America, Germany, Belgium, Canada, Japan, Australia, Spain, Italy, Korea, Portugal and China."一句知含有 A、B、C 三国,但没有 D、Russia。

**Questions 19 to 22 are based on the passage you have just heard.**

19. What was tested in the second experiment introduced in the passage?

【答案】B

【详解】此题要注意听题干,问的是第二次试验,注意这一点即可从"they tested to see what effects the speeches had on the listeners' attitudes toward the speakers"知答案选 B。

20. What do students want to learn most from their speech class?

【答案】C

【详解】由"In fact, when students explain what they hope to learn from their speech class, they almost always put the ability to organize their ideas more effectively at the top of the list."知答案为 C。

21. What are the basic parts of a speech according to the passage?

【答案】A

【详解】由"The first step in developing a strong sense of speech organization is … of a speech—introcluction, body, and conclution—and the strategic role of each."知演讲稿的基本要素是选项 A 里的三点。

22. What is the passage mainly about?

【答案】D

【详解】这是一道主旨题,要从整篇文章上把握,因为通篇都是围绕组织演讲稿谈的,故选 D。

**Questions 23 to 25 are based on the passage you have just heard.**

23. Which degree lies in the last place?

【答案】D

【详解】根据文章对各种学位的说明,Doctor's degree 是排在最后一位,故选 D。

24. What is a terminal degree?

【答案】A

【详解】根据"Not all associate degrees are designed for transfer. Some are technical degree that are called terminal degrees."可得出答案为 A。

25. When one plans to work at the master's level, which degree is normally required?

【答案】B

## Section C

The new Victoria Line was (26) opened in 1969. This new line is very (27) different from the others. The stations on the other lines need a lot of workers to sell (28) tickets, and to check and collect them when people (29) leave the trains.

This is all different on the Victoria Line. Here a machine checks and (30) collects the tickets, and there are no (31) workers on the (32) platform. On the train, there is only one worker. If (33) necessary, the man can drive the train. (34) But usually he just starts it, it turns and stops by itself. The trains are controlled by electrical signals which are sent by the so – called "command spots".

(35) The command spots are the same distance apart. Each sends a certain signal. (36) The train always moves at the speed that the command spots allow. If the command spots send no signals, the train will stop.

【详解】根据"The bachelor's is normally required before one can work at the master's level."可知答案为B。

## Section C

26. opened 27. different 28. tickets
29. leave 30. collects 31. workers
32. platform 33. necessary
34. But usually he just starts it, it turns and stops by itself
35. The command spots are the same distance apart
36. The train always moves at the speed that the command spots allow

# Unit 15

### 录音文字材料

## Section A

1. M: Is this school really as good as people say?
   W: It used to be even better.
   Q: How is the school?
2. M: What time does the train leave?
   W: Not until 9:15, But I'd like to get to the station by 8:30.
   Q: When will the train leave?
3. M: Make thirty copies for me and twenty copies for Mr. Brown.
   W: Certainly, sir. As soon as I make the final corrections on the original.
   Q: What is the probable relationship between the two speakers?
4. M: Hello, operator. I'd like to call Brazil. When is the best time?
   W: The cheapest times are before eight in the morning, after six at night, or at the weekends.
   Q: What does the man want to do?
5. M: You look nice with your hair down.
   W: Thanks. The barber was a good one.
   Q: What is the man saying?
6. M: What can I do for you, madam?
   W: Oh, yes, my husband bought this yellow skirt here yesterday. It is very nice, but it is a little small. Have you got any bigger ones?
   Q: What does the woman want to do?
7. W: Good morning. Your passport, please. Do you have anything to declare?
   M: Only these two cartons of cigarettes, a bottle of brandy and some jewelry. That's all.
   Q: With whom is the man speaking?

### 答案与详解

1.【答案】A
【详解】女士说它以前更好,即现在没有以前好了,选A。

2.【答案】C
【详解】女士说9:15以前不会开车,所以应是C.9:15。

3.【答案】D
【详解】男士要求女士复印文件,从这里可判断只有D项,老板和秘书的关系比较合适。

4.【答案】D
【详解】男士要打电话到巴西,询问最好的时间,故选D。

5.【答案】C
【详解】男士说女士头发垂下来很好看,女士说那个理发师不错,说明女士刚剪了个新发型,选C。

6.【答案】C
【详解】根据女士所说的话可知她是想帮丈夫换件大点的衬衣,选C。

7.【答案】D
【详解】女士向男士要护照也看出女士是customs official,选D。

8. M: I was surprised to see Mary using that record player you were going to throw away.

W: Yes it's very old. That she got it to work amazes me.

Q: What does the woman mean?

**Conversation One**

W: Hey, I hear you and Stephanie are really getting serious.

M: Yeah, I think she'll be impressed with my new exercise program.

W: What? What are you talking about? What exercise program? What did you tell her?

M: Well, you know, I enjoy staying in shape. [Right] First, I generally get up every morning at 5:30 a. m.

W: Oh, yeah. Since when? You don't roll out of bed until at least 7:30 p. m.

M: No, no, and on Mondays and Wednesdays,...

W: Ah, not another tall tale...

M: I almost always go jogging for about a half hour, you know, to improve my endurance.

W: Hey, jogging to the refrigerator for a glass of milk doesn't count.

M: Of course, before I leave, I usually make sure I do some stretches so I don't pull a muscle on my run.

W: Right. One jumping jack.

M: Then, I told her that I usually lift weights Tuesdays and Thursdays for about an hour after work.

W: Humph.

M: This really helps me build muscle strength.

W: A one-pound barbell.

M: Finally, I often go hiking on Saturdays with my dog [What dog!?], well, and I like hiking because it helps me burn off stress and reduce anxiety that builds up during the week.

W: Oh yeah, those lies.

M: Well, uh, as for Fridays, I sometimes just relax at home by watching a movie or inviting you over to visit.

W: If I buy the pizza.

M: But... bu... And on Sundays, I take the day off from exercising, but I usually take my dog for a walk.

W: Forget it. She'll never buy this story.

**Conversation Two**

W: Oh, hi. What was your name again? I can't keep straight all the students' names this being the second day of school.

M: It's okay. I have a hard time remembering names myself.

W: How, uh, Karen, right?

M: No, it's Nancy. My mom's name is Karen.

W: Nancy. Okay. I think I heard you were from England.

M: Well, I was born there, but my parents are American. I grew up in France.

W: Oh, a world traveller!

M: But then we moved here when I was nine.

W: So, what does your father do now?

M: Well, he's a college professor, and he is in Scotland at the moment.

W: How interesting. What does he teach?

M: Oh, I haven't a clue. Nah, just joking. He teaches chemistry.

8.【答案】D

【详解】根据男士说的对于 Mary 能让那台机子工作很令他吃惊,可知答案选 D。

**Questions 9 to 12 are based on the conversation you have just heard.**

9. What does the man usually do on Mondays and Wednesdays?

【答案】A

【详解】根据对话,Michael 说他在周一和周三会起得很早,去 go jogging,也就是跑步的意思,答案为 A。

10. What does the man do before the activity in Question 9?

【答案】C

【详解】before I leave, I usually make sure I do some stretches so I don't pull a muscle on my run. 明显,答案是 C。

11. Why does the man lift weights?

【答案】A

【详解】Michael 说为了增强肌肉的力量 This really helps me build muscle strength,答案为 A。

12. Why does the man go hiking on Saturdays?

【答案】A

【详解】I like hiking because it helps me burn off stress and reduce anxiety that builds up during the week. 这句话就是答案所在,关键是对 burn off 的理解,意思是蒸发掉,也就是消除的意思。答案为 A。

**Questions 13 to 15 are based on the conversation you have just heard.**

13. Where does the girl grow up before nine?

【答案】B

【详解】对话中出现了几个国家名,注意不要混淆。Nancy 的父母是美国人,她出生在英国,在法国长大。九岁的时候搬迁到这里。因此答案是 B。

14. What does the girl's father do for a living?

【答案】B

【详解】被问到她爸爸现在做什么工作时,女孩回答"he's a college professor",因此选 B。

15. What is one thing NOT mentioned about the girl's family?

【答案】A

【详解】对话中女孩提到了母亲在家做家庭主妇,她在法国长大,九岁的时候搬迁到这里,只有 A 选项中的为什么她的父母会在英国生活几年没有在对话中涉及到,因此选

W: Oh, chemistry, and uh, what about your mother?

M: She works full time at home.

W: Oh, and what, does she have her own business or something?

M: Nah, she takes care of me.

W: Well, being a homemaker can be a real hard, but rewarding job.

M: I think so too.

**Section B**

Passage One

Sometimes we say that someone we know is "a square peg in a round hole". This simply means that the person we are talking about is not suited for the job he is doing. He may be a bookkeeper who really wants to be an actor or a mechanic who likes cooking. Unfortunately, many people in the world are "square pegs", they are not doing the kind of work they should be doing, for one reason or another. As a result they probably are not doing a very good job and certainly they are not happy.

Choosing the right career is very important. Most of us spend a great part of our lives at our jobs. For that reason we should try to find out what our talents are and how we can use them. We can do this through aptitude tests, interviews with specialists, and study of books in our field of interest.

There are many careers open to each of us. Perhaps we like science, then we might prepare ourselves to be chemists, physicists, or biologists. Maybe our interests take us into the business world and such work as accounting, personnel management or public relations. Many persons find their place in government service. Teaching, newspaper work, medicine, engineering—these and many other fields offer fascinating careers to persons with talent and training.

Passage Two

For all its benefits, television's influence has been extremely harmful to the young. Children do not have enough experience to realize that TV shows present an unreal world; that commercials lie in order to sell products that are sometimes bad or useless. They believe and want to imitate what they see. They do believe that they will make more friends if they use a certain soap, or some other product. By the time they are out of high school, most young people have watched about 15,000 hours of television, and have seen about 18,000 killings or other acts of violence. How could they be choked to see the same in real life? All educators and psychologists agree that they "television generations" are more violent than their parents and grandparents. It is certain that television has deeply transformed our lives and our society. It is certain that, along with its benefits, it has brought enormous problems. To those problems we must soon find a solution because—whether we like it or not—television is here to stay.

Passage Three

About three hundred words in the English language come from the

A。

**Questions 16 to 18 are based on the passage you have just heard.**

16. What is implied in a "square peg in a round hole"?

【答案】C

【详解】"This Simply means that the person we are talking about is not suited for the job he is doing."一句可知这个词组的意思是什么,选 C。

17. Why is it important to find the right career?

【答案】C

【详解】由文中"As a result they probably are not doing a very good job and certainly they are not happy."一句可知合适的工作使人快乐,选 C。

18. What are some of the careers found in the scientific world according to the passage?

【答案】B

【详解】由"Perhaps we like science, then we might prepare ourselves to be chemists, physicists, or biologists."一句可知在科学领域,可以做化学、地理、生物学家,选 B。

**Questions 19 to 22 are based on the passage you have just heard.**

19. What does the paragraph before this passage most probably discuss?

【答案】D

【详解】根据文章的开头 for all its benefits,可知此段话前面应讲的是电视的好处。选 D。

20. What do we learn from the passage?

【答案】A

【详解】由"Children do not have enough experience to realize that TV shows present an unreal world."一句告诉我们 A 项正确,B、C、D 有违于文中的意思。

21. How many days of television have most young people watched when they graduate from high school?

【答案】A

【详解】根据文中"By the time they are out of high school, most young people have watched about 15,000 hours of television."一句计算,将 15000 除以 24 得 625,即选 A。

22. What can be inferred from the passage?

【答案】C

【详解】根据文章最后一句"because—whether we like it or not—television is here to stay",我们可以推测出选项 C。

**Qustions 23 to 25 are based on the passage you have just heard.**

names of people. Many of these words are technical words. When there is a new invention or discovery, a new word may be created after the inventor or scientist. It is interesting to observe how many common words have found their way into the language from the names of people. Lord Sandwich who lived from 1718 – 1792 used to sit at the gambling table eating slices of bread with meat in between. As the Lord was the only one among his friends who ate bread in that way, his friends began to call the bread "sandwich" for fun. Later on the word became part of the English language.

The word "boycott" means to refuse to have anything with somebody or something. It comes from a man called Captain Boycott. He was a land agent in 1880 and he collected rents and taxes for an English landowner in Ireland. But the Captain was a very hard man. He treated his poor tenants very badly. His tenants decided not to speak to him at all. Eventually word got back to the landowner and the Captain was removed. The word "boycott" became popular and was used by everyone to mean the kind of treatment that was received by Captain Boycott.

**Section C**

The history of the United States is quite (26) short. It began a little more than 200 years ago. In 1776, 13 (27) colonies located on the eastern (28) coast of North America (29) declared independence and fought a (30) revolution against the British. In 1783 the (31) colonists won the revolution. After the revolution, the United States (32) bought a large (33) section of country from Napoleon of France.

(34) Texas and most of what is now the Southwestern part of the United States belonged to Mexico. Later, the United States and Mexico went to war. (35) If Mexico had defeated the United States in that war, California and New Mexico would have been part of Mexico today.

In 1861 one-half of the United States went to war with the other half. (36) This was the Civil War. Following the Civil War, the United States bought Alaska from Russia. Alaska became the largest state in the country and is a very important one today.

23. What was Lord Sandwich fond of?

【答案】D

【详解】由"Lord Sandwich who lived from... eating slices of bread with meat in between."一句可找出答案,是 D、带肉的面包。

24. What is the word "Sandwich" created by?

【答案】C

【详解】由"his friends began to call the bread 'sandwich' for fun, Later on the word because part of the English leanguage."由这一句我们可知 Sandwich 这一单词是 Lord Sanwich 的朋友说起的。

25. When the landowner found out that the tenants were boycotting his land agent, what did he do?

【答案】A

【详解】由"Eventually word got back to the landowner and the Captain was removed."可知他解雇了 agent,选 A。

**Section C**

26. short 27. colonies 28. coast
29. declared 30. revolution 31. colonists
32. bought 33. section

34. Texas and most of what is now the Southwestern part of the United States belonged to Mexico
35. If Mexico had defeated the United States in that war, California and New Mexico would have been part of Mexico today
36. This was the Civil War. Following the Civil War, the United States bought Alaska from Russia. Alaska became the largest state in the country and is a very important one today

四级考试核心考点
**精确打击**

# Part 4

# 仔细阅读理解

Reading Comprehension
( Reading in Depth )

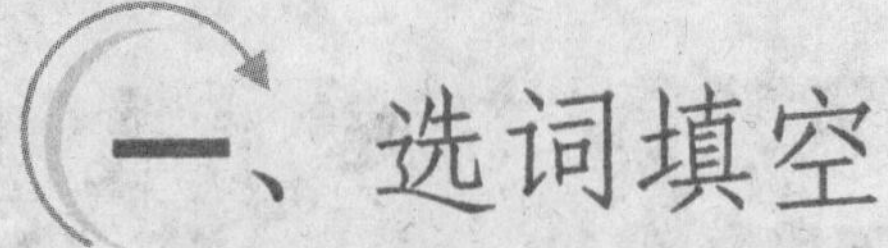

# 命题规律与应试技巧

## 1. 题型聚焦

选词填空是以前四级没有考过的题型。而阅读理解的这种考查题最早在雅思考试中就出现过。两者相比较，四级中的选词填空短一些，一般230词左右（样题219词，2006年6月真题244词），句子总共有12～13句；就难度而言，每句只会出现一个空，全文共10个空。文章下面提供一个词库，词库里有15个词备选。在文章中出现的词汇和词库里的词汇基本都属于四级大纲内词汇。但对于考生来说，这个新题型仍是个不小的挑战，因为我们绝大部分考生从小接触的就是像完形填空那样的题型，一个空有四个备选项，而现在的选词填空，从理论上讲，每个空有15个可能的答案。下面我们来看看2006年新四级的样题，对题型作更深一步的分析：

When Roberto Feliz came to the USA from the Dominican Republic, he knew only a few words of English. Education soon became a __47__. "I couldn't understand anything," he said. He __48__ from his teachers, came home in tears, and thought about dropping out.

Then Mrs. Malave, a bilingual educator, began to work with him while teaching him math and science in his __49__ Spanish. "She helped me stay smart while teaching me English," he said. Given the chance to demonstrate his ability, he __50__ confidence and began to succeed in school.

Today, he is a __51__ doctor, runs his own clinic, and works with several hospitals. Every day, he uses the language and academic skills he __52__ through bilingual education to treat his patients.

Roberto's story is just one of __53__ success stories. Research has shown that bilingual education is the most __54__ way both to teach children English and ensure that they succeed academically. In Arizona and Texas, bilingual students __55__ outperform their peers in monolingual programs. Calexico, Calif., implemented bilingual education, and now has dropout rates that are less than half the state average and college __56__ rates of more than 90%. In El Paso, bilingual education programs have helped raise student scores from the lowest in Texas to among the highest in the nation.

| | | | |
|---|---|---|---|
| A) wonder | B) acquired | C) consistently | D) regained |
| E) nightmare | F) native | G) acceptance | H) effective |
| I) hid | J) prominent | K) decent | L) countless |
| M) recalled | N) breakthrough | O) automatically | |

47.【答案】nightmare

【解析】结合前句 he knew only a few words of English（他对英语所知甚少），可以判断受教育对 Roberto Feliz 来说是很困难，很吃力的。词库中只有 nightmare（噩梦）能表达这方面的意思。

48.【答案】hid

【解析】本句缺谓语,应填动词,hide from 在此处指躲避。

49.【答案】native

【解析】从此句可知 Mrs. Malave 用西班牙语来教他,所以西班牙语应是 Roberto Feliz 的母语,故填 native。

50.【答案】regained

【解析】从本句后半部分 and began to succeed in school(并且开始在学习上取得成功)可以推知 Roberto Feliz 应该是又有了信心,而 regain 符合此意。

51.【答案】prominent

【解析】从文章前半段可知 Roberto Feliz 在学业方面顺利,那么成为医生的话,应该是一位成功的医生。prominent 指显著的,卓越的,突出的。

52.【答案】acquired

【解析】这一题实际考查的是语言(language),技巧(skill)这两个词之前一般使用什么动词。获得技能,掌握语言都可用 acquire。

53.【答案】countless

【解析】从 one of + 复数(题中是 success stories)结构可判断应填一个表数量的形容词,countless 最符合题意。

54.【答案】effective

【解析】纵观全文,作者对双语教学(bilingual education)十分推崇,所以本题应选 effective。

55.【答案】consistently

【解析】consistently 表示一贯地,一向,始终如一地,符合句意。

56.【答案】acceptance

【解析】前半句说辍学率(dropout rates)是少于全国平均水平的半数(less than half the state average),后半句可以看到有 more than,两厢对照,就应该是升学率上升了,而大学接受率(college acceptance rate)就是升学率。

从上面的分析可以看出,选词填空是一种综合考查学生阅读能力,词汇使用和语法知识的题型。考查考生对文章的全面理解;对词汇的色彩,内涵外延,搭配的熟练掌握;对语法知识的了解。

## 2. 解题三大步骤精确突破

### 步骤 1 快速阅读全文,了解全文大意。

由于全文并不长,因此在了解大意的时候,要特别注意那些表示逻辑关系的介词,连词和副词。

表示相同:and, as, as if /though, in the same way, like, likewise, similarly

表示不同:although, but, even though, however, in contrast, nevertheless, on the other hand, still

举例、重复、强调:as an example, for instance, specifically, such as, once again, once more, as a matter of fact, indeed, more important, to be sure

表示补充:also, and, in addition, as well, besides, furthermore, in addition, moreover, too

表示因果:as a result, because, hence, since, so that, then, therefore, thus

表示条件:as long as, as soon as, even if, in case, provided that, unless, when

这些连接词和短语虽不见得能直接对解题有什么帮助,但文章里面一旦出现的话,应该特别留意,有助于对文章整体脉络和逻辑的把握,间接有助于解题。

### 步骤 2 进入对填空的直接解答。

在选词之前,我们可以对词库中的词作一个分析,从样题和 2006 年 6 月考的一次新题型来看,词库中的词都是实义词,即名词、动词(含分词形式)、形容词和副词。我们可以看到样题中有 3 个名词,4 个动词,5 个形容词和 2 个副词,还有一个词 wonder,是既可作动词又可作名词。分布比较平均。让我们以 2006 年 6 月四级考试中的选词填空题为例,看看这样做有什么好处:

El Nino is the name given to the mysterious and often unpredictable change in the climate of the world. This strange __47__ happens every five to eight years. It starts in the Pacific Ocean and is thought to be caused by a failure in the *trade winds*(信风),which affects the ocean currents driven by these winds. As the trade winds lessen in __48__,the ocean temperatures rise, causing the Peru current flowing in from the east to warm up by as much as 5℃.

The warming of the ocean has far - reaching effects. The hot, *humid*(潮湿的) air over the ocean causes severe __49__ thunderstorms. The rainfall is increased across South America, __50__ floods to Peru. In the West Pacific, there are droughts affecting Australia and Indonesia. So while some parts of the world prepare for heavy rains and floods, other parts face drought, poor crops and __51__.

El Nino usually lasts for about 18 months. The 1982 - 83 El Nino brought the most __52__ weather in modern history. Its effect was worldwide and it left more than 2,000 people dead and caused over eight billion pounds __53__ of damage. The 1990 El Nino lasted until June 1995. Scientists __54__ this to be the longest El Nino for 2,000 years.

Nowadays, weather experts are able to forecast when an El Nino will __55__,but they are still not __56__ sure what leads to it or what affects how strong it will be.

| | | | |
|---|---|---|---|
| A) estimate | B) strength | C) deliberately | D) notify |
| E) tropical | F) phenomenon | G) stable | H) attraction |
| I) completely | J) destructive | K) starvation | L) bringing |
| M) exhaustion | N) worth | O) strike | |

该词库中有5个名词 strength, starvation, exhaustion, phenomenon, attraction;动词有 estimate, notify, bringing 共3个;既可做名词又可做动词的一个:strike;既可做名词又可做形容词的一个:worth;形容词有 destructive, tropical, stable 共3个;副词 completely, deliberately 共2个。这样可大大减少选择范围,因为不同词性的词总出现在句子的特定位置,下面看技巧一:

冠词+______; ______+of;______+动词或动词短语;介词+______;则空格内必为名词;

名词+______ +(名词、形容词、副词或动词不定式及分词形式);则空格内必为动词;

______+名词;则一般空格内为形容词;

______+形容词;则一般空格内为副词;

运用技巧一,我们开始解题:

47. ______+动词,故填名词。

【答案】phenomenon

【解析】根据技巧一,我们已知道要填一个名词,而根据空格前的 this,我们可知道填的名词在前面一句应该有所表现,而前一句也就是文章第一句讲的是厄尔尼诺现象,故应填 phenomenon。从这一题,我们可以总结出技巧二:结合上下文,注意文义的重现。

48. 介词+______

【答案】strength

【解析】根据技巧一,可判断是名词,根据技巧二,我们在上一句可看到 failure in the trade winds, 此处的 failure 显然不是指失败,而应是表示变得不足,不如以前的情况,由此可判断 trade winds 力量减弱,所以填 strength。

49. ______+名词

【答案】tropical

【解析】本句中已经有谓语动词,又根据技巧一,可知填形容词,已知的形容词已经有三个,destructive,tropical,stable;还有一个既可做名词又可做形容词的 worth,但 worth 和 stable 词义显然不符,可以排除,剩下的两个 destructive 和 tropical 对于语感一般的考生确实头痛,此时我们采用技巧三:无法判断的先放在一边。

50.【答案】bringing

【解析】此题完全是个句法题,根据句子前面已有的被动态 is increased 能判断出 flood 在此是做名词,又根据空格前的逗号,可知此处是一个分词引导的短语,而给出的15个词中只有 bringing 一个分词。所以这一题甚至不看全文都能做出来,用的是技巧四:根据语法做判断。

51. 出现 and 连词,是名词排比

【答案】starvation

【解析】既然是名词排比,就自然填名词,根据技巧一,看看这组排列前面的词 drought(干旱),poor crops(歉收),可知填 starvation(饥荒)。

52. ________ + 名词

【答案】destructive

【解析】不仅是________ + 名词，而且前面还有 most，肯定填形容词。已知的形容词已经有三个，destructive、tropical、stable；还有一个既可做名词又可做形容词的 worth，但 worth 和 stable 词义显然不符，可以排除，tropical 前面无法加 most，只能填 destructive，这是技巧五：排除法。此时我们再回到 49 题，不难得知 49 题填 tropical。

53.【答案】worth

【解析】前面是金额数量，则很自然用 worth。

54. 名词 + ________ + 代词

【答案】notify

【解析】根据技巧一可知填动词，已用掉一个动词 bringing，剩下的 3 个中 notify ... to... 是固定搭配，本题运用技巧六：借助固定搭配。

55. 名词 + ________

【答案】strike

【解析】名词 + ________ 且有助动词，肯定填动词，此时动词只剩两个，estimate 和 strike，根据句意，显然是选 strike，本题也是用技巧五排除法。

56. ________ + 形容词

【答案】completely

【解析】首先根据技巧一可判断填副词，然后本题同样用排除法，15 个词中只有两个副词，deliberately 表示故意的，从意思上无法修饰 sure，而 completely 可以。

## 步骤 3 检查。

检查分两步进行：第一步，带着选出的答案再一次通读文章，看是否通顺流畅，任何意思不清楚的地方都不能放过。第二步，审视选剩下的 5 个词。做到如下几个辨析：

**(1) 近义词的辨析**

例：So while some parts of the world prepare for heavy rains and floods, other parts face drought, poor crops and __51__.（2006 年 6 月真题）

我们填了 starvation，检查时会看到剩下 exhaustion。心中没有把握的同学应对这两个意思相近的词作出区分。两个词其实都可以表达缺乏，不足，耗尽，但 starvation 还可表示饥饿，饥荒，这正是文章需要的。

**(2) 词性相同的词的辨析**

词性相同，往往就意味着用法相近，比如样题中的 prominent 和 decent。这时就需要对词的内涵和外延的准确把握。

**(3) 拼写相近的词的辨析**

这个辨析其实就是防止自己粗心大意，导致失分。

# 二、短句回答

# 命题规律与应试技巧

## 1. 三大题型精确定位

短句回答是选词填空的备用题型，二者选其一交替出现在每次的四级考卷中，作为仔细阅读理解的 Section A。短句回答的命题主要有以下 3 种类型：①主题思想题，②事实细节题，③分析推理题，而以事实细节题考得最多，约占全部考题的 70%。

### (1) 主题思想题

此题型主要是用来检验学生高度概括的能力，是简答题的重要内容之一。每一篇文章都有它的主题思想，一般能在文章每一段第一句找到。有时，主题句也出现在文章某一段落的末尾或中间。常见提问方式有：

What does the passage mainly discuss?

What is the best title for the passage?

What is the main topic /subject of the passage?

What is the main idea expressed in the passage?

关键是找到主题句，通过对主题句进行"手术"，就能找到最佳答案。而主题句通常都在段首或段末，有时是一个疑问句或简单句，且具有语法上的独立性。

关键词有"in short"，"in my opinion"，"that is"，"therefore"，"I believe/suggest/think"，"in fact"等。

例如：What is the passage mainly about?

【分析】文章主要讲美国的偷车现象及其解决办法。阅读完毕，发现几个关键句子，如 Vehicle theft is a common phenomenon, which has a direct impact on over four million victims a year. How can you protect your car?

如果对这两个句子进行巧妙裁剪、嫁接，就能有许多满意答案，如：

①Vehicle theft and security system in U. S.

②Car thefts in the U. S. A.

③Theft of Car.

④Stolen－vehicle phenomenon and solution 等。

这些答案中关键词都可以在主题句或原文中找到，技巧只是如何把它们很好地结合起来。

### (2) 事实细节题

细节题是根据短文提供的信息和事实进行提问。此类题目涉及范围广泛，考试中所占比重大，提问方式多样。细节题的答案常常隐含在某些词语中，要求考生细读若干个句子，弄清题目意图，抓住关键词，对这些事实和根据作出评估，然后再作综合概括。本题型是考试中使用最多的题型，它们常围绕以下方面的内容展开：

Why was it. . . ?　　When did . . . ?　　Where did . . . ?　　Who was . . . ?　　How . . . ?

According to the passage, ...?

此类题型是简答题中考查得最多的题型,考查形式多样,下面一一剖析。

**①描述型题目:能直接从原文中找到答案。**

此类细节题答案一般是句子谓语或宾语等主干成分,略做变化即能写出答案。

例如:

How serious did the author predict the annual vehicle theft could be in the United States in 1989?

【分析】答好此题的关键,是要看到文中的一句话,即:

In 1987,... If current trends continue, experts predict annual vehicle thefts could exceed two million by the end of the decade.

前面提到了1987,可知 the end of the decade 指80年代末,即1989年;exceed two million 是关键词。因此答案可以为:Over two million vehicles could be stolen.

**②因果型题目:短文中能直接找到原因或结果。**

通常表原因的关键词有:for that reason, for, as, because, since, as a result of, owing to, thanks to 等。表结果的关键词有:as a result, therefore, consequently, thus, accordingly, so。阅读时要注意这些词后面的内容。

例如:

Why are newspaper considered as an important medium according to the passage?

【分析】此题用 why 来提问,短文中似乎没有 because... 回应句。但考生在第一段应该会看到这样一句话:Of all the media, television is clearly dominant, with newspapers a close second, at least as a source of news and other information. 此句中直接谈到"newspapers"的作用"as a source of news and other information",as 是重要标志,因而题干答案即为:Because they are a source of news and other information.

例如:Why was it easy for boats to tumble over in the Colorado?

【分析】此题在短文第一段中没有出现任何表因果关系的连词,但文中有两处关键的话,即:all of us naturally set aside any pretenses(矫饰) and put out backs into every stroke to keep the boat from tumbling over.

此处要知道 keep ... from ... 之意,防止……被……,working together to cope with the unpredictable twists and turns of the river. 前一句话暗示出小船极易"tumble over",后一句话表明河流有:"the unpredictable twists and turns",故答案为:Because the river is full of twists and turns.

**③范例型题目:需要概括答案。**

表示举例的关键词有:for example, such ... as, for instance, that is, as follow 等。对这些词后面的内容要注意。

例如: What caused the sharp conflict in the GM plant in the late 1970s?

【分析】此题的答案需要从文中所举例子的前后经过的描述中才能概括出来。文中用 For example 道出事情经过:For example, in the late 1970s a General Motor plant in Fremont, Calif., was the scene of constant warfare between labor and management. Distrust ran so high that the labor contract was hundreds of pages of tricky legal terms...

关键词有"distrust","high","tricky"。此外,上一段"the teamwork is the key to making dreams come true"也很重要,因为所举例子是论证此观点的,故可概括出答案:Distrust and lack of teamwork.

**④对照比较型题目:对照比较所涉及的两个事物之间的不同或相似之处,进而说明主题。**

表对照的关键词有:how ever, nevertheless, in contrast, on the other hand, but, yet, while 等。

表比较的词有:likewise, in the same way, as if, as 等。

例如:

Developing children's self-confidence helps bring them up to be ______.

【分析】此题答案在短文最后一句能直接找到。Giving children the opportunity to develop new resources to enlarge their horizons and discover the pleasures of doing things on their own is, on the other hand, a way to help children develop a confident feeling about themselves as capable and intersting people.

句中 on the other hand 表对照,暗示出与前面相反的结果,故答案为:capable and interesting people.

**⑤描写叙述型题目:需要变换词法或句型。**

此类题的答案内容和表达词在原文中能直接找到,关键是进行时态和句型的转换。在替换过程中切勿因粗心犯错误,如时态、单复数和动名词等。

例如:

What does the author think Joe Jempler should be blamed for?

【分析】此题答案可在文中找到大部分词句,但需要作些变换。

原句有:Joe Templer should have known better:... It won't hurt to leave the key in the truck this once,... 应知道:"should + have + V-ed"用法,其次"leave the key in the truck"是关键词。通过这些知识点可知答案为:Having left his key in the truck.

## (3)分析推理题

此类题旨在考查学生归纳、演绎、与综合分析等逻辑推理能力,其答案一般根据已知信息来推理。要回答这类题不仅要求考生弄懂文章字面的意思,还要求考生领会文章潜在的含义和作者所给的提示。解题的关键是:靠推断而不是原文照搬,把握文章的主题思想和每段的内容;抓住作者的观点;分析文章的有关信息,用自己的话来叙述。推理题一般有以下方式:

It can be inferred from the passage that...

What does the passage imply about?

According to the author, what does the sentence suggest?

The author implies that...

What is the author's main purpose in the passage?

例如:What made it possible for the TIME reporters to come up with so many interesting stories about pets?

本题要求考生仔细阅读、进行合理的推断,对文章第二段一、二句有透彻的理解。文章说:每一位宠物的主人都有类似的故事并渴望听的人也有同感,《时代周刊》的记者们一下就给出了25个动物的故事,每个故事都说自己的宠物是世界上最聪明的动物。从这里我们可以推断,正是因为宠物的主人急于把故事说给别人听,记者才有机会收集这么多的故事。所以本题的答案可以是:

Pet owners want others to share their stories.

Pet owners are willing to share their stories (with others).

Pet owners want to show that their pets are smart.

Pet owners are proud of raising clever pets.

这类题型主要包括两种:描述事实基础上的推理和逻辑推理。

**①在描述事实基础上的推理题。**

考生只要在描述细节句子范围内进行推理,不必顾及整个文章主旨,以防干扰。

例如:

What did Newton seem puzzled about?

【分析】此题文中第一段只描写了一个事实:Whenever I tossed out a Frisbee(飞碟) for him to chase, he'd take off in hot pursuit but then seem to lose track of it. Moving back and forth only a yard of two from the toy, Newton would look all around, even up into the trees. He seemed genuinely puzzled.

通过对这几句描述的理解,应明白 Newton 似乎不知上哪儿去找飞碟,故答案为:Not knowing the Frisbee's track.

**②通过前后句以及上下文的内在逻辑推理。**

此类题一般是针对主题思想、作者意图而设计的。考生要注意首句、段尾句和表示转折或因果关系的一些词,如:but, however, yet, in short, as, although, as a result, because, since, therefore, thus, so 等。

例如:

Why does the author say Newton had unique sense of humor?

【分析】短文第一段先描写了 Newton 似乎不知道该上哪儿找飞碟,但当作者准备去帮他时,"he would run invariably straight over the Frisbee, grab it and start running like mad, looking over his shoulder with what looked suspiciously like a grin."

从这前、后句可以判断 Newton 是在挑逗、欺骗作者。故答案为:Because Newton intended to deceive him.

# 2. 应试技巧精确突破

(1)答案尽量用短语或词组,句子要简洁,一般都规定了不能超过10个单词,多了要扣分。

(2)尽量用短文中出现的词组或短语,句子的用词也应以短文中出现的关键词为先。若非用自己的语言组织、表达不可,则应注意句子结构的精炼、完整和时态的正确。

(3)注意大小写正确及书写的工整。

# 三、多项选择

# 命题规律与应试技巧

## 1. 五大题型命题规律精确定位

多项选择阅读理解题,实际上就是传统的阅读理解。所不同的是,传统四级阅读为四篇文章,现改为两篇。这两篇 300 多词的阅读文章和选词填空(或短句回答)构成了仔细阅读理解部分。多项选择选用的阅读理解文章,如根据试题类型划分,可大致分为以下五类:

### (1)主旨大意题

这类题型主要是用来考查考生是否理解了文章或段落的主旨和大意,其提问方式常有以下几种:

①What is the main idea of the passage?

②What is the author's purpose in writing the passage?

③What is the passage mainly about?

④Which of the following statements best expresses the main idea of the passage?

⑤Which of the following can be the best title of the passage?

⑥Which of the following best summarizes the author's opinion?

主旨是一篇文章或一个段落的核心。就四级考试而言,文章或段落的主旨通常以主题句(topic sentence)的形式出现。

### (2)事实细节题

细节(detail)或事实(fact)是相对主题而言的。段落中的主题要靠以不同形式联结在一起的事实和细节来进行深化和阐述。在篇章中找出深化主题的重要事实和细节是一项重要的阅读技能。事实、细节题一般主要考查文章中作者提到的有关事件的时间、地点、过程等。一般来说,这些事件都是用来支持作者的主要论点或者帮助作者阐述主题的。从以往的四级阅读题来看,细节题的设问表达方式有:

①Which of the following is included in the article / passage?

②Which of the following is mentioned in the article / passage?

③We learn from the first (the second, third ...) paragraph that ______.

④It can be inferred from the passage that ______.

⑤According to the passage, ______.

⑥Which of the following words can best describe ...?

⑦Which of the following is TRUE according to the passage?

⑧It is suggested in the article / passage that ______.

⑨In the article / passage the author advocates all of the following except ______.

做这类试题应把握的原则是“身在其外而意在其中”，即我们通常所说的“实际是说……”。总之，不能脱离该细节的上下文来想象、推理。

## (3)词汇理解题

词汇(Vocabulary)题是四级阅读理解测试中重要的一项。词汇类其实是就细节进行提问，词汇题往往要求对文章中的某个单词、短语甚至句子等找出近义词或最合适的解释。因此考生可以利用上下文的特定语境推断生词的意思。阅读理解中词汇类问题的常见提问方式有下列几种：

①According to the author ,the word “...”means ______.

②What does the author probably mean by“...”in...paragraph?

③Which of the following is nearest in meaning to “...”?

④From the passage, we can infer that the word“...”is ______.

⑤The term “...”in paragraph... can be best replaced by ...

⑥What does the author probably mean by“...”in...paragraph?

⑦What's the meaning of “...”in line ... of paragraph...?

⑧As used in the line ..., the word “...”refers to ______.

## (4)分析推理题

在四级英语阅读题型中，有一种较难的题——推理题。它要求考生根据语篇的已知信息和事实推断出某个合乎逻辑的结论，要求考生有较强的分析、综合、推理、判断的能力。在主旨题中，我们已涉及了推理这一问题，因为在综合篇章的内容时，实际上是在进行归纳。但主旨题中的归纳都是在语篇提供了较明确和直接的信息和线索(如主题句)的基础上进行的，而在做推理题时，考生应能在语篇提供的信息很隐秘、线索较间接的条件下进行推断。因此，推理题要求考生有较强的逻辑思维能力。如果说前几种题型在语篇中一般能发现直接的提示和线索的话，在推理题中考生找不到上下文中的这种直接的提示，能找到的只是间接的提示和条件。推理题型(或引申题型)主要分为判断、推理和预测三大类。

**①判断题**

判断题是指对文章中的有关事实和观点进行分析和研究，按照事实发展的逻辑次序，总结出合情合理的结论的过程。

**②推理题**

所谓推理，即以已知的事实为依据来获得未知的信息的整个过程。

在阅读理解测试中，除了以上所介绍的题型外，一般还有以下几种形式。

A. The author of the passage implies that...

B. It is implied in the passage that...

C. Which of the following is an inference from the passage?

D. It can be inferred from the passage that...

E. The passage intends to say...

F. It can be concluded from the passage that...

G. We may conclude from reading that...

H. Which of the following is TRUE according to the passage you have just read?

I. Which of the following statements is true according to the passage you have just read?

**③预测题**

预测，即通过阅读文章，凭借文章中的知识推测出下文将会讲什么，或者判断所选文章的出处。

预测的方法多种多样。首先可以根据逻辑上的意义进行预测。其次可以通过语法、连接词等来预测。再次可以通过有关自己所掌握的背景知识及常识进行预测。还有可以通过主题句进行预测。

该题型的一般形式有：

A. The paragraph that follows this one may be about ______.

B. What kind of book do you think this passage is selected from?

### (5)态度倾向题

判断作者的观点和倾向题实质上是判断作者对文章的主观态度。在这类题中,一般没有明显的解题线索,要通过个别带有主观色彩的词来进行推断,透过作者叙述的客观事实来判断隐藏在词句后面的作者的意图,因此,这类题在很大程度上也是一种推理题,属于较难的题型。

判断作者的观点和倾向题的设问方式通常有:

A. According to the author ______.

B. The author's attitude towards ______.

C. The author suggests that ______.

D. In the author's eyes ______.

E. In the author's opinion ______.

F. The tone of the author is ______.

G. What is the author's opinion concerning ______?

H. What is the author's viewpoint on ______?

## 2. 解题技巧精确突破

### (1)利用文章的结构特点应试

大学英语四级考试中的阅读理解文章的体裁主要有三类:叙述文、说明文和议论文。下面我们结合大学英语四级考试真题来具体说明如何利用文章的结构特点达到阅读的目的。

**①叙述文**

叙述文一般以讲述个人生活经历为主,对于经历的陈述通常由一定的时间概念贯穿其中,或顺序或倒序。但是四级考试中一般不出现单纯的叙述文,因为单纯的叙述文比较简单、易懂。所以四级考试中的叙述文大多是夹叙夹议的文章。这类文章的基本结构模式是:

A. 用一段概括性的话引入要叙述的经历(话题)

B. 叙述先前的经历(举例1)及其感悟或发现

C. 叙述接下来的经历(举例2)及其感悟或发现

D. 做出总结或结论

有一次四级考试阅读理解的第二篇就是这样的结构。我们可以将其结构简化为:

A. 总括性的话

Engineering students are supposed to be example of practicality and rationality, but when it comes to my college education I am an idealist and a fool.

B. 先前的经历或想法

In high school I wanted to be, but I didn't choose a college with a large engineering department.

C. 往后的经历

I chose to study engineering at a small liberal - arts university for a broad education.

D. 接下来的经历

I headed off for sure that I was going to have an advantage over others.

E. 再下来的经历

Now I am not so sure... I have learned the reasons why few engineering students try to reconcile engineering with liberal - arts courses in college.

F. 结论

I have realized that the struggle to reconcile the study of engineering and liberal - arts is difficult.

只要理解了这类文章的结构特点解答问题就相当简单,因为这类文章后的阅读理解试题大多是和文章的内容先后顺序一致的细节题。

**②说明文(描述文)**

说明文的一般结构模式和叙述文的结构模式有相似之处，即：提出问题（以一个事例引出问题）——（专家）发现直接原因——分析深层原因——得出结论或找到出路。

某年四级考试阅读理解就有这样的一篇文章。

A. Priscilla Ouchida's "energy - efficient" house turned out to be a horrible dream... a strange illness.（事例）

B. Experts finally traced the cause of her illness.（直接原因）

C. The Ouchidas are victims of indoor air pollution,...（深层原因）

D. The problem appears to be more troublesome in newly constructed homes rather than old ones.（得出结论）

知道了类似的文章结构特点，就可以据此来进行考题预测。比如，我们看出了该篇文章属于这种结构类型，就能判断出几个问题中肯定有一个要问原因，还有可能要出现推断题。

**③议论文**

大家最容易辨认出来的议论文模式是主张一反主张模式。在这一模式中，作者首先提出一种普遍认可的观点或某些人认可的主张或观点，然后进行澄清，说明自己的主张或观点，或者提出相反的主张或真实情况。有一年大学英语四级考试阅读理解就有这样的结构。

文章的开始提出某 college teacher 认为："High school English teachers are not doing their jobs." 因为 His students have a bad command of English.

作者的反观点是：

A. It is inevitable for one generation to complain the one immediately following it. And it is human nature to look for reasons for our dissatisfaction.

B. The people who criticize the high school teachers are not aware that their language ability has developed through the years.

最后的结论是：The concern about the decline and fall of the English language is a generation, and is not new and peculiar to today's young people.

议论文的这种结构特点决定了它的主要题型是作者观点态度题，文章主旨题以及推理判断题。只要发现了这种结构特点，解答问题的主要任务就变成了到段落内找答案，基本上不存在任何困难。

通过研究以上的文章结构特点，我们不难发现，在四级考试阅读理解中无论任何体裁的文章都遵循着这样一个共同的模式：提出话题（观点或事例）——用事例分析原因（或批驳观点）——得出结论。对文章结构特点的把握有助于读者更加自觉地关注文章的开始和结尾，分清观点和事例，从而在四级考试的阅读理解中准确定位，快速答题。

## (2)阅读理解词汇题应试技巧

**①词义猜测8种方法**

一般来说，在文章的阅读中解决释义的最好的办法是猜测词义。猜测词义也需要一定的技巧，主要包括上下文之间意义的联系；同义关系，反义关系；词的定义；对词的解释和举例；构词法知识猜测词义等等。

A. 利用上下文词语意义的互相联系猜测词义

文章的作者本人也意识到文章中的某些词十分生僻或不常使用而故意在同位语、修饰性从句等中给出一定的提示或进行一些解释。这些用来解释或进行提示的词，意义与生僻词基本相同，基本可以互换。也就是说，通过阅读解释部分，生僻词的意义便明晰了。

B. 利用文章中词与词的同义和反义关系猜测词义

可以根据上下文所提供的信息进行有机的联想及推断，进而达到理解生词含义的目的。这种类型的题，上下文的解释一般较为具体，而生词则是文章中较为概括的部分。

C. 利用比较关系

例如：

The Asian monkey like other apes, is specially adapted for life in trees.

如果不认识 ape，但认识 monkey，这里用 like 把 ape 和 monkey 进行比较，还用了一个 other 说明 monkey 可能是 ape 的一种，即 ape 也就是"猿类"的意思了。

D. 利用同位替代关系

例如：

Many famous scientists are trying to understand the problems modern people suffer from, but never these eminent scholars are confused about what causes them.

在句中,为避免重复,"these eminent scholars"替换"many famous scientists",既然 scholars 和 scientists 同义,eminent 也就和 famous 同义,为"著名的"。

E. 利用比喻猜测词义

当作者作比较的时候,一般是强调二者的相似之处。比喻则更是如此。因此,两个相比较的东西只要认出其中之一,便可大致猜出另一物的实质,从而了解全句的含义。

例如:

She sat there for the moment, quiet and silent, suddenly, the meek lamb burst in bad temper, as ferocious as a lion.

在示例中,也许考生对"meek"和"ferocious"两词都很生疏。从文中分析,作者首先把"她"比作"绵羊",后来又比作"狮子"。绵羊一般很温顺,狮子一般很凶残。因此,"meek"大约指"温顺","ferocious"的意义可解释为"凶残"。

F. 利用常识猜测词义

很多词的词义,放在某一类词汇中间,读者可以很容易凭借自己的生活经验或生活常识来猜词义。

G. 利用信号猜测词义

作者在行文过程中使用的一些标点符号、单词或短语,常充当着"信号"的指示作用,提供了有关作者思路及篇章结构的线索,阅读者可利用这些线索推测词义。此为"信号法"。

H. 利用构词法知识猜测词义

英语的构词方法很多,大致可以分为两种,一种是词缀辨认,另一种是词汇复合。

**②绕开生词另辟蹊径**

绕开生词的方法和上面分析文章结构特点的思路是统一的,也就是说,只要我们从总体上把握了文章,不用认识每一个单词也能照样理解整篇文章。

A. 英语文章中不是所有的词的功能都是同等的,有些词担负着传达主要信息的功能,而有些词主要起语法作用,它所传达的信息和下文的其他信息没有联系。这类词有:表示人名、地名、机构名等专有名词。遇到这些词,只要我们能辨认出它是专有名词,就能理解文章而不必知道它的意思。比如在下面的句子中:"In fact", says David Dinges, a sleep specialist at the University of Pennsylvania School of Medicine, "there's even a prohibition against admitting we need sleep."两个引号之间的部分就不必去管它。类似的还有:" We have to totally change our attitude toward napping," says Dr. William Dement of Stanford University, the godfather of sleep research.

B. 我们不用弄清上面某些部分的原因是,它们的后面往往有一个同位语来解释说明它们的意思。也就是说如果我们对文章中的某一个单词不熟悉,我们还可以根据同一篇文章中的其他信息来帮助判断。这类信息有:同位语、下定义、解释、举例、同义词、反义词、上下词以及标点符号(如破折号、冒号都表示解释和说明)等。

**③抓住"第三词汇"**

语法中的功能词对理解句子十分重要,同样文章中那些起组织作用的实义词对理解文章也是非常重要的,因为掌握了它们就可以大大增强阅读理解中的预知下文的能力。我们把这些词称作"第三词汇"(区别于仅起语法作用的功能词和一般实义词)。抓住了它们,就抓住了文章的核心意思。这类词有很多,其中最常见的有:

achieve, addition, attribute, cause, change, consequence, deny, effect, explanation, fact, form, grounds, instance, kind, manner, matter, method, opposite, point, problem, reason, respect, result, same, situation, thing, way 等等。

另外,有人认为"第三词汇"主要是一些"照应名词",其中包括:

abstraction, approach, belief, classification, doctrine, dogma, evaluation, evidence, insight, investigation, illusion, notion, opinion, position, supposition, theory, viewpoint 等等。

还有一些"第三词汇",比如:在"问题—解决"文章模式中,这些"第三词汇"就更加固定和明确。它们有:

问题:concern, difficulty, dilemma, drawback, hamper, hinder(hindrance), obstacle, problem, snag 等。

反应:change, combat, come up with, develop, find, measure, respond, response 等。

解决或结果:answer, consequence, effect, outcome, result, solution, (re)solve 等。

评价:(in)effect, manage, overcome, succeed, (un)successful, viable, work 等。

# 四、仔细阅读理解

# 核心考点精确打击

## Unit 1

### Section A

**Directions:** *In this section, there is a passage with ten blanks. You are required to select one word for each blank from a list of choices given in a word bank following the passage. Read the passage through carefully before making your choices. Each choice in the bank is identified by a letter. Please mark the corresponding letter for each item on* ***Answer Sheet*** *2 with a single line through the centre.* ***You may not use any of the words in the bank more than once.***

**Questions 1 to 10 are based on the following passage.**

During the summer session there will be a __1__ schedule of services for the university __2__. Specific changes for __3__ bus services, cafeteria summer hours, infirmary, recreational, and athletic facilities will be posted on the bulletin board outside the cafeteria. Weekly movie and concert schedules, which are in the process of being __4__, will be posted each Wednesday outside the cafeteria.

Intercampus buses will leave the main hall every hour on the half hour and make all of the regular stops on their __5__ around the campus. The cafeteria will __6__ breakfast, lunch, and early dinner from 7 a. m. to 7 p. m. during the week and from noon to 7 p. m. on weekends. The library will __7__ regular hours during the week, but shorter hours on Saturdays and Sundays. The weekend hours are from noon to 7 p. m.

All students who want to use the library borrowing services and the recreational, athletic, and entertainment facilities must have a __8__ summer __9__ card. This announcement will also __10__ in the next issue of the student newspaper.

| | | | | |
|---|---|---|---|---|
| A) identification | B) arranged | C) society | D) standard | E) revised |
| F) valid | G) maintain | H) way | I) prepared | J) route |
| K) keeps | L) serve | M) intercampus | N) appear | O) community |

### Section B

**Directions:** *There are 2 passages in this section. Each passage is followed by some questions or unfinished statements. For each of them there are four choices marked A, B, C and D. You should decide on the best choice and mark the corresponding letter on* ***Answer Sheet*** *2 with a single line through the centre.*

## Passage One

**Questions 11 to 15 are based on the following passage.**

The development of Jamestown in Virginia during the second half of the seventeenth century was closely related to the making and use of bricks. There were several practical reasons why bricks became important to the colony. Although the forests could initially supply sufficient timber, the process of lumbering was extremely difficult, particularly because of the lack of roads. Later, when the timber on the peninsula had been depleted, wood had to be brought from some distance. Building stones were also in short supply. However, as clay was plentiful, it was inevitable that the colonists would turn to brick-making.

In addition to practical reasons for using brick as the principal construction material, there was also an ideological reason. Brick represented durability and permanence. The Virginia Company of London instructed the colonists to build hospitals and new residences out of brick. In 1662, the Town Act of the Virginia Assembly provided for the construction of thirty-two brick buildings and prohibited the use of wood as a construction material. Had this law ever been successfully enforced, Jamestown would have been a model city. Instead, the residents failed to comply fully with the law. By 1699, Jamestown had collapsed into a pile of rubble with only three or four habitable houses.

11. What is the subject of this passage?

A. The reasons for brick-making in Jamestown.　　B. The cause of the failure of Jamestown.

C. The laws of the Virginia colonists.　　D. The problems of the early American colonies.

12. Which of the following was NOT a reason for using bricks in construction?

A. Wood had to be brought from some distance.　　B. There was considerable clay available.

C. The lumbering process depended on good roads.　　D. The timber was not of good quality.

13. It can be inferred from the passage that Jamestown was established on ______.

A. a rocky peninsula with a small forested area　　B. a barren peninsula near other towns

C. an uninhabitable peninsula with few natural resources　　D. a wooded peninsula with clay soil

14. It can be inferred that the Virginia Assembly, by passing a law regarding building construction, hoped to ______.

A. increase the manufacture of bricks　　B. prevent the destruction of trees in the area of Jamestown

C. establish a city that would be an example for the future　　D. discourage people from settling in Jamestown

15. According to the passage, what eventually happened to Jamestown?

A. It was practically destroyed.　　B. It became a model city.

C. It remained the seat of government.　　D. It was almost completed.

## Passage Two

**Questions 16 to 20 are based on the following passage.**

The manner in which desert locust plagues develop is very complex. The two most important factors in that development are meteorology and the gregariousness of the insect. Since locusts breed most successfully in wet weather, rain in the semiarid regions inhabited by the desert locust provides ideal breeding conditions for a large increase in population. This increase must be repeated several times in neighboring breeding areas before enough locusts crowd together to form a swarm. As the supply of green, palatable food plants decreases toward the end of the rainy season, the locusts become even more concentrated. They move on to other warm, damp, verdant places where they settle, feed, and reproduce. As this process is repeated a swarm eventually develops. Plagues are unpredictable and irregular because the meteorological patterns favorable to crowding are themselves irregular.

16. The passage deals primarily with desert locust swarms and their ______.

A. plagues　　B. reproduction　　C. formation　　D. unpredictability

17. According to the author, which of the following is NOT an important factor in the growth of desert locust swarms?

A. An abundant food supply.　　B. Large population increases.

C. Rain in the semiarid regions.　　D. The gregarious nature of locusts.

18. According to the passage, a decrease in the number of palatable food plants causes ______.

A. a decrease in the number of locusts　　B. death to the local locust population

C. a heavier concentration of locusts　　D. increased breeding of locusts

19. It can be inferred from the passage that in periods of little or no rain the locust population becomes ______.

A. smaller　　B. denser　　C. more active　　D. more unpredictable

20. The next paragraph would most probably deal with ______.

A. the gregarious nature of locusts
B. rain patterns in the desert
C. the type of food plants preferred by locusts
D. the control of locust plagues

# Unit 2

## Section A

**Directions:** *In this part there is a short passage with five questions or incomplete statements. Read the passage carefully. Then answer the questions or complete the statements in less than* 10 *words.*

**Questions 1 to 5 are based on the following passage.**

When Lynne Waihee, wife of Hawaii's former governor, heard Jim Trelease speak, she was inspired. She soon persuaded Rotary Clubs, libraries, schools and several corporations in her state to develop the "Read to Me" campaign. Its goal: to see that every child in Hawaii is read to for at least ten minutes every day. "For years our literacy program had targeted the adult population," Waihee says, "but we realized that if we could focus attention on raising a literate population instead of fixing up an illiterate one, our chance of success would be much greater."

"Read to Me" promotes its message through advertisements, including radio and TV spots. In addition, every elementary school and library in Hawaii has received a bibliography of recommended children's books and a ten-minute videotape on the whys and hows of reading aloud.

Hawaii's enormously popular program has been adopted in Colorado, Wyoming, Alasks and Texas. Several other states are also planning to launch it. On the national level several trade organizations have recently begun a similar campaign called "The Most Important 20 Minutes of Your Day."

Meanwhile, Trelease goes on planting the seeds of reading. He is walking about the auditorium in St. Helena now, gesturing to nobody in particular. "You, sir, had time to watch your favorite ball team yesterday. You, ma'am, had time to go to the mall. You had time to run to the corner store to play that lottery ticket, get cigarettes, rent a video. You had time to chase dust balls under the couch. But you didn't have time to read to your child? I can tell you this unequivocally: 20 years from now the dust balls will still be under the couch, but your little boy or girl will no longer be your little boy or girl."

The message takes. Two hours after he began, a hundred people go home to sleeping children. And tomorrow, for reasons they will not understand, a lot of kids will hear their parents read to them, perhaps for the first time in years.

1. "Read to me" campaign in Hawaii persuades that ______________________________.
2. The phrase "fix up" in the first paragraph probably means to ______________________________.
3. We can infer from the fourth paragraph that parents read to their children will ______________________________.
4. The title of this passage should be ______________________________.
5. What is the author's attitude toward this passage?

______________________________.

## Section B

**Directions:** *There are 2 passages in this section. Each passage is followed by some questions or unfinished statements. For each of them there are four choices marked A, B, C and D. You should decide on the best choice and mark the corresponding letter on* **Answer Sheet** 2 *with a* *single line through the centre.*

### Passage One

**Questions 6 to 10 are based on the following passage.**

A revolution in our understanding of the Earth is reaching its climax as evidence accumulates that the continents of today are not venerable landmasses but amalgams of other lands repeatedly broken up, juggled, rotated, scattered far and wide, then crunched together into new configurations like ice floes swept along the shore of a swift-flowing stream.

After considerable modification this became the new largely accepted concept of "plate tectonics", explaining much of what is observed regarding our dynamic planet. Some oceans, such as the Atlantic, are being split apart, their opposing coasts carried away from one another at one or two inches per year as lava wells up along the line of separation to form new seafloor. Other oceans, such as the Pacific, are shrinking as seafloor descends under their fringing coastlines off shore areas of island.

The Earth's crust, in this view, is divided into several immense plates that make up the continents and seafloors, and that all float on a hot, plastic, subterranean "mantle". What causes these plates to jostle each other, splitting apart or sliding under one another at their edges, is still a mystery to geologists: it may be friction from circulating rock in the Earth's mantle or it may be an effect produced by gravity.

6. What's the author's main purpose in the passage?

A. To explain the theory of plate tectonics.

B. To compare and contrast the Atlantic and Pacific Oceans.

C. To praise geologists for their explanations and discoveries.

D. To display any misconceptions about the rotation of the Earth.

7. The author implies that people used to believe the continents were ______.

A. immobile bodies of land　　B. frozen chunks of ice

C. rotating masses of rock　　D. hardened crusts of lava

8. The word "swept" in the last sentence of paragraph one could best be replaced by ______.

A. won　　B. cleaned　　C. carried　　D. removed

9. According to the passage, the Pacific Ocean is changing in which of the following ways?

A. It is getting smaller.　　B. It is growing warmer.　　C. It is being split apart.　　D. It is filling up with lava.

10. According to the passage, one possible cause of the movement of the tectonic plates is ______.

A. wave motion　　B. the expansion of the oceans

C. gravitational pull　　D. the position of the Moon

### Passage Two

**Questions 11 to 15 are based on the following passage.**

Dam is a barrier constructed across a stream or river to impound water and raise its level. The most common reasons for building dams are to concentrate the natural fall of a river at a given site, thus making it possible to generate electricity; to direct water from rivers into canals and irrigation and water-supply systems; to increase river depths for navigational purposes; to control water flow during times of flood and drought; and to create artificial lakes for recreational use. Many dams fulfill several of these functions. In the United States the network of dams under the Tennessee Valley Authority is an outstanding example of a multipurpose dam development.

The first dam of which record exists was built about 4000 BC to divert the Nile in Egypt in order to provide a site for the city of Memphis. Many ancient earth dams, including a number built by the Babylonians, were part of elaborate irrigation systems that transformed unproductive regions into fertile plains capable of supporting large populations. Because of the ravages of periodic floods, very few dams more than a century old are still standing. The construction of virtually indestructible dams of appreciable height and storage capacity became possible after the development of portland cement concrete and the mechanization of earth-moving and materials-handling equipment.

Controling and using water by means of dams profoundly affects the economic prospects of vast areas. One of the first stages in the progress of developing countries usually involves gaining the ability to use water for power generation, agriculture, and flood protection.

11. What is the main idea of this passage?

A. To tell the main functions of dams.　　B. To date the existence of the first dam.

C. To emphasize the economic effects of dam.　　D. To give a brief introduction to dam.

12. Which of the following is not mentioned as one of the common reasons for building dams?

A. People build dams to generate electricity.

B. Dams are used to control water flow in time of flood and drought.

C. People build dams to divert rivers.

D. Dams serves as part of the irrigation system.

13. The word "ravage" in the second paragraph may mean ______.

A. destruction B. occurrence C. destroy D. savage

14. Many ancient dams mainly served the need to ______.

A. generate electricity B. create artificial lakes for recreational use

C. irrigate fields D. protect people from the suffering of floods

15. According to the passage, when did the construction of virtually indestructible dams become possible?

A. After many ancient earth dams were destroyed.

B. Not until the development of cement concrete and earth moving machines.

C. About 4000 BC.

D. Before the turn of last century.

# Unit 3

## Section A

**Directions:** *In this part there is a short passage with five questions or incomplete statements. Read the passage carefully. Then answer the questions or complete the statements in less than* 10 *words.*

**Questions 1 to 5 are based on the following passage.**

In what now seems like the prehistoric times of computer history, the early post-war year, there was a quite wide-spread concern that computers would take over the world from man one day. Already today, less than forty years later, as computers are relieving us of more and more of the routine tasks in business and in our personal lives, we are faced with a less dramatic but also less foreseen problem. People tend to be over-trusting of computers and are reluctant to challenge their authority. Indeed, they behave as if they were hardly aware that wrong buttons may be pushed, or that a computer may simply malfunction.

Obviously, there would be no point in investing in a computer if you had to check all its answers, but people should also rely on their own internal computers and check the machine when they have the feeling that something has gone wrong. Questioning and routine double checks must continue to be as much a part of good business as they were in pre-computer days. Maybe each computer should come with the following warning: for all the help this computer may provide, it should not be seen as a substitute for fundamental thinking and reasoning skills.

1. The main purpose of this passage is to warn against ____________________.

2. The author advises those dealing with computers to ____________________.

3. An "internal computer" is a person's ____________________.

4. The author suggests that the present-day problem with regard to computers is ____________________.

5. It can be inferred from the passage that the author would disapprove of companies which depend exclusively on computers for ____________________.

## Section B

**Directions:** *There are 2 passages in this section. Each passage is followed by some questions or unfinished statements. For each of them there are four choices marked A, B, C and D. You should decide on the best choice and mark the corresponding letter on* ***Answer Sheet*** *2 with a single line through the centre.*

### Passage One

**Questions 6 to 10 are based on the following passage.**

"It hurts me more than you," and "This is for your own good." These are the statements my mother used to make years ago when I had to learn Latin, clean my room, stay home and do homework.

That was before we entered the permissive period in education in which we decided it was all right not to push our children to achieve their best in school. The schools and the educators made it easy on us. They taught that it was all right to be parents who take a let-alone policy. We stopped making our children do homework. We gave calculators, turned on the television, left the teaching to the teachers and went on vacation.

Now teachers, faced with children who have been developing at their own pace for the past 15 years, are realizing we've made a terrible mistake. One such teacher is Sharon Klompus who says of her students "so passive" and wonders what happened. Nothing was demanded of them, she believes. Television, says Klompus, contributes to children's passivity. "We're not training kids to work any more." says Klompus. "We're talking about a generation of kids who've never been hurt or hungry. They have learned somebody will always do it for them. Instead of saying 'go look it up', you tell them the answer. It takes greater energy to say no to a kid."

Yes, it does. It takes energy and it takes work. It's time for parents to end their vacation and come back to work. It's time to take the car away, to turn the TV off, to tell them it hurts you more than them but it's for their own good. It's time to start telling them no again.

6. Children are becoming more inactive in study because ______.
   A. they watch TV too often
   B. they have done too much homework
   C. they have to fulfill too many duties
   D. teachers are too strict with them
7. To such children as described in the passage ______.
   A. it is easier to say no than to say yes
   B. neither is easy—to say yes or to say no
   C. it is easier to say yes than to say no
   D. neither is difficult—to say yes or to say no
8. We learn from the passage that the author's mother used to lay emphasis on ______.
   A. learning Latin
   B. natural development
   C. discipline
   D. education at school
9. By "permissive period in education" the author means a time ______.
   A. when children are allowed to do what they wish to
   B. when everything can be taught at school
   C. when every child can be educated
   D. when children are permitted to receive education
10. The main idea of the passage is that ______.
    A. parents should leave their children alone
    B. kids should have more activities at school
    C. it's time to be more strict with our kids
    D. parents should always set a good example to their kids

### Passage Two

**Questions 11 to 15 are based on the following passage.**

There are two basic differences between the large and the small enterprise. In the small enterprise you operate primarily through personal contacts. In the large enterprise you have established "publics", "channels" of organization, and fairly rigid procedures. In the small enterprise you have, moreover, immediate effectiveness in a very small area. You can see the effect of your work and of your decisions right away, once you are a little above the ground floor. In the large enterprise even the man at the top is only a *cog*(嵌齿) in a big machine. To be sure, his actions affect a much greater area than the actions and decisions of the man in the small organization, but his effectiveness is remote, indirect, and difficult to see at first sight. In a small and even in a middle-sized business you are normally exposed to all kinds of experiences, and expected to do a great many things without too much help or guidance. In the large organization you are normally taught one thing thoroughly. In the small one the danger is of becoming a Jack-of-all-trades and master of none. In the large one it is of becoming the man who knows more and more about less and less.

There is one other important thing to consider: do you derive a deep sense of satisfaction from being a member of a well-known organization—general Motors, the Bell Telephone System, the government? Or is it more important to you to be a well known and important figure within your own small pond? There is a basic difference between the satisfaction that comes from being a member of a large, powerful, and generally known organization, and the one that comes from being a member of a family; between impersonal *grandeur*(伟大) and personal—often much too personal—*intimacy*(亲密); between life in a small office on the top floor of a skyscraper and life in a crossroads gas station.

11. In a large enterprise, ______.
    A. new technology is employed quickly
    B. all people work efficiently
    C. one's effectiveness is felt very slowly
    D. one can get promotion easily
12. In the first paragraph, a "jack-of-all-trades" means ______.
    A. a person who doesn't know anything about business
    B. a person who is very capable as a businessman
    C. a person who knows a little bit of everything
    D. a person who is very knowledgeable about trade
13. We can conclude from the first paragraph that the writer ______.
    A. prefers to work for a large enterprise
    B. does not mention his own preference
    C. prefers to work for a small enterprise
    D. is against anything that goes to its extreme
14. In the second paragraph, the contrast between the organization and the family is employed to show ______.
    A. how necessary a deep sense of satisfaction is
    B. what satisfaction means to different types of people
    C. how families may differ from one another
    D. what large enterprises can offer to ordinary families
15. It seems that the writer ______.
    A. is giving advice to applicants for jobs
    B. is commenting on the country's industry
    C. has written the passage from an economist's view
    D. has been working for many enterprises

# Unit 4

## Section A

**Directions:** *In this section, there is a passage with ten blanks. You are required to select one word for each blank from a list of choices given in a word bank following the passage. Read the passage through carefully before making your choices. Each choice in the bank is identified by a letter. Please mark the corresponding letter for each item on* **Answer Sheet** *2 with a single line through the centre.* ***You may not use any of the words in the bank more than once.***

**Questions 1 to 10 are based on the following passage.**

Pub etiquette is concerned mainly with the form of your conversation, not the content. The regular __1__ to each greeting, usually addressing the greeter by name or nickname. No one is __2__ of obeying a rule or following a formula, yet you will hear the same greeting ritual in every pub in the country.

Pub etiquette does not dictate the actual words to be used in this exchange and you may hear some inventive and *idiosyncratic*(别具风格的) __3__. The words may not even be particularly polite. When you first enter a pub, don't just order a drink-start by saying "good evening" or "good morning", with a friendly nod and a smile, to the bar staff and the regulars at the bar counter. By greeting before ordering, you have communicated friendly __4__. Although this does not make you an "instant regular", it will be noticed, and your __5__ attempts to initiate contact will be received more favorably.

You may well hear a lot of arguments in pubs—arguing is the most popular pastime of regular pub goers—and some may seem to be quite heated. But pub-arguments are not like arguments in the real world. They are conducted in __6__ with a strict code of etiquette. This code is based on the first commandment of pub law: "thou salt not take things too seriously". The etiquette of pub-arguments __7__ the principles enshrined in the unwritten "constitution" governing all social interaction in the pub: the constitution prescribes equality, *reciprocity*

(互惠), the pursuit of intimacy and a unspoken non-aggression *pact*(默契).

Rule number one: the pub-argument is an __8__ game—no strong views or deeply held convictions are necessary to engage in a __9__ dispute. In the end, everyone may have forgotten what the argument was __10__ to be about. Opponents remain the best of mates, and a good time has been had by all.

| | | | | |
|---|---|---|---|---|
| A) responds | B) subsequent | C) lively | D) superior | E) conscious |
| F) accordance | G) supposed | H) expresses | I) variations | J) reflects |
| K) resorts | L) suggested | M) intentions | N) enjoyable | O) various |

## Section B

**Directions:** *There are 2 passages in this section. Each passage is followed by some questions or unfinished statements. For each of them there are four choices marked A, B, C and D. You should decide on the best choice and mark the corresponding letter on* **Answer Sheet** *2 with a single line through the centre.*

### Passage One

**Questions 11 to 15 are based on the following passage.**

Lucinda Childs' spare and orderly dances have both mystified and mesmerized audiences for more than a decade. Like other so-called "postmodern" choreographers, Childs sees dance as pure form. Her dances are mathematical explorations of geometric shapes, and her dancers are expressionless, genderless instruments who etch intricate patterns on the floor in precisely timed, repetitive sequences of relatively simple steps. The development of Childs' career, from its beginning in the now legendary Judson Dance Theater, paralleled the development of minimalist art, although the choreographer herself has taken issue with those critics who describe her work as minimalist. In her view, each of her dances is simply "an intense experience of intense looking and listening." In addition to performing with her troupe, the Lucinda Childs Dance Company, Childs has appeared in the avant-garde opera Einstein on the Beach, in two off-Broadway plays, and in the films Jeanne d'Iman by Marie Jimenez and 21:12 Piano Bar.

As a little girl, Childs had dreamed of becoming an actress. She appeared regularly in student productions throughout her school years, and when she was about eleven she began to take drama lessons. It was at the suggestion of her acting coach that the youngster, who was, by her own admission, "clumsy, shapeless, and on the heavy side," enrolled in a dancing class. Among her early teachers were Hanya Holm, the dancer and choreographer who introduced the Wigman system of modern dance instruction to the United States, and Helen Tamiris, the Broadway choreographer. Pleased with her pupil's progress, Ms. Tamiris eventually asked the girl to perform on stage. After that exhilarating experience, Lucinda Childs "wasn't sure [she] even wanted to be an actress anymore."

11. What is the passage mainly about?
A. Minimalist art. B. Mathematical forms. C. A choreographer. D. Broadway plays.

12. The word "its" in Line 4 refers to ______.
A. career B. development C. steps D. the Judson Dance Theater

13. The work of Lucinda Childs has been compared to which of the following?
A. Avant-garde opera. B. The Wigman system. C. Realistic drama. D. Minimalist art.

14. In which artistic field did Childs first study?
A. Painting. B. Dance. C. Drama. D. Film.

15. Where in the passage does the author mention how Childs regards her own work?
A. Lines 1-2. B. Lines 6-7. C. Lines 10-11. D. Lines 15-16.

### Passage Two

**Questions 16 to 20 are based on the following passage.**

The difference between a liquid and a gas is obvious under the conditions of temperature and pressure commonly found at the surface of the Earth. A liquid can be kept in an open container and fills it to the level of a free surface. A gas forms no free surface but tends to diffuse throughout the space available; it must therefore be kept in a closed container or held by a gravitational field, as in the case of a planet's atmosphere. The distinction was a prominent feature of early theories describing the phases of matter. In the nineteenth century, for example, one theory maintained that a liquid could be "dissolved" in a vapor without losing its identity, and another theory held that the two phases are made up of different kinds of molecules: liquidons and gasons. The theories now prevailing take a quite different approach by emphasizing what liquids and gases have in common. They are both forms of matter that have no permanent structure, and they both flow readily. They are fluids.

The fundamental similarity of liquids and gases becomes clearly apparent when the temperature and pressure are raised somewhat. Suppose a closed container partially filled with a liquid is heated. The liquid expands, or in other words becomes less dense; some of it evaporates. In contrast, the vapor above the liquid surface becomes denser as the evaporated molecules are added to it. The combination of temperature and pressure at which the densities become equal is called the critical point. Above the critical point the liquid and the gas can no longer be distinguished; there is a single, undifferentiated fluid phase of uniform density.

16. Which of the following would be the most appropriate title for the passage?

A. The Properties of Gases and Liquids.　　B. High Temperature Zones on the Earth.

C. The Beginnings of Modern Physics.　　D. New Containers for Fluids.

17. According to the passage, the difference between a liquid and a gas under normal conditions on Earth is that the liquid ______.

A. is affected by changes in pressure　　B. has a permanent structure

C. forms a free surface　　D. is considerably more common

18. It can be inferred from the passage that the gases of the Earth's atmosphere are contained by ______.

A. a closed surface　　B. the gravity of the planet　　C. the field of space　　D. its critical point

19. According to the passage, in the nineteenth century some scientists viewed liquidons and gasons as ______.

A. fluids　　B. dissolving particles　　C. heavy molecules　　D. different types of molecules

20. According to the passage, which of the following is the best definition of the critical point?

A. When the temperature and the pressure are raised.　　B. When the densities of the two phases are equal.

C. When the pressure and temperature are combined.　　D. When the container explodes.

# Unit 5

## Section A

**Directions:** *In this section, there is a passage with ten blanks. You are required to select one word for each blank from a list of choices given in a word bank following the passage. Read the passage through carefully before making your choices. Each choice in the bank is identified by a letter. Please mark the corresponding letter for each item on* ***Answer Sheet*** *2 with a single line through the centre.* ***You may not use any of the words in the bank more than once.***

**Questions 1 to 10 are based on the following passage.**

Newspapers are not nearly as popular today as they were in the past. There are not very many people who __1__ read a newspaper every day. Most people read only the sports pages, the advice or gossip columns, the comics, and perhaps the __2__ advertisements. Most people don't take the time to read the real news. Newspaper editors say that their readers are lazy. They say they have to __3__ people into reading the news. They __4__ to catch the reader's interest with pictures and exciting headlines. These __5__ are used on the front page because it is the first thing you see when you pick up the paper. The first page __6__ attention and __7__ the reader to look through the rest of the paper. This is why editors always look for a good first page story and __8__ that make you stop and look. If the headline is horrible enough or frightening enough or wild enough, perhaps you will go on to read the rest of the story. Just the same, there are a lot of people who do not even read the front page __9__. They may read the headlines, but that is all. Then they turn to the sports page, or comics, or advertisements. It seems that people do not want the news from a newspaper anymore. They say they get the news on the television now.

More people watch television news because it is easier and more interesting than reading a newspaper. What about you? Do you read news from a newspaper? Do you watch the news on television? Do you think it easier to get the news from television? Do you listen to the radio? Or do you even __10__ about news at all? Would you mind if there were no news?

| | | | | |
|---|---|---|---|---|
| A) classified | B) care | C) seriously | D) headlines | E) demand |
| F) encourages | G) subject | H) attempt | I) want | J) concern |
| K) strictly | L) trick | M) techniques | N) attracts | O) anymore |

## Section B

**Directions:** *There are 2 passages in this section. Each passage is followed by some questions or unfinished statements. For each of them there are four choices marked A, B, C and D. You should decide on the best choice and mark the corresponding letter on **Answer Sheet** 2 with a single line through the centre.*

### Passage One

**Questions 11 to 15 are based on the following passage.**

When a new movement in art attains a certain vogue, it is advisable to find out what its advocates are aiming at, for however far-fetched and unreasonable their tenants may seem today, it is possible that in years to come they may be regarded as normal. With regard to Futurist poetry, however, the case is rather difficult, for whatever Futurist poetry may be—even admitting that the theory on which it is based may be right—it can hardly be classed as literature.

This, in brief, is what the Futurist says: for a century, past conditions of life have been continually speeding up, till now we live in a world of noise and violence and speed. Consequently, our feelings, thoughts and emotions have undergone a corresponding change. This speeding up of life, says the Futurist, requires a new form of expressions. We must speed up our literature too, if we want to interpret modern stress. We must pour out a cataract of essential words, unhampered by stops, or qualifying adjectives, or finite verbs. Instead of describing sounds we must make up words that imitate them: we must use many sizes of types and different colored inks on the same page, and shorten or lengthen words at will.

Certainly their descriptions of battles are vividly chaotic. But it is a little disconcerting to read in the explanatory notes that a certain line describes a fight between a Turkish and a Bulgarian officer on a bridge off which they both fall into the river—and then to find that the line consists of the noise of their falling and the weights of the officers: "Pluff! Pluff! A hundred and eighty five kilograms."

This, though it fulfills the laws and requirements of Futurist poetry, can hardly be classed as literature. All the same, no thinking man can refuse to accept their first proposition: that a great change in our emotional life calls for a change of expression. The whole question is really this: have we essentially changed?

11. The main idea of this selection is best expressed as ______.

A. the past versus the future
B. changes in modern life
C. merits of the Futurist movement
D. an evaluation of Futurist poetry

12. When novel ideas appear, it is desirable, according to the writer, to ______.

A. discover the aims of their adherents
B. ignore them
C. follow the fashion
D. regard them as normal

13. The Futurists claim that we must ______.

A. increase the production of literature
B. look to the future
C. develop new literary forms
D. avoid unusual words

14. The writer believes that Futurist poetry is ______.

A. too emotional
B. too new in type to be acceptable
C. not literature as he knows it
D. essential to a basic change in the nature mankind

15. The Futurist poet uses all the following devices EXCEPT ______.

A. imitative words
B. qualifying adjectives
C. different colored inks
D. a stream of essential words

### Passage Two

**Questions 16 to 20 are based on the following passage.**

Unlike the carefully weighed and planned compositions of Dante, Goethe's writings always have a sense of immediacy and enthusiasm.

He was a constant experimenter with life, with ideas and with forms of writing. For the same reason, his works seldom have the qualities of finish or formal beauty which distinguish the masterpieces of Dante and Virgil. He came to love the beauties of classicism, but these were never an essential part of his make-up. Instead, the urgency of the moment, the spirit of the thing, guided his pen. As a result, nearly all his works have serious flaw of structure, of inconsistencies, of excesses and redundancies.

In a large sense, Goethe represents the fullest development of the romanticism. It has been argued that he should not be so designated because he so clearly matured and outgrew the kind of romanticism exhibited by Wordsworth, Shelley, and Keats. Shelley and Keats died young; Wordsworth lived narrowly and abandoned his early attitudes. In contrast, Goethe lived abundantly and developed his faith in the spirit, his understanding of nature and human nature, and his reliance on feelings as man's essential motivating force. The result was an all-encompassing vision of reality and a philosophy of life broader and deeper than the partial visions and attitudes of other romanticists. Yet the spirit of youthfulness, the impatience with close reasoning or "logic chopping", and the continued faith in nature remained his to the end, together with an occasional waywardness and impulsiveness and a disregard of artistic or logical propriety which savor strongly of romantic individualism. Since so many twentieth century thoughts and attitudes are similarly based on the stimulus of the Romantic Movement. Goethe stands as particularly the poet of the modern man as Dante stood for medieval man and as Shakespeare for the man of the Renaissance.

16. The title that best expresses the ideas of this passage is ______.
   A. Goethe and Dante
   B. The Characteristics of Romanticism
   C. Goethe, the Romanticist
   D. Goethe's Abundant Life

17. Goethe's work shows a lack of ______.
   A. a vision of reality
   B. repetitions
   C. formal polish
   D. knowledge of Shakespeare

18. A characteristic of romanticism not mentioned in this passage is ______.
   A. interest in nature
   B. modernity of ideas
   C. youthful attitude
   D. simplicity of language

19. Goethe is called the poet of the modern man because ______.
   A. he developed his faith
   B. he lived longer than Shelley and Keats
   C. he presents many twentieth-century ideas
   D. his work has serious flaws

20. According to this passage, Goethe ______.
   A. stimulated many modern ideas
   B. disliked Dante and Virgil
   C. should be called a classicist
   D. was illogical

# Unit 6

## Section A

**Directions:** *In this section, there is a passage with ten blanks. You are required to select one word for each blank from a list of choices given in a word bank following the passage. Read the passage through carefully before making your choices. Each choice in the bank is identified by a letter. Please mark the corresponding letter for each item on* ***Answer Sheet*** *2 with a single line through the centre.* ***You may not use any of the words in the bank more than once.***

**Questions 1 to 10 are based on the following passage.**

There are various ways in which individual economic units can __1__ with one another. Three basic ways may be __2__ as the market system, the administered system, and the traditional system.

In a market system individual economic units are free to interact each other in the marketplace. It is possible to buy commodities from other economic units or sell commodities to them. In a market, __3__ may take place via barter or money exchange. In a barter economy, real goods such as automobiles, shoes, and pizzas are traded against each other. __4__, finding somebody who wants to trade my old car in

5 for a sailboat may not always be an easy task. Hence, the introduction of money as a 6 of exchange eases transactions considerably. In the modern market economy, goods and services are bought or sold for money.

An alternative to the market system is administrative controlled by some agency over all transactions. This agency will issue 7 or commands as to how much of each kind of goods and services should be produced, exchanged, and consumed by each economic unit. Central planning may be one way of administering such an economy. The central plan drawn up by government, shows amounts of each commodity produced by the various firms and 8 to different households for consumption. This is an example or complete planning of production, consumption, and exchange for the whole economy.

In a traditional society, production and consumption patterns are governed by tradition: every person's place within the economic system is fixed by parentage, religion, and custom. Transactions take place on the basis of tradition, too. People 9 to a certain group of caste may have an obligation to care for other persons, provide them with food and shelter, care for their health, and provide for their education. Clearly, in a system where every decision is made on the basis of tradition alone, progress may be difficult to achieve, a 10 society may result.

| | | | | |
|---|---|---|---|---|
| A) stagnant | B) belonging | C) wholesale | D) interact | E) medium |
| F) charge | G) transactions | H) Possibly | I) described | J) involving |
| K) Obviously | L) exchange | M) prosperous | N) edicts | O) allocated |

## Section B

**Directions:** *There are 2 passages in this section. Each passage is followed by some questions or unfinished statements. For each of them there are four choices marked A, B, C and D. You should decide on the best choice and mark the corresponding letter on* **Answer Sheet** *2 with a single line through the centre.*

### Passage One

**Questions 11 to 15 are based on the following passage.**

The exact number of English words is not known. The large dictionaries have over half a million entries, but many of these are compound words (schoolroom, sugar bowl) or different derivatives of the same word (rare—rarely, rarefy), and a good many are obsolete words to help us read older literature. Dictionaries do not attempt to cover completely words that we can draw on: the informal vocabulary, especially slang, localism, the terms of various occupations and professions; words use only occasionally by scientists and specialists in many fields; foreign words borrowed for use in English; or many new words or new senses of words that come into use every year and that may or may not be used long enough to warrant being included. It would be conservative to say that there are over a million English words that any of us might meet in our listening and reading and that we may draw on in our speaking and writing.

Professor Seashore concluded that first-graders enter school with at least 24,000 words and add 5,000 each year so that they leave high school with at least 80,000. These figures are for recognition vocabulary, the words we understand when we read or hear them. Our active vocabulary, the words we use in speaking and writing, is considerably smaller.

You cannot always produce a word exactly when you want it. But consciously using the words you recognize in reading will help get them into your active vocabulary. Occasionally in your reading pay particular attention to these words, especially when the subject is one that you might well write or talk about. Underline or make a list of words that you feel a need for and look up the less familiar ones in a dictionary. And then before very long find a way to use some of them. Once you know how they are pronounced and what they stand for, you can safely use them.

11. In the author's estimation, there are ______ words in English.

A. more than half a million B. at least 24,000 C. at least 80,000 D. more than a million

12. The word "obsolete" most probably means ______.

A. no longer in use B. profound C. colorful or amusing D. common

13. One's recognition vocabulary is ______.

A. less often used than his active vocabulary B. smaller than his active vocabulary

C. as large as his active vocabulary D. much larger than his active vocabulary

14. The author does not suggest getting recognition vocabulary into active vocabulary by ______.

A. making a list of words you need and looking up the new ones in a dictionary

B. everyday spending half an hour studying the dictionary

C. consciously using the words you recognize in reading

D. trying to use the words you recognize

15. From this passage we learn that ______.

A. dictionaries completely cover the words we can make use of

B. "schoolroom" is used in the passage as an example of a specialized term

C. once you know how a word is pronounced and what it represents, you have turned it into your active word

D. active vocabulary refers to words we understand when we read and hear them

## Passage Two

**Questions 16 to 20 are based on the following passage.**

In the past, it was believed that depression was more prevalent among the poorer and less educated people, but that is not the case. The truth is, depression can afflict people from all walks of life, and often it hits the most ambitious, creative and conscientious. It is fallacious to think that people on the corporate rung of the social ladder are not *prone*(易于……) to this *malady*(疾病). In fact, executives and professionals who are burdened with mounting pressures in their work may succumb to all these pressures and become depressed. The suicide of Vincent Foster, Jr, a noted American lawyer and White House official is a case in point. Despite the glamour of his position, which many people thought enviable, he felt overburdened with pressures and decided to take the easy way out.

Studies show that people born later in this century have experienced much more depression than those born earlier. In fact the rate of depression over the last two generations has increased tenfold. Experts theorize that it could be due to the fact that the younger generations have higher expectations from life and are therefore more likely to suffer from failure, disappointment and hence, depression.

Depression is easily recognizable. The depressed person feels sad or down in the dumps, and loses interest in even the most pleasurable activities. Moreover, he suffers from either significant gain or loss of weight, sleeplessness or over-sleeping, sluggish movement and thinking, fatigue, feelings of guilt and worthlessness, impaired concentration and forgetfulness, and in extreme cases, the afflicted person may have suicidal tendencies. It is often believed that depression runs in the family, but this is not conclusively so, since there are cases where depressed persons do not have a history of depression in their families. Depression is often work-related although at times it has its roots in family relations.

People who suffer from depression need not stay in the closet, since it is not a sin or a shame to be depressed. A prompt visit to a psychologist means prompt treatment and hence prompt recovery. Experts guarantee that depression is easily treatable, and in nip-in-the-bud cases, the patient fully recovers in a few days, thanks to the variety of effective treatment available.

16. According to the author, depression ______.

A. affects the poor and poor educated people more than the rich and successful ones

B. seldom attacks executives

C. is not limited to any particular class of people

D. is caused by one's ambition

17. Studies show depression ______.

A. is more common today than in the past decades

B. was more common in the old generations

C. increased ten times in the days of our parents and grandparents

D. afflicts only young people

18. Which of the following statements is NOT TRUE?

A. Very high expectations for the individual make him more prone to depression.

B. A depressed person feels sad and dejected.

C. People need not attempt to hide their depression because it's not a shameful thing.

D. Depression is hereditary.

19. Depression can be easily cured ______.

A. if the patient is young
B. at the early stage of the disease
C. when plants are in bud
D. if the patient goes to see a psychologist

20. The passage is mainly a ________.

A. description of the symptoms of depression
B. brief introduction of depression
C. summary of the causes of depression
D. list of treatment of depression

# Unit 7

## Section A

**Directions:** *In this part there is a short passage with five questions or incomplete statements. Read the passage carefully. Then answer the questions or complete the statements in less than 10 words.*

**Questions 1 to 5 are based on the following passage.**

Students who score high in achievement needs tend to make higher grades in college than those who score low. When degree aptitude for college work, as indicated by College Entrance Examination Board Tests, is held constant, engineering students who score high in achievement needs tend to make higher grades in college than the aptitude test scores would indicate.

We can define this need as the habitual desire to do useful work well. It is a salient influence characteristic of those who need little supervision. Their desire for accomplishment is a stronger motivation than any stimulation the supervisor can provide. Individuals who function in terms of this drive do not "bluff" in regard to a job that they fail to do well.

Some employees have a strong drive for success in their work; others are satisfied when they make a living. Those who want to feel that they are successful have high aspiration for themselves. Thoughts concerning the achievement drive are often prominent in the evaluations made by the typical employment interviewer who interviews college seniors for executive training. He wants to find out whether the senior has a strong drive to get ahead or merely to hold a job. Research indicates that some who do get ahead have an even stronger drive to avoid failure.

1. What is the main subject of this passage?

________________________________________.

2. How can individuals obtain high achievement scores?

________________________________________.

3. According to the passage, individuals with a strong drive to succeed will ________________

________________.

4. What quality do employment interviewers look for in college seniors for executive training?

________________________________________.

5. What motivates some seniors to succeed?

________________________________________.

## Section B

**Directions:** *There are 2 passages in this section. Each passage is followed by some questions or unfinished statements. For each of them there are four choices marked A, B, C and D. You should decide on the best choice and mark the corresponding letter on* ***Answer Sheet*** *2 with a single line through the centre.*

### Passage One

**Questions 6 to 10 are based on the following passage.**

It is common knowledge that drug abuse leads to harmful consequences. Why then do people—particularly youngsters—continue to use drugs? Psychologists claim that there are three basic motivations that influence people to take drugs: curiosity, stress and environmental factors. Sometimes, youngsters take drugs simply because they are curious. Taking drugs seems to be the "in-thing" for their generation, so they want to know what drugs are like. The trouble is that they do not know that taking soft and seemingly *innocuous*(无害的) drugs can develop into *cravings*(渴望) for stronger stuff later on. In some cases, youngsters are depressed or frustrated because of problems related to parents, school or the opposite sex. They take drugs to escape from the stress brought on by all these problems. In other cases, the environment is conductive to taking drugs. If, for instance, a youngster belongs to a community, school, or peer group where other youngsters take drugs, he may soon be tempted to follow suit, for fear of ostracism or non-acceptance.

There is a growing consensus nowadays among social workers and psychologists that the best possible approach to the problem of drug addiction among the young is for school authorities, social workers and the Police Narcotics Division to work together to provide young people with much-needed education on the effects and dangers of drug abuse. Moreover, parents can do a great job in leading children away from drugs. They should spend more time with their children, listening and talking to them. Most importantly, parents should show them attention, concern and love. Parents who always scream at their children and *nag*(唠叨) them about their failings and weaknesses are regarded as unwitting drug pushers. As far as young people are concerned, a warm and happy family, wherein members share both joys and sorrows and where children get maximum encouragement and support, is the best bulwark against the onslaughts of drugs. It is no exaggeration to say that a happy home is a drug-free home.

6. The expression "in-thing" in the first paragraph most probably means ______.
   A. curiosity　B. fashion　C. demand　D. pressure
7. Which of the following is NOT MENTIONED as a reason why some youngsters take soft drugs?
   A. They think that soft drugs are not harmful.　B. They wonder what drugs are like.
   C. They are disturbed by problems.　D. Their parents are drug-takers.
8. Social workers and psychologists hold a common belief that ______.
   A. the Police Narcotics Division should take sole responsibility for the problem of drug addiction among the young
   B. parents ought to be educated about the effects and danger of drug abuse
   C. young people tend to be addicted to drugs
   D. the concerned authorities should join efforts to educate youngsters about the evil consequences of drug addiction
9. A youngster grows in a community where people around him take drugs ______.
   A. may also take drugs to adapt to the trend
   B. may run away from home for fear to be involved in it
   C. may be very cautious in his choice of friends
   D. may be tempted into doing the same thing to be accepted
10. The best way to prevent youngsters from taking soft drugs is ______.
   A. to issue a ban on the sale of drugs　B. to punish the drug addicts
   C. to give them a warm and loving family　D. to teach them principles

## Passage Two

**Questions 11 to 15 are based on the following passage.**

Centuries ago, man discovered that removing moisture from food helps to preserve it, and that the easiest way to do this is to expose the food to sun and wind.

All foods contain water—cabbage and other leaf vegetables contain as much as 93% water, potatoes and other root vegetables 80%, lean meat 75% and fish ranging from 80% to 60% depending on how fatty it is. If this water is removed, the activity of the *bacteria*(细菌) which cause food to go bad is checked.

Nowadays most foods are dried mechanically. The <u>conventional</u> method of such *dehydration*(脱水) is to put food in chambers through which hot air is blown at temperatures of about 110℃ at entry to about 43℃ at exit. This is the usual method for drying such things as vegetables, minced meat, and fish.

Liquids such as milk, coffee, tea, soups and eggs may be dried by pouring them over a heated horizontal steel cylinder or by spraying them into a chamber through which a current of hot air passes. In the first case, the dried material is scraped off the roller as a thin film which is then broken up into small, though still relatively coarse flakes. In the second process it falls to the bottom of the chamber as a fine powder. Where recognizable pieces of meat and vegetables are required, as in soup, the ingredients are dried separately and then mixed.

Dried foods take up less room and weight less than the same food packed in cans or frozen, and they do not need to be stored in special conditions. For these reasons they are invaluable to climbers, explorers and soldiers in battle, who have little storage space. They are also popular with housewives because it takes so little time to cook them. Usually it is just a case of replacing the dried-out moisture with boiling water.

11. The chief point of the second paragraph is about ______.

A. the comparison of lean meat and fish B. the removal of water in food

C. the water content in food D. the relationship between water and food

12. It can be inferred from the passage that ______.

A. the fattier fish contain as much water as the lean ones

B. the fattier the fish is the more water it may contain

C. a fatty fish holds less water than a lean one

D. the water content of fish has nothing to do with the content of their fat

13. The word "conventional" in Paragraph 3 can most probably be replaced by ______.

A. traditional B. scientific C. usual and acceptable D. not common

14. Which of the following statements is NOT TRUE about drying food?

A. The removal of water in food helps prevent it from going rotten.

B. The open-air method of drying food has been known for hundreds of years.

C. In the course of dehydration, the temperature of hot current coming from entry to exit is gradually going up.

D. The process of drying liquids is much more complex than that of drying solid food.

15. The last paragraph tends to discuss ______.

A. the reason why housewives like dried food B. the general convenience of dried food

C. the methods of storing food D. the advantages of dried, canned and frozen food

# Unit 8

## Section A

**Directions:** *In this part there is a short passage with five questions or incomplete statements. Read the passage carefully. Then answer the questions or complete the statements in less than 10 words.*

**Questions 1 to 5 are based on the following passage.**

One big step in transport technology will be automated roads: regulating vehicles in *convoys*(车队) on motorways so that they are safer and can be packed closer together. Sensors would establish what is around each vehicle and electronic control systems would keep them moving in the right direction, at the right safe speed, with maximum comfort and economy. The technology is massively expensive now, but eventually it will become a reality. You would just pay tolls, couple your car into an electronic convoy and sit back to enjoy the journey.

Motorways will gradually become more like railways, with freight vehicles electronically coupled in trains running at relatively high speeds. At suitable intervals, they would *uncouple*(分离) to travel the remainder of the journey with their own driver. That's almost certainly going to happen. It would make better use of the roads and be safer, cheaper and greener, as well as making driving more pleasant for everyone.

That sort of combination of personal and centralized control is the direction we're going in road transport, probably first of all for freight. Any rail system has in the end to be inflexible; it doesn't go where you want, especially in rural communities, where the nearest station can be 30 miles away. We are *wedded to*(与……结下不解之缘) private cars, because of their flexibility and the pride people take in ownership—not to mention the huge sums we've spent on the road network.

So cars aren't going to go away. But under electronic control they will become greener and safer.

1. How will vehicles travel on motorways in the future?

______________________________________________.

2. What should you do before you enjoy the driving on motorways?

______________________________________________.

3. What's the main idea of paragraph 2?

______________________________________________.

4. What's the disadvantage of rail system?

______________________________________________.

5. Why aren't cars going to go away?

______________________________________________.

## Section B

**Directions:** *There are 2 passages in this section. Each passage is followed by some questions or unfinished statements. For each of them there are four choices marked A, B, C and D. You should decide on the best choice and mark the corresponding letter on* ***Answer Sheet*** *2 with a single line through the centre.*

### Passage One

**Questions 6 to 10 are based on the following passage.**

The advantages and disadvantages of a large population have long been a subject of discussion among economists. It has been argued that the supply of good land is limited. To feed a large population, inferior land must be cultivated and the good land worked intensively. Thus, each person produces less and this means a lower average income than could be obtained with a smaller population. Other economists have argued that a large population gives more scope for specialization and the development of facilities such as ports, roads and railways, which are not likely to be built unless there is a big demand to justify them.

One of the difficulties in carrying out a world-wide birth control program lies in the fact that official attitudes to population growth vary from country to country depending on the level of industrial development and the availability of food and raw materials. In the developing country where a vastly expanded population is pressing hard upon the limits of food, space and natural resources, it will be the first concern of government to place a limit on the birthrate, whatever the consequences may be. In the highly industrialized society the problem may be more complex. A decreasing birthrate may lead to unemployment because it results in a declining market for manufactured good. When the pressure of population on housing declines, prices also decline and the building industry is weakened. Faced with considerations such as those, the government of a developed country may well prefer to see a slowly increasing population, rather than one which is stable or in decline.

6. A small population may mean ______.

A. higher productivity, but a lower average income
B. lower productivity, but a higher average income
C. lower productivity, and a lower average income
D. higher productivity, and a higher average income

7. According to the passage, a large population will provide a chance for developing ______.

A. agriculture
B. transport system
C. industry
D. national economy

8. In a developed country, people will perhaps go out of work if the birthrate ______.

A. goes up
B. is decreasing
C. remains stable
D. is out of control

9. According to the passage slowly rising birthrate perhaps is good for ______.

A. a developing nation
B. a developed nation
C. every nation with a big population
D. every nation with a small nation

10. It is no easy job to carry out a general plan for birth control throughout the world because ______.

A. there are too many underdeveloped countries in the world
B. underdeveloped countries have low level of industrial development
C. different governments have different views of the question

D. even developed countries may have complex problems

## Passage Two

**Questions 11 to 15 are based on the following passage.**

The full influence of mechanization began shortly after 1850, when a variety of machines came rapidly into use. The introduction of these machines frequently created rebellions by workers who were fearful that the machines would rob them of their work. Patrick Bell, in Scotland, and Cyrus McCormick, in United States, produced threshing machines. Ingenious improvements were made in plows to compensate for different soil types. Stream power came into use in 1860s on large farms. Hay rakes, hay-loaders, and various special harvesting machines were produced. Milking machines appeared. The internal-combustion *engine*（内燃机）run by gasoline became the chief power source for the farm.

In time, the number of certain farm machines that came into use skyrocketed and changed the nature of farming. Between 1940 and 1960, for example, 12 million horses and mules gave way to 5 million tractors. Tractors offer many features that are attractive to farmers. There are, for example, numerous attachments: cultivators that can penetrate the soil to varying depths, rotary hoes that chop needs; spray devices that can spray pesticides in bands 100 feet across, and many others.

A piece of equipment has now been invented or adapted for virtually every laborious hand or animal operation on the farm. In the United States, for example, cotton, tobacco, hay, and grain are planted, treated for pests and diseases, fertilized, cultivated and harvested by machine. Large devices shake fruit and nut from trees, grain and blend feed, and dry grain and hay. Equipment is now available to put just the right amount of fertilizer in just the right place, to spray an exact row width, and to count out, space, and plant just the right number of seeds for a row.

Mechanization is not used in agriculture in many parts of Latin America, Africa. Agriculture innovation is accepted fastest where agriculture is already profitable and progressive. Some mechanization has reached the level of plantation agriculture in parts of the tropics, but even today much of that land is laboriously worked by people leading draft animals pulling primitive plows.

The problems of mechanization some areas are not only cultural in nature. For example, tropical soils and crops differ markedly from those in temperate areas that the machines are designed for, so adaptations have to be made. But the greatest obstacle to mechanization is the fear in underdeveloped countries that the workers who are displaced by machines would not find work elsewhere. Introducing mechanization into such areas requires careful planning.

11. The first paragraph uses several examples to convey the ideas that ______.

A. the introduction of machines into agricultural work created rebellions on the part of the farmers

B. the use of internal-combustion engine as a chief power source for the farm produced great influence

C. the mechanization of agricultural work after 1850 gradually robbed many farmers of their work

D. ingenious improvements were made in farming machines in the 1860s to yield production

12. In the first sentence of the second paragraph, the word "skyrocketed" probably means ______.

A. became various　　B. was updated　　C. increased rapidly　　D. remained the same

13. In tropical areas, ______.

A. mechanization is not yet used in agriculture　　B. agriculture is accepted fastest

C. a lot of farm work is still done in the old way　　D. mechanization is avoided to save primitive forest

14. By saying that "the problems of mechanizing some areas are not only cultural in nature", the author means ______.

A. mechanization is not yet introduced in some areas for economic reasons

B. human and animal labor in some areas are less expensive

C. culture is not a factor in obstacling the introduction of mechanization

D. different kinds of mechanized farming tools are used in different cultures

15. To introduce mechanization into underdeveloped areas, ______ should be required.

A. force　　B. more technicians　　C. careful planning　　D. more money

# Unit 9

## Section A

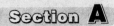

**Directions:** *In this part there is a short passage with five questions or incomplete statements. Read the passage carefully. Then answer the questions or complete the statements in less than 10 words.*

**Questions 1 to 5 are based on the following passage.**

Most Americans receive at least two or three credit card applications in the mail every month. Why have credit cards become so popular? For a merchant, the answer is obvious. By *depositing charge slips*(收款存条) in a bank or other financial institutions, the merchant can convert credit card sales into cash. In return for processing the merchant's credit card transactions, the bank charges a fee that ranges between 1.5 and 5 percent.

It is important to choose a credit card carefully because terms and conditions vary widely. Annual fees range from $15 to $75 a year, but some credit card companies charge no annual fee at all. If you will be one of the growing numbers of people who don't pay off their credit card transactions in full each month, look for the card with the lowest interest rate. Interest rates generally range from 12 to 18 percent, though it is possible to find cards with lower rates.

A credit card can be your friend because it can get you through unexpected emergencies. And if there is a problem with the products or service you purchase with your credit card, you have an opportunity to withhold payment by asking the credit card company to "charge back" to the retailer until the dispute is settled. Monthly credit card statements can also help you keep your records in order. Finally, if you make payments on time, the card helps you to establish a good credit history.

A credit card can be your enemy because it is an invitation to purchase items you really do not need. The credit card companies continuous offers of low minimum payments, cash advances, and even months without payments may seem like a way to skate through a *money crunch*(财政困境). In reality, your finance charges and fees only increase, and you go deeper into debt.

If you do find yourself in trouble, do not ignore the bills. Contact your creditors to explain your problem and express your desire to pay down your card balance. If that fails, a nonprofit organization like Consumer Credit Counseling Service can assist you in getting back on your financial feet.

1. How does the merchant convert credit card sales into cash in a bank?

______________________________.

2. If people will possibly not pay off the credit card transactions in full each month, they had

______________________________.

3. What should people do when there is some problem with the products they buy with the credit card?

______________________________.

4. What will happen to consumers at last with some measures being taken by the credit card companies to stimulate them to purchase items?

______________________________.

5. What's the main idea of Paragraph 5?

______________________________.

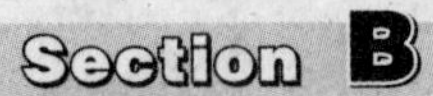

## Section B

**Directions:** *There are 2 passages in this section. Each passage is followed by some questions or unfinished statements. For each of them there are four choices marked A, B, C and D. You should decide on the best choice and mark the corresponding letter on* ***Answer Sheet*** *2 with a single line through the centre.*

### Passage One

**Questions 6 to 10 are based on the following passage.**

In our day of the automobile and paved highway few people ever encounter quicksand. Yet quicksand is still common in many parts of the country. It may be more dangerous for being less familiar.

Quicksand is usually found along the shores and in the beds of rivers. It is simply sand saturated with water from beneath, as from a spring. The water flowing into the sand separates the grains. The suspended grains give rather easily, and a heavy object placed on the surface is likely to sink. How fast it sinks depends on its weight and surface area.

How does one detect quicksand? It cannot be done by the eye alone, since sand which looks firm may suddenly collapse and trap anyone who ventures out on it. The only way to be sure is to test the sand before walking on it. For test probing a pole or long stick should be used. If the pole sinks more than six inches, the sand is probably quicksand.

A traveler who stumbles into quicksand will soon sink to the depth of his knees. If he stands still or struggles wildly, he will sink even further. He should at once lie on his back and stretch out his arms. Contrary to popular notion, quicksand does not suck objects down, and will support more weight than water alone. While the trapped person "floats" on the surface of the sand, rescuers should build a platform with boards or branches. Then they can pull him out slowly.

If the trapped person is alone, he can rescue himself. When he is in the floating position, he should begin rolling towards solid ground. Rolling is the only way of getting free. It should be done with frequent rests, so that the trapped person does not tire himself. When he reaches solid ground, he should swing his legs to safety, and quickly scramble out of the quicksand.

6. This selection can best be titled ________.

A. All about Quicksand　　B. How to Detect Quicksand

C. How to Escape from Quicksand　　D. Where Quicksand is Found

7. The main idea of this passage is ______.

A. that today few people ever encounter quicksand

B. that quicksand is still common in many parts of the country

C. that quicksand cannot be detected by eye alone

D. expressed by none of the above

8. According to the passage, quicksand is usually found ______.

A. on hillside　　B. near ravines　　C. far inland　　D. near water

9. For detecting quicksand the author recommends ______.

A. good eyesight　　B. fast thinking

C. the use of a long pole　　D. testing the surface with your shoes

10. To escape from quicksand you should be ______.

A. calm　　B. excited　　C. strong　　D. daring

## Passage Two

**Questions 11 to 15 are based on the following passage.**

In old times there lived a King, who was so cruel and unjust towards his subjects that he was always called the Tyrant. So heartless was he that his people used to pray night and day that they might have a new king. One day, much to their surprise, he called his people together and said to them: "My dear subjects, the days of my tyranny are over. Henceforth you shall live in peace and happiness, for I have decided to try to rule henceforth justly and well."

The King kept his words so well that soon he was known throughout the land as The Just King. By and by one of his favorites came to him and said: "Your Majesty, I beg of you to tell me how it was that you had this change of heart towards your people?"

And the King replied: "As I was galloping through my forests one afternoon, I caught sight of a hound chasing a fox. The fox escaped to his hole, but he had been bitten by the dog so badly that he would be lame for life. The hound, returning home, met a man who threw a stone at him, which broke his leg. The man had not gone far when a horse kicked him and broke his leg. The horse, starting to run, fell into a hole and broke his leg. Hence I came to my sense, and decided to change my rule. 'For surely,' I said to myself, 'he who does evil will sooner or later be overtaken by evil.'"

11. The people used to wish to have a new king because ______.

A. the present king was too old　　B. the present king was too foolish

C. the present king was too cruel and unjust　　D. a new king would be extremely just

12. The King declared to his subjects that ______.

A. he would do what he liked
B. he would strengthen his tyranny
C. he would live in peace and happiness
D. he would rule better

13. We can see that the King ______.
A. went back on his words
B. did what he had promised
C. still ruled unjustly
D. still wasn't considered a just king

14. One afternoon the king saw in his forests ______.
A. a dog chasing a fox
B. a man throwing a stone at a dog
C. a horse kicking a man
D. a horse fallen into a hole

15. The King's conclusion is that ______.
A. evil people will remain evil
B. he who does bad things will never do anything good
C. if a person does bad things, something bad will sooner or later be done to him
D. sooner or later an evil thing will surely happen

# Unit 10

## Section A

**Directions:** *In this part there is a short passage with five questions or incomplete statements. Read the passage carefully. Then answer the questions or complete the statements in less than 10 words.*

**Questions 1 to 5 are based on the following passage.**

Twenty years ago, when only the lowly tadpole had been cloned, bioethicists raised the possibility that scientists might someday advance the technology to include human beings as well. They wanted the issue discussed. But scientists assailed the moralists' concerns as alarmist. Let the research go forward, the scientists argued, because cloning human beings would serve no discernible scientific purpose. Now the cloning of human is within reach, and society as a whole is caught with its ethical pants down.

Today the sheep—tomorrow the shepherd? Whether the cloning of human beings can be ethically justified is now firmly, perhaps permanently, on the nation's moral agenda. President Clinton has given an advisory panel of experts just 90 days to come up with proposals for government action. The government could prohibit the cloning of human beings or issue regulations limiting what researchers can do. But the government cannot control the actions of individuals or private groups determined to clone humans for whatever purpose. And science has a way of outdistancing all ethical restraints. "In science, the one rule is that what can be done will be done," warns Robbi Moses Tendler, professor of medical ethics at Yeshiva University in New York.

Some ethicists regard the cloning of humans as inherently evil, a morally unjustifiable intrusion into human life. Others measure the morality of any act by the intentions behind it, still others are concerned primarily with the consequences—for society as well as for individuals. Father Richard McCormick, a veteran Jesuit ethicist at the University of Notre Dame, represents the hardest line: any cloning of human is morally repugnant. A person who would want a clone of himself, says McCormick, "is overwhelmingly self-centered. One Richard McCormick is enough." But why not clone another Einstein? Once you program for producing superior beings, he says, you are into eugenics. "and eugenics of any kind is inherently discriminatory." What's wrong with duplicating a sibling whose bone marrow could save a sick child? That, he believes, is using another human being merely "as a source for replaceable organs." But why shouldn't an infertile couple resort to cloning if that is the only means of having a child? "Infertility is not an absolute evil that justifies doing any and every thing to overcome it," McCormick insists.

1. Scientists think the moralists' warning is ______________________________.

2. The phrase "be caught with pants down" in the first paragraph probably means ______________________________ ______________________________.

3. The main idea in the second paragraph is ________________________________________.

4. What the government can do is to set regulations to ________________________________________.

5. Should human be cloned? McCormick insists that ________________________________________.

## Section B

**Directions:** *There are 2 passages in this section. Each passage is followed by some questions or unfinished statements. For each of them there are four choices marked A, B, C and D. You should decide on the best choice and mark the corresponding letter on* ***Answer Sheet*** *2 with a single line through the centre.*

### Passage One

**Questions 6 to 10 are based on the following passage.**

The science of *meteorology*(气象学) is concerned with the study of the structure, state, and behavior of the atmosphere. The subject may be approached from any vantage point. Different views must be combined to give perspective to the whole picture.

One may consider the condition of the atmosphere at a given moment and attempt to predict changes from that condition over a period of a few hours to a few days ahead. This approach is covered by the branch of the science called synoptic meteorology.

Synoptic meteorology is the scientific basis of the technique of weather forecasting by means of the preparation and analysis of weather maps and serological diagrams. In serving the needs of shipping, aviation, agriculture, industry, and many other interests and fields of human activity with accurate weather warnings and professional forecast advice, great benefits are obtained in the form of the saving of human lives and property and in economic advantages of various kinds. One important purpose of the science of meteorology is constantly to make great efforts, through study and research, to increase our knowledge of the atmosphere with the aim of improving the accuracy of weather forecasts.

The tools needed to advance our knowledge in this way are disciplines of mathematics and physics applied to solve meteorological problems. The use of these tools forms that branch of the science called dynamic meteorology.

6. Which of the following could be the best title for the passage?

A. Limitation of Meteorological Forecasting  B. New Advances in Synoptic Meteorology

C. Approaches to the Science of Meteorology  D. The Basic of Dynamic Meteorology

7. The predictions of synoptic meteorologists are directly based on the ______.

A. application of the physical sciences  B. preparation and study of weather maps

C. anticipated needs of industry  D. observations of commercial airline pilots

8. The author implies that increased accuracy in weather forecasting will lead to ______.

A. more funds given to meteorological research  B. greater protection of human life

C. a higher number of professional forecasters  D. less-specialized forms of synoptic meteorology

9. Which of the following statements best describes the organization of the third paragraph of the passage?

A. A scientific theory is explained.

B. Two contrasting views of a problem are presented.

C. Recent scientific advancements are outlined in order of importance.

D. A problem is examined and possible solutions are given.

10. In the sentence of the last passage the phrase "the tools" refers to ______.

A. weather forecasts  B. meteorological problems  C. mathematics and physics  D. economic advantages

### Passage Two

**Questions 11 to 15 are based on the following passage.**

Whenever any new invention is put forward, those for it and those against it can always find medical men to approve it or condemn it. The same is true of the railway.

In the early days, people who rejected the railway produced well-known doctors who said that tunnels would be most dangerous to public health; they would produce colds and some other diseases. The deadening noise of the engine fire would have a bad effect on the nerves. Further, being moved through the air at a high speed would do grave harm to delicate lungs. In those with high blood pressure, the movement of the train might produce a stroke. The sudden plunging of the train into the darkness of a tunnel, and the equally sudden rush into full daylight, would cause great damage to eyesight. However, those who wanted railways were of course able to produce equally **eminent** medical men to say just the opposite. They said that the speed and the swing of the train would equalize the flow of blood, promote digestion, calm the nerves, and insure good sleep. More than a rapid and comfortable means of transport, they actually saw the railway as a factor in world peace. They did not expect that the railway would be just one movement for the rapid movement of hostile armies. None of them expected that the more we are together, the more chances there are of war. Any boy or girl who is one of a large family knows that.

11. Those who did not want railways ______.

    A. tried to show that tunnels were certain to cause colds

    B. said that tunnels would be cold

    C. produced doctors who would show the colds they had caught in tunnels

    D. would show people the colds and diseases they had got in tunnels

12. The word "eminent" (in Paragraph 2) most probably means ______.

    A. optimistic　　B. outstanding　　C. desperate　　D. ordinary

13. Those who welcomed the railway did so because ______.

    A. it was a convenient way of making a change

    B. they expected more than just a quicker way of traveling

    C. they realised it would not get faster or more comfortable for a very long time

    D. they thought it would enable armies to be moved rapidly

14. All boys and girls in large families know that ______.

    A. the faster hostile armies were moved, the more chances there are of war

    B. we are more together than used to be

    C. a lot of people being together makes fights likely

    D. whenever there is a new invention there are always people to condemn it

15. Which of the following statements is NOT true?

    A. The author's attitude toward the railway is a negative one.

    B. Every new invention is open to both approval and disapproval.

    C. War is the inevitable result of the railway.

    D. Those who wanted railways said people would sleep well if they traveled in trains.

# Unit 11

## Section A

**Directions:** *In this section, there is a passage with ten blanks. You are required to select one word for each blank from a list of choices given in a word bank following the passage. Read the passage through carefully before making your choices. Each choice in the bank is identified by a letter. Please mark the corresponding letter for each item on* ***Answer Sheet*** *2 with a single line through the centre.* ***You may not use any of the words in the bank more than once.***

**Questions 1 to 10 are based on the following passage.**

Praise is like sunlight to the human spirit; we cannot flower and grow without it. And yet, while most of us are only too ready to __1__ to others the cold wind of criticism, we are somehow __2__ to give our fellows the warm sunshine of praise. Perhaps it's because few of us know how to accept compliments __3__. Instead, __4__ and *shrug off*(对……不屑一顾) the words we are really so glad to hear. Because of this defensive reaction, direct compliments are surprisingly difficult to give. That is why some of the most valued compliments are those which come to us indirectly, in a letter or passed on by a friend. When one thinks of the speed with which spiteful remarks are conveyed, it seems a pity that there isn't more effort to relay pleasing and flattering comments. Praise is particularly appreciated by those doing __5__ jobs: gas station attendants, waitresses—even housewives. Do you ever go into a house and say, "what a tidy room!"? Hardly anybody does. That's why housework is considered such a boring job.

Behavioral scientists have done countless experiments to prove that any human being __6__ to repeat an act which has been immediately followed by a pleasant result. In one such experiment, a number of schoolchildren were divided into three groups and given arithmetic tests daily for five days. One group was __7__ praised for its previous performance; another group was criticized; the third was ignored. Not __8__, those who were praised improved dramatically. Those who were criticized improved also, but not so much. And the scores of the children who were ignored hardly improved at all.

To give praise costs the giver nothing but a moment's thought and a moment's effort—perhaps a quick phone call to pass on a compliment, or five minutes __9__ writing an appreciative letter. It is such a small investment—and yet consider the results it may produce. So, let's be __10__ to the small excellences around us—and comment on them.

| | | | | |
|---|---|---|---|---|
| A) apply | B) reluctant | C) gracefully | D) embarrassed | E) routine |
| F) tends | G) Consistently | H) considerately | I) fortunately | J) surprisingly |
| K) common | L) spent | M) remarkably | N) ordinary | O) alert |

## Section B

**Directions:** *There are 2 passages in this section. Each passage is followed by some questions or unfinished statements. For each of them there are four choices marked A, B, C and D. You should decide on the best choice and mark the corresponding letter on* ***Answer Sheet*** *2 with a single line through the centre.*

### Passage One

**Questions 11 to 15 are based on the following passage.**

It has always been common for students to work to earn money, not only in vacations but also, when practicable, by doing part-time jobs during term-time. As the total cost of study and living may be $ 2,000 to $ 3,000 a year these earnings are useful and often essential. Mostly students do rather unskilled work. Some students do paid work for the university at which they study, in the library or the restaurant, or even by acting as lifeguards at a bathing-place. Others work outside. One popular occupation is that of porter at a supermarket, carrying housewives' groceries out to their cars.

Since 1958 the financial position of students has been improved by the provision of loans by the Federal government. The National Defense Education Act of 1958 enabled students to borrow money to help with expenses, provided that they needed the money and had a good academic record after a period of study, and by 1965, 750,000 students had received loans, amounting to up to $ 1,000 a year per student. The Higher Education Act of 1965 was an important new development, allowing students to receive loans in their first year at college, on the basis of need alone. Students may take up to eleven years to repay the loans, though those who themselves become teachers in public schools only have to repay a portion of the loan. Those who teach in depressed areas are specially favored and each year of depressed-area teaching wipes out fifteen per cent of the loan received.

11. The reason why students earn money while studying in universities is that ______.

A. they have to learn how to do so
B. they are asked to do so
C. the tuition is high
D. their parents don't pay their tuition

12. There are many jobs for the students to do, but the popular one is ______.

A. to work as librarians
B. to help housewives at supermarkets
C. to work as waiters or waitresses
D. to act as lifeguards

13. It can be inferred from the passage that ______.

A. the American government attaches great importance to education
B. the American government attaches great importance to education only to higher education
C. a student who receives high marks doesn't pay tuition
D. a student who is really poor doesn't need to repay the loans

14. A student doesn't need to repay the loans if he teaches in a depressed area ______.

A. for 11 years　　B. for 15 years　　C. for about 7 years　　D. all his life

15. Which of the following statements is TRUE according to the passage?

A. Some students are forced to earn their tuition in American colleges.

B. Only unskilled work is suitable for college students to do.

C. Loans by the government are the last but not the least way for one's education.

D. Teaching in public schools and depressed areas is encouraged by the government.

## Passage Two

**Questions 16 to 20 are based on the following passage.**

Tests conducted at the University of Pennsylvania's Psychological Laboratory showed that anger is one of the most difficult emotions to detect from facial expression. Professor Dallas E. Buzby confronted 716 students with pictures of extremely angry persons, and asked them to identify the emotion from the facial expression. Only two percent made correct judgments. Anger was most frequently judged as "pleased". A typical reaction of one student confronted with the picture of a man who was hopping mad was to classify his expression as either "bewildered", "*quizzical*"(好奇的)or simply "amazed". Other studies have shown that it is extremely difficult to tell whether a man is angry or not, just by looking at his face. The investigation found further that women are better at detecting anger from facial expression than men are. Paradoxically, they found that psychological training does not sharpen one's ability to judge a man's emotions by his expressions but appears actually to hinder it. In the university tests, the more courses the subject had taken in psychology, the poorer judgement scores he turned in.

16. The information in this passage centers about ______.

A. the relation between anger and other emotions

B. the findings of Professor Dallas E. Buzby

C. the differences between men and women with respect to emotion

D. the detection of anger from facial expression

17. Based on the information of the passage, the main thought of this passage is that anger ______.

A. is difficult to detect by looking at a person's face

B. is frequently confused with other emotions

C. is detected by women better than by men

D. cannot be detected by a psychologically trained person

18. According to the passage, the students with psychological training who were tested ______.

A. marked less than two percent of their possible choices correctly

B. were less able to judge correctly than the average student

C. did better than the average student in the group

D. did as well as the women students

19. To achieve the greatest success in detecting anger from facial expression, it would be best to ______.

A. use adults rather than students as judges

B. ask women in fields rather than psychologist to judge

C. ask women rather than men to judge

D. ask psychologists to judge

20. "Subject" in the last sentence most probably refers to ______.

A. a person whose emotions are judged

B. a person who conducts tests like Professor Dallas E. Buzby

C. a person who judges a man's emotions by his expressions

D. none of the above

# Unit 12

## Section A

**Directions:** *In this part there is a short passage with five questions or incomplete statements. Read the passage carefully. Then answer the questions or complete the statements in less than 10 words.*

**Questions 1 to 5 are based on the following passage.**

"We're more than halfway now; it's only two miles farther to the inn," said the driver.

"I'm glad of that!" answered the stranger in a more sympathetic mood. He meant to say more but the east wind blew clear down a man's throat if he tried to speak. The girl's voice was something quite charming, however, and presently he spoke again.

"You don't feel windy so much at twenty below zero out in the Western Country. There's none of this damp chill," he said, and then it was a disagreeable day, and he began to be conscious of a warm hopefulness of spirit, and a sense of pleasant adventure under all the woolen clothes.

"You'll have a cold drive going back," he said anxiously, and put up his hand for the twentieth time to see if his coat collar was as close to the back of his neck as possible. He had wished a dozen times for the warm old hunting clothes in which he had many a day confronted the worst of weather in the Northwest.

"I shall not have to go back!" exclaimed the girl, with eager pleasantness. "I'm on my way home now. I drove over early to meet you at the train. We had word that someone was coming to the inn."

1. How far was the drive from the train station to the inn?

______________________________.

2. What was the driver, a boy or girl?

______________________________.

3. From the passage we gather that the two speakers are in ______________________________.

4. According to the passage, how are the weather in the West?

______________________________.

5. Where did most probably the driver live?

______________________________.

## Section B

**Directions:** *There are 2 passages in this section. Each passage is followed by some questions or unfinished statements. For each of them there are four choices marked A, B, C and D. You should decide on the best choice and mark the corresponding letter on **Answer Sheet** 2 with a single line through the centre.*

### Passage One

**Questions 6 to 10 are based on the following passage.**

One phase of the business cycle is the expansion phase. This phase is a two-fold one, including recovery and prosperity. During the recovery period there is ever-growing expansion of existing facilities and new facilities for production are created. More businesses are created and older ones expanded. Improvements of various kinds are made. There is an ever increasing optimism about the future of economic growth. Much capital is invested in machinery or heavy industry. More labor is employed. More raw materials are required. As one part of the economy develops, other parts are affected. For example, a great expansion in automobiles results in an expansion of the steel, glass, and rubber industries. Roads are required; thus the cement and machinery industries are stimulated. Demand for labor and materials results in greater prosperity for workers and suppliers of raw materials, including farmers. This increases purchasing power and the volume of goods bought and sold. Thus prosperity is diffused among the various segments of the population. This prosperity period may continue to rise and rise without an apparent end. However, a time comes when this phase reaches a peak and stops spiraling upwards. This is the ending of the expansion phase.

6. We may assume that in the next paragraph the writer will discuss ______.

A. cyclical industries B. the status of the farmer C. the higher cost of living D. the recession period

7. The title below that best expresses the ideas of this passage is ______.

A. The Business Cycle B. The Recovery Stage C. An Expanding Society D. The Period of Good Times

8. Prosperity in one industry ______.

A. reflects itself in many other industries B. will spiral upwards

C. will affect the steel industry D. will end abruptly

9. Which of the following industries will probably be a good indicator of a period of expansion?

A. Toys. B. Machine tools. C. Foodstuffs. D. Farming.

10. During the period of prosperity, people regard the future ______.

A. cautiously B. in a confident manner C. opportunity D. indifferently

## Passage Two

**Questions 11 to 15 are based on the following passage.**

The Ordinance of 1784 is most significant historically because it embodied the principle that new states should be formed from the western region and admitted to the Union on an equal basis with the original commonwealths. This principle, which underlay the whole later development of the continental United States, was generally accepted by this time and cannot be properly credited to any single man. Thomas Jefferson had presented precisely this idea to his own state of Virginia before the Declaration of Independence, however, and if he did not originate it he was certainly one of those who held it first. It had been basic in his own thinking about the future of the Republic throughout the struggle for independence. He had no desire to break from the British Empire simply to establish an American one—in which the newer region should be subsidiary and tributary to the old. What he dreamed of was an expanding union of self-governing commonwealths, joined as a group of peers.

11. Which of the following proposals did the Ordinance of 1784 incorporate?

A. New states should be admitted to the Union in numbers equal to the older states.

B. The Union should make the western region into tributary states.

C. New states should share the same rights in the Union as the original states.

D. The great western region should be divided into twelve states.

12. According to the passage, what was the general attitude toward the principle underlying the Ordinance of 1784?

A. It was considered the most important doctrine of the day.

B. It received wide support at that time.

C. It was more popular in Virginia than elsewhere.

D. It was thought to be original and creative.

13. According to the passage, one of Thomas Jefferson's political goals was to ______.

A. maintain strong ties with the British Empire B. fight for more territory for his country

C. found the Republican Party D. guarantee the status of new states

14. The author implies which of the following happened to new lands that became part of the British Empire?

A. They were not considered equal to Britain itself.

B. They established a separate empire of their own.

C. They had an equal share in the government of the empire.

D. They were ruled by a group of peers.

15. The paragraph following this passage most probably would discuss ______.

A. Jefferson's home in eastern Virginia B. the implementation of the Ordinance of 1784

C. British colonial expansion outside North America D. the economic development of Virginia

# Unit 13

## Section A

**Directions:** *In this part there's a short passage with five questions or incomplete statements. Read the passage carefully. Then answer the questions or complete the statements in less than 10 words.*

**Questions 1 to 5 are based on the following passage.**

As early as 1710 the iron industry in English complained of increasing competition from the American colonies. The American iron industry developed rapidly from that date until, by 1750, numerous furnaces, forges, and mills were operating in New England, the middle colonies, and Virginia. When large quantities of pig iron from the American colonies first entered England in 1735, the product proved to be of such excellent quality that English iron – makers became involved in a bitter argument over the future of the colonial iron industry. The English iron smelters, who changed native English iron ore into pig iron, insisted that American pig iron be kept out of England by means of high import taxes and, in fact, that the whole colonial iron industry be suppressed. In agreement with the iron smelters were owners of English mines and even forests, whose wood was used to fuel the furnaces which smelted the iron ore.

On the other side of the issue were the English iron manufacturers who desired more cheap pig iron to make into nails, tools, and other iron wares. The iron manufacturers therefore encouraged the production of pig iron in the American colonies. They wanted it to enter England tax free, but, at the same time, demanded that the colonists be prevented from working their crude iron into finished products. In addition to the iron manufacturers, English merchant ship-owners were in favor of receiving American pig iron, for they looked forward to transporting the crude iron from America to England and the manufactured iron products from England to the colonies. The English wool industry supported the iron manufacturers, also, in the belief that the Americans would use the money received for shipments of crude iron to buy cloth made in England, thus discouraging the growth of wool manufacturing in America.

1. English iron smelters and English iron manufacturers were both opposed to ______________________________.

2. The passage implies that American pig iron was ______________________________.

3. The passage suggests that the American wool industry would have developed rapidly if ______________________________.

4. What is the main topic of this passage?

______________________________.

5. It was believed that the colonists would use profits from the sale of their pig iron to ______________________________.

## Section B

**Directions:** *There are 2 passages in this section. Each passage is followed by some questions or unfinished statements. For each of them there are four choices marked A, B, C and D. You should decide on the best choice and mark the corresponding letter on **Answer Sheet** 2 with a single line through the centre.*

### Passage One

**Questions 6 to 10 are based on the following passage.**

For many years, scientists couldn't figure out how atoms and molecules on the Earth combined to make living things. Plants, fish, dinosaurs, and people are made of atoms and molecules, but they are put together in a more complicated way than the molecules in the primitive ocean. What's more, living things have energy and can reproduce, while the chemicals on the Earth 4 billion years ago were lifeless.

After years of study, scientists figured out that living things, including human bodies, are basically made of amino acids and nucleotide bases. These are molecules with millions of hydrogen, carbon, nitrogen, and oxygen atoms. How could such complicated molecules have been formed in the primitive soup? Scientists were stumped.

Then, in 1953, two scientists named Harold Urey and Stanley L. Miller did a very simple experiment to find out what had happened on the Primitive Earth. They set up some tubes and bottles in a closed loop, and put in some of the same gases that were present in the atmosphere 4 billion years ago: water vapor, ammonia, carbon dioxide, methane, and hydrogen.

Then they shot an electric spark through the gases to simulate bolts of lightning on the ancient Earth, circulated the gases through some water, sent them back for more sparks, and so on. After seven days, the water that the gases had been bubbling through had turned brown. Some new chemicals were dissolved in it. When Miller and Urey analyzed the liquid, they found that **it** contained amino acids—the very kind of molecules found in all living things.

6. When did scientists come to realize how the atoms and molecules on the Earth combined to make living thing?

A. 4 billion years ago. B. In 1953. C. After seven days. D. Many years later.

7. Scientists figured out that human bodies are basically made of ______.

A. amino acids
B. molecules
C. hydrogen, carbon, nitrogen and oxygen atoms
D. water vapor, ammonia, carbon dioxide, methane and hydrogen

8. Harold Urey and Stanley L. Miller did their experiment in order to ______.

A. find out what had happened on the Earth 4 billion years ago
B. simulate bolts of lightning on the ancient Earth
C. dissolve some new chemicals
D. analyze a liquid

9. At the end of the last paragraph, the word "it" refers to ______.

A. a closed loop B. an electric spark C. water D. the liquid

10. According to the writer, living things on the Earth include ______.

A. atoms and molecules
B. chemicals
C. plants, fish, dinosaurs and human beings
D. the primitive soup

## Passage Two

**Questions 11 to 15 are based on the following passage.**

TOEFL TEACHERS

Summer Posts

Once again we require 10 excellent TOEFL teachers for our summer program. Large thriving Arels-Felco school offers a special package to qualified, TOEFL experienced teachers. $1,500 and free accommodation for 200 hours teaching from 2 July to 24 August. Overtime available.

Good possibility of longer term and permanent posts. Shorter contracts available.

Letters of application and C. V. to Teacher Recruitment Dept. E), Churchill House School, 40-42 Spencer Square, Ramsgate, Kent (T11 9LD) Fax: (0843) 584827.

Established 20 years.

Recognized by the British Council and a member of Arels-Felco

11. Most likely, this advertisement is for ______ teachers of English as a foreign language.

A. employing B. part-time C. permanent D. newly qualified

12. What does "package" in the advertisement refer to?

A. The salary.
B. The number of teaching hours.
C. The free accommodation provided.
D. All the above.

13. Some teachers may be able to ______.

A. accomplish the job ahead of schedule
B. continue working at the school after the summer
C. enjoy free accommodation for a longer time
D. quit the job when they choose to do so

14. If you apply for the summer posts, which agency should you write to?
A. Arels-Felco.
B. British Council.
C. Spencer Square.
D. Churchill House School.

15. Arels-Felco is probably ______.
A. a company
B. an educational organization
C. the name of a school
D. a housing agency

# Unit 14

## Section A

**Directions:** *In this section, there is a passage with ten blanks. You are required to select one word for each blank from a list of choices given in a word bank following the passage. Read the passage through carefully before making your choices. Each choice in the bank is identified by a letter. Please mark the corresponding letter for each item on* ***Answer Sheet*** *2 with a single line through the centre.* ***You may not use any of the words in the bank more than once.***

**Questions 1 to 10 are based on the following passage.**

Loneliness is a curious thing. Most of us can remember feeling most lonely when we are not in fact __1__ at all, but when we were surrounded by people. Everyone has __2__, at some time, that nuttier sense of isolation that comes over you when you are at a party or in an audience at a lecture. It suddenly seems to you as if everybody knows everybody else; everybody is sure of himself; everybody, that is, except you.

This feeling of loneliness which can overcome you when you are in a crowd is very difficult to get rid of. People living alone are advised to __3__ their loneliness by joining a club or a society, by going out and meeting people. Does this really help?

There are no easy __4__. Your first day at work, or at a new school or university, is a __5__ situation in which you are likely to feel lonely. You feel that everybody else is full of __6__ and knows what to do, but you are __7__ and helpless. The fact of the matter is that, in order to survive, we all put on a show of self-confidence to hide our __8__ and doubts.

In a big city it is particularly easy to get the feeling that everybody except you is leading a full, rich, busy life. Everybody is going somewhere, and you tend to __9__ that they are going somewhere nice and interesting, whereas your __10__ is less exciting and fulfilling.

| | | | | |
|---|---|---|---|---|
| A) tackle | B) confidence | C) assume | D) common | E) alone |
| F) retain | G) destination | H) satisfaction | I) uncertainties | J) worry |
| K) experienced | L) typical | M) solutions | N) avoid | O) adrift |

## Section B

**Directions:** *There are 2 passages in this section. Each passage is followed by some questions or unfinished statements. For each of them there are four choices marked A, B, C and D. You should decide on the best choice and mark the corresponding letter on* ***Answer Sheet*** *2 with a single line through the centre.*

### Passage One

**Questions 11 to 15 are based on the following passage.**

Men usually want to have their own way. They want to think and act as they like. No one, however, can have his own way all the time. A man cannot live in society without considering the interests of others as well as his own interests. "Society" means a group of people with the same laws and the same way of life. People in society may make their own decisions, but these decisions ought not to be unjust or harmful to others. One man's decisions may so easily harm another person. For example, a motorist may be in a hurry to get to a friend' s house. He sets out, driving at full speed like a competitor in a motor race. There are other vehicles and also *pedestrians*(步行者) on the road. Suddenly there is a crash. There are screams and confusion. One careless motorist has struck another car. The collision has injured two of the passengers and killed the third. Too many road accidents happen through the thoughtlessness of selfish drivers.

We have governments, the police and the law courts to prevent or to punish such criminal acts. But in addition, all men ought to observe certain rules of conduct. Every man ought to behave with consideration for other men. He ought not to steal, cheat, or destroy the property of others. There is no place of this sort of behavior in a civilized society.

11. A man cannot have his own way all the time because ______.
   A. he may have no interest in other people
   B. he has to share the same interest with the people in the same society
   C. his decisions are always unjust
   D. his decisions may harm other people

12. According to the passage, people in a civilized society should usually ______.
   A. be honest to each other
   B. be cautious in doing everything
   C. behave in a responsible way
   D. punish criminal acts

13. The purpose of this passage is to ______.
   A. tell people how to behave in society
   B. illustrate the importance of laws
   C. teach people how to prevent criminal
   D. persuade people not to make their own decision

14. It is implied that there will be fewer road accidents if ______.
   A. the drivers are more considerate of other people
   B. there are fewer cars or walkers in the street
   C. the motorists are not always in a hurry
   D. the passengers are calm but not confused before the accidents

15. We can draw a conclusion that ______.
   A. the government should contribute more efforts
   B. the criminals should be more severely punished
   C. man should be more strict with himself
   D. man should have more and more similar interests

## Passage Two

**Questions 16 to 20 are based on the following passage.**

In the past, the concept of marketing emphasized sales. The producer or manufacturer made a product he wanted to sell. Marketing was the task of figuring out how to sell the product. Basically, selling the product would be accomplished by sales promotion, which included advertising and personal selling. In addition to sales promotion, marketing also involved the physical distribution of the product to the places where it was actually sold. Distribution consisted of transportation, storage, and related services such as financing, standardization and grading, and the related risks.

The modern marketing concept encompasses all of the activities mentioned, but it is based on a different set of principles. It subscribes to the notion that production can be economically justified only by consumption. In other words, goods should be produced only if they can be sold. Therefore, the producer should consider who is going to buy the product—or what the market for the product is—before production begins. This is very different from making a product and then thinking about how to sell it.

16. Marketing used to be mainly concerned with ______ the product.
   A. making  B. distributing  C. selling  D. advertising

17. The two main aspects of traditional marketing are ______.
   A. selling and distributing  B. advertising and selling  C. producing and selling  D. financing and grading

18. How many aspects does distribution involve?
   A. Three.  B. Four.  C. Five.  D. Six.

19. While traditional marketing is mainly concerned with sales of a product, modern marketing ______.
   A. caters for selling justified by production
   B. excludes the sales activities involved in traditional marketing
   C. puts more emphasis on economy in production
   D. aims to achieve a balance between production and sales

20. The producer is advised to first consider how to ______.
   A. advertise the product
   B. distribute the product
   C. meet the needs of the consumer
   D. make the product

# Unit 15

## Section A

**Directions:** *In this section, there is a passage with ten blanks. You are required to select one word for each blank from a list of choices given in a word bank following the passage. Read the passage through carefully before making your choices. Each choice in the bank is identified by a letter. Please mark the corresponding letter for each item on* **Answer Sheet** *2 with a single line through the centre.* ***You may not use any of the words in the bank more than once.***

**Questions 1 to 10 are based on the following passage.**

In the traditional marriage, the man worked at a job to earn money for the family. Most men worked in an office, a factory, or some other place away from the home. Since the man earned the money, he paid the __1__. The money was used for food, clothes, a house, and other family needs. The man made most of the __2__. He was the boss.

In the traditional marriage, the woman __3__ worked away from the house. She stayed at home to care for the children and her husband. She cooked the meals, cleaned house, washed the clothes, and did other household work. Her job at home was very __4__.

In recent years, many __5__ continue to have a traditional relationship of this kind. The man has a job and earns the money for the family. The woman stays at home and cares for the children and the house. Many Americans are happy with this kind of marriage. But some other Americans have a different impression of marriage and family __6__.

There are two important differences in male and female roles now. One is that both men and women have many more __7__. They may choose to marry or to stay __8__. They may choose to work or stay at home. Both men and women may choose roles that are __9__ for them.

A second difference in male and female roles is that within marriage many decisions and responsibilities are __10__. The husband and wife may choose to have children, or they may not. If they have children, the man may take care of them some of the time, all of the time, or not at all. The woman may want to stay at home and take care of the children. Or she may want to go to work. Men and women now decide these things together in a marriage. Many married people now share these decisions and the responsibilities of their families.

| | | | | |
|---|---|---|---|---|
| A) comfortable | B) decisions | C) choices | D) debates | E) bills |
| F) seldom | G) hard | H) single | I) shared | J) important |
| K) responsibilities | L) impossible | M) forced | N) often | O) couples |

## Section B

**Directions:** *There are 2 passages in this section. Each passage is followed by some questions or unfinished statements. For each of them there are four choices marked A, B, C and D. You should decide on the best choice and mark the corresponding letter on* **Answer Sheet** *2 with a single line through the centre.*

### Passage One

**Questions 11 to 15 are based on the following passage.**

Nuclear power is obtained from the energy which can be released from the nucleus of an atom. Until the twentieth century man used water, wood and the *fossil*(化石) fuels(coal, oil and gas) as sources of power. During the first quarter of the twentieth century physicists investigated the structure of the atom.

In 1919 Rutherford split the atom artificially. Thirteen years later the neutron was discovered. In 1939 Hahn and Strassman investigated the action on neutrons of uranium-235. They found that it was split into two equal pieces. This process is known as *fission*(分裂). It released great amounts of energy. The neutrons that are released in fission produce fission in other atoms. This is known as a chain reaction. On 2nd December 1942 Enrico Fermi and his colleagues produced the first controlled nuclear chain reaction.

Since then atomic energy has been used in war and peace. In 1951 electricity was first produced by using the heat from a nuclear reactor. More recently nuclear energy has been used to power submarines. Nuclear batteries are now being used in *cardiac pacemakers*(心脏起搏器). More and more countries are building nuclear power stations to produce electricity.

11. What was done 23 years before the first controlled nuclear reaction?

A. The structure of the atom was investigated.
B. The atom was split artificially.
C. The reaction of neutrons was studied.
D. The neutrons are released in fission.

12. What can we know about "a chain reaction" from the passage?

A. A process of change in which a product of the change leads to another further change.
B. A series of events in which one happens after another.
C. A number of things which are joined to each other.
D. An amount of energy which are released in fission.

13. How many years passed between the discovery of fission and the use of nuclear energy to produce electricity?

A. About 12 years. B. About 13 years. C. About 23 years. D. About 32 years.

14. The last paragraph is mainly about ______.

A. the gradual development of atomic energy
B. the production of electricity by nuclear power
C. the wide uses of nuclear power
D. the tendency of building nuclear power stations

15. Which statement is NOT true according to the passage?

A. Water, wood and the fossil fuels were used as sources of power before the 20th century.
B. After 1942, atomic energy has been used in war and peace.
C. The structure of atom was investigated from 1900 to 1925.
D. Nuclear power is released from the nuclear of an atom.

## Passage Two

**Questions 16 to 20 are based on the following passage.**

One of the most fascinating things about television is the size of the audience. A novel can be on the "best seller" list with a sale of fewer than 100,000 copies, but a popular TV show might have 70 million viewers. TV can make anything or anyone well-known overnight.

This is the principle behind "quiz" or "game" shows, which put ordinary people on TV to play a game for prizes and money. A quiz show can make anyone a star, and it can give away thousands of dollars just for fun. But all of this money can create problems. For instance, in the 1950s, quiz shows were very popular in the U. S. and almost everyone watched them. Charles Van Doren, an English instructor, became rich and famous after winning money on several shows. He even had a career as at a television personality. But one of the losers proved that Van Doren was cheating. It turned out that the show's producers who were pulling the strings, gave the answers to the most popular contestants beforehand. Why? Because if the audience didn't like the person who won the game, they turned the show off. The result of this cheating was a huge scandal. Based on his story, a movie under the title "Quiz show" is on 40 years later.

Van Doren is no longer involved with TV. But game shows are still here, though they aren't taken as seriously. In fact, some of them try to be as ridiculous as possible. There are shows that send strangers on vacation trips together, that try to cause newly-married couples to fight on TV, or that punish posers by humiliating them. The entertainment now is to see what people will do just to be on TV. People still win money, but the real prize is to be in front of an audience of millions.

16. The sale of novels is talked about in comparison with ______.

A. the size of TVs
B. the number of TV viewers
C. the sale of "best sellers"
D. the number of TV audiences

17. In Charles quiz scandal, who, according to the passage, is to be blamed most?

A. One of the losers.
B. The shows producers.
C. Charles Van Doren himself.
D. His audience.

18. Charles Van Doren was mentioned in order to prove that ______.

A. a quiz show could make anyone a star
B. game shows cheat audiences
C. the cheating of his quiz show results in a huge scandal
D. he became rich after winning money on several shows

19. In the last paragraph, the word "seriously" can most probably be replaced by ______.

A. importantly B. honestly C. solemnly D. formally

20. The writer looks upon game shows on TV ______.

A. unconcernedly B. hopefully C. seriously D. critically

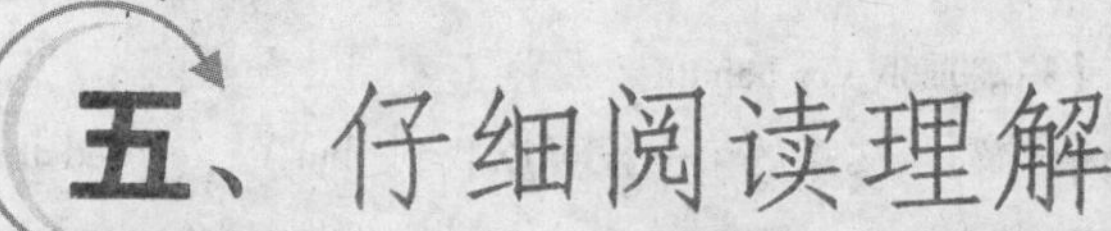

# 答案详解

## Unit 1

### Section A

1.【答案】E

【详解】根据文义,应是一形容词,后面的 changes 一词暗示此处填 revised。

2.【答案】O

【详解】浏览选项只有 community 和 society 可选,但 society 不合文义,故选 community。

3.【答案】M

【详解】根据前面的 university 可知是在校园,易找到 intercampus。

4.【答案】B

【详解】此空的干扰项有 prepared,但对于 schedule 还是 arranged 比较合适。

5.【答案】J

【详解】此空的干扰项有 way,但 route 含有规定好的路线之意,way 只指一般意义上的道路。

6.【答案】L

【详解】前有 cafeteria 后有 breakfast,lunch,明显应选 serve。

7.【答案】G

【详解】maintain 和 keep 两个词这里接上下文意思都可选,但因为这里是将来时,所以 keeps 不能选。

8.【答案】F

【详解】修饰 card 可选用 valid 和 standard,相比较 valid 要更合适一些。

9.【答案】A

【详解】根据文义,这里是一张证明身份的卡片,所以选 identification。

10.【答案】N

【详解】根据上下文很容易选出 appear。

### Section B

11.【答案】A

【详解】纵观全文可以看出,本篇短文共分两个部分来阐述为什么在 Jamestown 制砖并用砖来建筑房屋。第一部分讲实用方面的原因(practical reasons),第二部分讲思想观念方面的原因(ideological reason)。A 项点明了短文的主题,因此为正确答案。B,C 项只是文章中所涉及的两个细节,不能做主题。D 项意义范围太大,不适合本题。

12.【答案】D

【详解】通过阅读短文,可知 D 项叙述并没有在原文中出现。原文只说"伐木极端困难"(... the process of lumbering was extremely difficult,...)并未说"木料质量不好"。故 D 项符合题意,为正确答案。

13.【答案】D

【详解】运用排除法做本题。A 项不正确,因为在 Jamestown 用砖建房前,都是用木料做建筑材料。据此推断,其森林面积不会太小;如果多岩石,就不会出现石料短缺。B 项也不正确,因为 Jamestown 所在的半岛(peninsula)上有森林,因此不会是光秃秃的(barren)。C 项也不正确,因为既然 Jamestown 有人居住,就不是一个不能居住的地方(uninhabitable)。剩下的 D 项为正确答案。从文中可以推断,Jamestown 所在的半岛多树而且黏土丰富(... clay was plentiful...)。

14.【答案】C

【详解】原文中说:"Had this law ever been successfully enforced, Jamestown would have been a model city."作者在这里用虚拟语气告诉读者,因为这项法律没有得到贯彻,所以 Jamestown 也就没有成为城市的典范(model city)。据此,我们可以推断,弗吉尼亚州议会(Virginia Assembly)通过这项法令的初衷是希望能把 Jamestown 建成一座未来城市的典范(an example for the future)。因此 C 项为正确答案。

15.【答案】A

【详解】见原文最后一句"By 1699, Jamestown had collapsed into a pile of rubble..."。

16.【答案】C

【详解】通读全文,可以看出,本文重点描述的不是蝗虫集结成群带来的后果(plague),因此,A 被排除。蝗虫的再繁衍(reproduction)与蝗灾的不可预见性(unpredictability)只是作者所涉及到两个细节,显然不能作为该短文主题。本文作者着重描述了蝗灾的形成过程,因此 C 项为正确答案。

17.【答案】A

【详解】半干旱区的降雨(rain in the semiarid regions)为蝗虫的大量繁殖提供了必要的条件;蝗虫数量增加(population increases)为蝗虫大量集结提供了前提条件,蝗虫喜群居(the gregarious nature)是蝗虫大量集结的基础条件。这三点对蝗虫大量集结非常重要,因此 B,C,D 项与题意不符。A 丰富的食源(abundant food supply),不能成为蝗虫集结的必要条件,因为蝗虫只有在感到食源不足时才集结迁移。故 A 项符合题意,为正确答案。

18.【答案】C

【详解】见原文"As the supply of green, palatable food plants decreases toward the end of the rainy season, the locusts become even more concentrated"。

19.【答案】A

【详解】原文中说:"Since locusts breed most successfully in wet weather."可以推出干旱的地方蝗虫数量肯定更少。因此 A 项符合题意。

20.【答案】D

【详解】本段重点讨论的是蝗灾的形成原因,据此可以推知本文的主题应是与蝗灾有关的话题。根据文章结构安排应遵循向高潮推进的顺序(climactic order)这一原则,下一段讨论的内容应是如何控制蝗灾。故 D 项为正确答案。如选其他任一项都偏离主题。

# Unit 2

## Section A

1.【答案】parents should read to children every day

【详解】根据"Read to Me"的目标"to see that every child in Hawaii is read to for at least ten minutes every day",可以得到本题的答案。

2.【答案】solve or handle

【详解】从文章的第一段可以找到"but we realized that if we could focus attention on raising a literate population instead of fixing up an illiterate one..."。这段话的意思是:把重点放在提高有文化人群的素质而不仅仅是解决文盲问题……因此可以推断出 fix up 是"解决"的意思。

3.【答案】influence the children's lives

【详解】从文章的倒数第二段 Trelease 的话中可以得出答案:"20 years from now the dust balls will still be under the couch, but your little boy or girl will no longer be your little boy or girl."由此可以看出,给孩子朗读故事将会对孩子的人生产生影响。

4.【答案】Read to Your Children

【详解】全文围绕着一个中心,即父母应该给孩子朗读故事。作者以美国各地发起的"读给我听"运动作为例子更加突出了这个中心,因此毫无疑问它应该是文章的标题。

5.【答案】Neutral.

【详解】作为一篇报道,内容要真实、准确。不要被 Trelease 的积极态度所误导。作者写这篇文章完全是从事实出发,并不包含个人情绪。

## Section B

6.【答案】A

【详解】文章第一段说明了"板块构造地质学(plate tectonics)"的定义。然后作者应用该理论来解释地球构成的地质现象,以便读者从一个更加具体的角度来理解该理论。由以上分析可知,本文写作的目的是为了向读者解释板块构造地质学理论。A 答案正确说明了作者这一写作目的,符合题意,是正确答案。

7.【答案】A

【详解】A. 固定的陆块。原文第一段中说:... the continents of today are not venerable landmasses but amalgams of other lands repeatedly broke up, jugged, rotated, scattered far and wide, then crunched together into new configurations...。……我们今天的大陆并不(一直)是这样大面积的陆地,而是由其他陆地的混合物不断地分裂,碰撞,旋转,……,然后相互挤压形成新

的构造……。作者在此处是暗示过去人们并不是这样认为的,他们认为大陆是从古就有,固定不变的大陆块。A 项与文义相符,是正确答案。B. 冻结的冰块,文章第一段结尾处是在拿冰块作比喻,而不是暗示大陆是冰块。从文中可以看出,C,D 不是过去人们的观点。

8.【答案】C

【详解】原文第一段结尾处... like ice floes swept along the shore of a swift-flowing stream。联系上下文,此处意为:……就像激流中携带的浮冰,沿着岸边(碰撞,挤压中)流动。可以看出,swept 在此处意为"携带";C. carried 有此意,符合题意,是正确答案。A. won 赢得;B. cleaned 把……弄干净;D. removed 移动,搬开。均不符合题意。

9.【答案】A

【详解】见原文第二段结尾处:Other oceans, such as the Pacific, are shrinking as seafloor descends under their fringing coastlines off shore areas of island. 其他的大洋,比如太平洋,都在缩小,由于其海底下降……这说明,太平洋的面积渐渐缩小,因此,A 项正确。

10.【答案】C

【详解】C. 地心引力。见原文最后部分的论述:What causes these plates to jostle each other, ... or it may be an effect produced by gravity. 导致这些板块相互挤压的原因是什么……或许是地心引力所产生的结果。C 答案与文义相符,是正确答案。A. 波状运动;B. 海洋的扩张;D. 月亮的位置,文中均未提及。

11.【答案】D

【详解】本文首先解释了什么叫水坝,然后说明了建设水坝的原因及其功用;接着介绍了水坝建设的发展历程,最后说明了水坝对经济的深远影响。据以上分析,可以看出,作者写作本文的意图在于对水坝做一个简要的介绍。D 答案正确表达了文章的这一主题,是正确答案。

12.【答案】C

【详解】阅读文章第一段,就可以看出,"C. 人们修建水坝是为了让河流改道"并没有被作者列为修建水坝的普通原因。C 答案符合题意,因此是正确答案。其他各项均可在第一段中找到相应的内容。

13.【答案】A

【详解】见原文第二段:Because of the ravages of periodic floods, very few dams more than a century old are still living. 联系上下文,此处意为:因为周期性洪水的破坏,一个世纪前修建的水坝现在已经很少见了,可以看出,ravage 为名词,在此处意为"破坏,毁坏"。A. destruction 毁坏,破坏,名词,与 ravage 意思相同,因此 A 项正确。B. occurrence 发生,出现;与题意不符。C. destroy 破坏,动词,不可用来解释名词;D. savage 残酷的,凶残的,与题意不符。

14.【答案】C

【详解】见原文第二段:Many ancient earth dams, ..., were part of elaborate irrigation systems... 很多古代修建的土筑大坝,……是复杂的灌溉系统的一部分……这说明古代的水坝很多是为了灌溉用的。因此 C 答案是正确的。

15.【答案】B

【详解】原文第二段直接给出了本题答案 The construction of virtually indestructible dams ... became possible after the development of portland cement concrete and the mechanization of earth-moving ... equipment.

## Unit 3

### Section A

1.【答案】a mentally lazy attitude toward computers

【详解】见文章第一段"People tend to be over – trusting of computers and are reluctant to challenge their authority."可知人们过份依赖电脑。

2.【答案】be reasonably skeptical about them

【详解】根据文章第一段最后一句"Indeed, they behave..., or that a computer may simply malfunction."可知不能尽信电脑,有时也需要持合理的怀疑态度。

3.【答案】store of knowledge and the ability to computer

【详解】根据文章第二段"but people should also rely on their own internal computer..."可知 internal computer 指的是人的大脑。

4.【答案】psychological

【详解】综合全文内容上看作者认为现如今人们应克服心理上对电脑的依赖,要注意自身的思考,持合理的怀疑态度。

5.【答案】decision-making

【详解】从文章讲述我们可以推断作者是反对公司完全依赖电脑来做决定的。

### Section B

6.【答案】A

【详解】见原文"Television, says Klompus, contributes to children's passivity."电视是造成孩子上课无精打采的原因。

7.【答案】C

【详解】见原文第二段结尾:"It takes greater energy to say no to a kid"对孩子说"不"需要更大的勇气。

8.【答案】C

【详解】作者在第一段中回忆了多年前,当他(她)还是个孩子时,母亲对他(她)的生活、学习常常重复的要求。由此,可以看出作者的母亲更重视对孩子的要求和锻炼。因此 C 项是正确答案。

9.【答案】A

【详解】见原文第二段“... the permissive period in education in which we decided it was right not to push our children to achieve best in school”作者通过一个主语从句说明在“permissive period in education”这个阶段里,父母不要求孩子在学校里做到最好。其含义是孩子们可以自己做主。据此分析 A 答案正确表达了这一涵义,是正确答案。其他各项均不能准确表达这一涵义。

10.【答案】C

【详解】作者发现,他们不再对孩子提任何要求,完全把孩子交给学校去管理时,孩子们的情况就变得越来越糟糕。他们的学习积极性越来越差。因此,作者呼吁“It's time for parents to end their vacation and come back to work... It's time to start telling them no again”。做父母的再也不能对孩子放任不管,是该向他们提出要求,约束他们对他们说不,对他们更严厉一点的时候了。这正是本文的主旨所在。C 项概况了这一主题,为正确答案。A. 父母应对孩子听之任之,与文章主题背道而驰。B. 孩子们在学校的表现应该更积极,只是说明孩子们的一种表现,不能作为文章主题。D. 父母应常为孩子们树立良好榜样,不是本文所探讨的问题。

11.【答案】C

【详解】A. 新技术可以迅速得到应用。文中并无提及。B. 所有成员工作效率很高。文中也没能提到。D. 员工很容易得到提升。也没有在文中出现。C. 要花很长时间才能感觉到个人的影响力。可以在文中得到证实。:In the large enterprise...“but his effectiveness is remote, indirect, and difficult to see at first sight”。

12.【答案】C

【详解】参考上文“In a small and even in a mid-sized business you are normally exposed to all kinds of experiences, and expected to do a great many things without too much help or guidance”在一个中小型公司里,一个员工通常要做各种各样的事情,在没有帮助和指导的情况下要完成大量的事务。这说明在小公司做事情的人要了解很多事情。C 项:一个对各种事情都了解一点的人,正确表述了这一概念,为正确答案。

13.【答案】B

【详解】B. 没有提及自己的偏爱。作者在第一段对比了在大公司与小公司工作的不同,分别指出了各自的优缺点,但并未表明自己的偏爱。因此 B 项是正确的。既然 B 项是正确的,A,C 两项就不是正确答案。D. 不喜欢任何走向极端的事物,作者虽然指出了在两种类型的公司工作各有危险性(danger),但并未表明作者对此的态度。

14.【答案】B

【详解】在第二段中,作者通过描述在社会组织(大企业或政府里)供职和呆在家里(或一个小机构)对于人的不同意义,说明了不同类型的人对满足的不同理解。B 答案阐述了这一观点,因此是正确答案。

15.【答案】A

【详解】本题考查考生的观察思维能力,要考生通过文章内容了解作者。作者在文章第一段中对在大小公司工作的不同观点做了详细的陈述和对比,在第二段中作者说明了不同类型的人适合不同的工作,关键是源于对满足感的不同理解。这有助于求职的人了解自己适合于何种工作。因此这篇文章很可能是给求职者提供建议的。A 项符合这一观点,是正确答案。B. 评论国家的工业形势;D. (作者)曾在很多公司工作过。这两项在文中没有提及,不属文章讨论内容。C. (作者)是从经济学家的角度来写这篇文章的。文章并没有大量运用经济学的知识,由此可见作者不一定是个经济学家。

# Unit 4

**Section A**

1.【答案】A

【详解】这句话的意思是:常客对每一句问候进行回答。respond 意为“回答”。

2.【答案】E

【详解】这句话的意思是:没有人意识到服从了规则或遵守了惯例。

3.【答案】I

【详解】这句话的意思是:你会听到一些有创意、随心所欲的不同的语言。

4.【答案】M

【详解】这句话的意思是:你已传达了友好的意图。

5.【答案】B

【详解】这句话的意思是:你随后主动接触的举动会更好地被接受。

6.【答案】F

【详解】这句话的意思是:他们做事要依照一套严格的规则。in accordance with 意为“与……一致,按照,依据”。

7.【答案】J

【详解】这句话的意思是:这个规则反应了不成文的原则。

8.【答案】N

【详解】这句话的意思是:酒吧争论是一种快乐的游戏。

9.【答案】C

【详解】这句话的意思是:热烈的争论中不一定包括……。

10.【答案】G

【详解】这句话的意思是:争论本应是关于什么的。

**Section B**

11.【答案】C

【详解】本文主要记述了卢西恩达·蔡尔兹,一个舞蹈动作设计师的有关情况。A. 抽象派艺术;B. 精确的(舞蹈)形式;D. 百老汇戏剧。都是文章在叙述过程提及的与卢西恩达有关的要点,但都不是主要内容。C. A choreographer 舞蹈动作设计师。反映了本文是把卢西恩达作为一个舞蹈动作设计师来进行描述的。因此 C 项是正确答案。

12.【答案】A

【详解】见原文"The development of Childs' career, from its... beginning...""蔡尔兹艺术事业的发展,从它的开端……"it 代指 career,符合逻辑关系,因为事业是有开头的,而它的发展已经不是开始了。所以说 it 不能代指 development。其他两项很明显与 it 没有指代关系。因此 A 项为正确答案。

13.【答案】D

【详解】本题问卢西恩达·蔡尔兹的艺术作品属于何种流派。见原文"The development of Childs' Career,... paralleled the development of minimalist art"很明地把她的作品定格为抽象派艺术。因此 D 项是正确答案。

14.【答案】C

【详解】蔡尔兹十一岁时就开始学习话剧(见原文第二段)。而 A. painting(绘画)文中并无叙述。B. Dance(跳舞)与 D. Film(演电影)都是在学习话剧之后发生的。因此话剧是她最早学习的艺术。答案 C 正确。

15.【答案】B

【详解】本题问本文何处提到蔡尔兹是如何看待自己的艺术的。见原文第 6 至第 7 行:In her view, each of her dances is simply an intense experience of intense looking and listening. "In her view"明确告诉了这是她自己的看法。其他各项提到的地方都是作者或作者引用别人的说法或观点。而不是她本人的看法。故此 B 项是正确答案。

16.【答案】A

【详解】本文通过比较和对比的方式说明了气体和液体各自所具有的特征。A. 气体和液体的特征,概况了本文讨论的主要内容。可用作本文的题目。其他各项概括范围太窄,均不适合用作本文题目。

17.【答案】C

【详解】见原文第一段"The difference between a liquid and a gas is obvious under the conditions of temperature and pressure commonly found at the surface of the Earth. A liquid can... and fills it to the level of a free surface. A gas forms no free surface but..."在正常的温度和压力条件下,液体可盛在敞口容器内,并在容器内一定的高度处形成一个游离的表面。而气体却不会形成这样的表面,它总是在空间内不断地膨胀。因此答案 C 是正确选项。A. 受压力变化的影响,这是气体和液体共有的特征。B. 有耐久的组织结构;这是气体和液体都不具有的特征。D. 被认为更常见,在文中并无提及。

18.【答案】B

【详解】见原文"... it (gas) must therefore be or held by a gravitational field, as in the case of a planet's atmosphere。"气体必须……,或者被存放在一个引力场内,就像行星周围的气体一样。暗示了行星周围空气之所以不扩散也是因为受到行星引力作用的结果。因此 B 项是正确的。

19.【答案】D

【详解】见原文第一段"In the nineteenth century...; and another theory held that the two phases are made up of different kinds of molecules: liquidons and gasons。"十九世纪,……有一种理论认为这两种物质(液体和气体)是由不同种类的微粒子组成,液体分子和气体分子。这么说,当时的科学家们认为"liquidons 和 gasons 是不同的微粒"。因此 D 答案是正确的。

20.【答案】B

【详解】见原文第二段"The combination of temperature and pressure at which the densities become equal is called the critical point。"

## Unit 5

### Section A

1.【答案】C

【详解】根据前后文这里应填副词,选项里有 seriously 和 strictly 可选,比较分析后,可知 seriously 有"认真地"的意思,故选 C。

2.【答案】A

【详解】空前都为报纸各个栏目的名称,依据生活常识可知这里应为 classified advertisements 为分类广告。选 A。

3.【答案】L

【详解】此空的难度很大,因为很容易选 encourage 和 attract,但由于选项里 encourage 和 attract 都加了 s,所以不可选,trick 一词似乎是贬义,但由于前面说读者很懒,所以这里可以说哄骗。

4.【答案】H

【详解】此空干扰项有 want,但由于后文已经有了行动,所以这里用 attempt 比较合适。

5.【答案】M

【详解】根据上下文易选出 techniques。

6.【答案】N

【详解】attract attention 为固定搭配。

7.【答案】F

【详解】根据后文选出 encourages。

8.【答案】D

【详解】前文已经提到好的头版故事来使你阅读,此空则为

headlines,干扰项 subject 没有 headlines 合适。

9.【答案】O

【详解】not... anymore 为固定搭配。

10.【答案】B

【详解】干扰项 concern 是"担心,焦急"之意,care 用在这里是"关心"之意。

**Section B**

11.【答案】D

【详解】本题考查文章大意。通读全文,不难看出,作者是在对 Futurist Poetry(未来主义诗歌)做出评价,与 D 项相同。A. 过去和未来,本文中没有涉及。B. 现代社会的变化,文中仅提到了变化的一个方面"... a great change in our emotional life"(第四段),显然不能做文章的主题。C. 未来主义运动的功绩,文中并无讨论。

12.【答案】A

【详解】注意:题目中的"novel idea",意为"新思想、新观点"。答案见原文第一段"When a new movement in art attains a certain vogue, it is advisable to find out what its advocates are aiming at."

13.【答案】C

【详解】见原文第二段"This speeding up of life, says the futurist, requires a new form of expressions. We must speed up our literature too..."生活的变化要求文学表现形式变化。

14.【答案】C

【详解】作者前后两次提到这一点。第一段"...,for whatever Futurist poetry may be...—it can hardly be classed as literature"和最后一段"This though it fulfills the laws and requirement of Futurist poetry, can hardly be classed as literature."。

15.【答案】B

【详解】见原文第二段"We must pour out a cataract of essential words, unhampered by stops, or qualifying adjectives,... we must use many sizes of type and different colored inks on the same page..."从这里可以看出。D. a stream of essential words, A. imitative words 和 C. different colored inks 都属于未来主义诗人表达应用手段。而 qualifying adjectives(限定性形容词)则不可夹在"一长串基本词"(a cataract of essential words)中间。故 C 项不在未来主义诗人采用之列,为正确答案。

16.【答案】C

【详解】本文阐述的重点是试图通过分析歌德(Goethe)诗歌的艺术特点来说明他是浪漫主义诗人的代表(Romanticist)。C 项题目很好地体现本文讨论的主题,可用作本文题目。A. Goethe and Dante 歌德和但丁。文章开头提到了他们诗歌特点的不同,但文章重点不是讨论二人的不同之处。故用此作题目不合适。B. The Characteristics of Romanticism 浪漫主义的特点。文章提到了浪漫主义的一些特点,但那是为了说明歌德的诗歌特点,不是文章的主题,因此 B 项题目也不合适。D. Goethe's Abundant Life 歌德的丰裕的生活。同样是为了说明歌德的诗歌特点,不能做文章题目。

17.【答案】C

【详解】见原文第一段末尾"As a result, nearly all his works have serious flaw of structure, of inconsistency, of excess and redundancies."。

18.【答案】D

【详解】A. 对自然的兴趣。作为浪漫主义的一个特点,在第二段第 6 行提到"and the continued faith in nature"。B. 观点的现代性。作为浪漫主义的特点之一在第二段第 4 行提到"The result was an all-encompassing vision of reality... and of other romanticists"。C. 年青的精神。在文章第二段第 5 行提到"Yet the spirit of youthfulness"。D. 语言的简洁。在文中第二段作为浪漫主义的特点并没有被提及。因此 D 项是正确答案。

19.【答案】C

【详解】见原文第二段"Since so many twentieth-century thoughts and attitudes are similarly based on the stimulus of romantic movement. Goethe stands, as particularly the poet of the modern man..."因为二十世纪的思想和态度是建立在(以歌德为代表的)浪漫主义的促进基础之上,所以歌德被看做是(浪漫主义诗人)的代表。

20.【答案】A

【详解】原文第二段中末尾"Since so many twentieth century thoughts and attitudes are similarly based on the stimulus of Romantic Movement."说明很多现代的思想观点都是在以歌德为代表的浪漫主义运动的刺激下产生的。也就是说,在某种程度上说是歌德促进了这些现代思想的产生。A 答案表达了这一含义,为正确答案。B. 歌德不喜欢但丁和维吉尔,文中没有提到。C. 歌德应被称作古典主义者。与文章的观点相反。D. 歌德(的诗歌)缺乏逻辑性。文章并没有提到这一点。

# Unit 6

**Section A**

1.【答案】D

【详解】interact with 意为"相互作用"。

2.【答案】I

【详解】be described as 意为"被描述为"。

3.【答案】G

【详解】市场上通过以物易物或交换金钱方式进行的应是交易。

4.【答案】K

【详解】找到一个想用帆船换取我旧车的人很显然不总是件容易事。

5.【答案】L

【详解】in exchange for 意为“交换”。

6.【答案】E

【详解】引入钱作为交换的中介使交易简单得多。

7.【答案】N

【详解】政府所颁布的当然是法令。

8.【答案】O

【详解】allocate to 是“分配”的意思。

9.【答案】B

【详解】belonging to 是现在分词作定语修饰 people。

10.【答案】A

【详解】前面一句话说取得进步,也许会很困难,那么社会也可能停滞不前。

**Section B**

11.【答案】D

【详解】在第一段中明确写道:保守估计,在听说读写各方面可能会遇到的英语单词超过一百万个。从这一句我们可以看出D项“一百多万个”是正确的。A项不正确,而B项是指高中毕业生的词汇量,C项是指一年级学生的词汇量。

12.【答案】A

【详解】对这一题我们可以通过上下文的意思来判断,文中写道 obsolete words 是用来帮助我们阅读古代文学的,由此我们可推断出,obsolete words 是用于古代文学中的,即老式词,所以A项“不再使用的”与原文相近,B,C,D都不正确。

13.【答案】D

【详解】在第二段末尾,明确提到 recognition vocabulary(认识词汇)是用于读或听的词汇,active vocabulary(会用的词汇)是说和写时用的词汇,后者比前者小得多,即会用的词汇小于认识的词汇,这样不难看出D项是正确的,而A项“比会用的词汇用得少”,这在原文中未曾提及,B,C两项与原文意思不符。

14.【答案】B

【详解】这一题要求我们对原文充分理解后,结合排除法来做。文中第三段讨论了这个问题,我们可以在其中发现A项:把你需要用的词列成一张单子,然后在字典中查出生词的涵义,C项:有意识地多用你在阅读中认识的新词,D项:尝试使用你认识的词。在排除以上三项之后只有B项,每天读半小时的字典符合题意。

15.【答案】C

【详解】这一题要求对全文通篇有所理解后才能决定出答案,并且这种题无法从文中找到一个可用来判断的确切依据,须用排除法结合大意来做。

A项认为“字典完全包括我们所用的词汇”,而原文中提到“字典并不企图完全包含我们所用的词汇”,因此A是不正确的。

B项认为“schoolroom 在本文中用于说明特定术语”,而文中提到 schoolroom 时是作为复合词的例证,因此B是不正确的。

C项认为“一旦你知道一个词如何发音且代表什么,你就掌握了这一单词”,这与原文中的“一旦你知道一个词如何发音且代表着什么,你就能安全地运用它”的意思相符,所以C项是正确的。

D项认为“会用的词汇指我们在理解阅读材料时所遇到的词汇和所听到的词汇”,而原文写道“会用的词汇是我们在说和写时所用的词汇。”这显然与原文不符,所以D是不正确的。

16.【答案】C

【详解】这一题主要是考查读者对第一段的理解。文中一开始就陈述了以往人们的错误观点:认为抑郁症在穷人和教育程度不高的人中更为普遍。并且强调事实上各行各业的人都有可能患该病,尤其是那些雄心勃勃,有创造力,尽责之人更易受此病的折磨。同时,文中还强调行政官员和专业人士由于承受的压力过大也会患此病。所以,A项、B项以及D项都与原文不符。答案应为C项。

17.【答案】A

【详解】文中第二段指出,一项研究结果表明,本世纪后期的人比早期的人更易患抑郁症。并举出后两代人患此病的比率增加了10倍的例子来证明。所以我们可判断最佳答案为A项。而B项认为“抑郁症在上辈们中更为普遍”,这显然与原文相反。C项认为:“该病的发生率在父母辈和祖父母辈期间增加了10倍”。这与文中所说的“最近的两代人,即年轻人及他们的父母辈”不相一致。D项认为“抑郁症仅折磨年轻人。”答案片面,因为父母辈也属于受该病折磨的一代人。所以D项也不正确。

18.【答案】D

【详解】这一题主要考查读者对整体的理解。且需用排除法来选择答案。A项认为“极高的期待使人易患抑郁症”。文中第二段曾提到年青一代易患此病是由于他们对生活有太高的期待从而容易失败进而患上抑郁症,所以A项符合原文,应排除。B项认为“一个患抑郁症的人常感到悲伤、沮丧”这在文中第三段第二句也提到,所以B项符合原文应排除。C项认为“人们无需隐瞒他们患抑郁症的事实,因为这并非一件羞耻的事。”这在文中第四段首句中也明确提到,C项也应排除。D项认为“抑郁症是遗传的”这与文中第三段“抑郁症并不一定是遗传的”这一提法相反。文中还强调有许多例子表明许多抑郁症患者的上辈并不曾患过此病。因而D项与原文不符,应选D。

19.【答案】B

【详解】这一题也是对细节的考查。文中提到“nip-in-the-bud case”即治疗及时的病例,患者就能在几天内康复。该短语的原意为“扼杀在萌芽状态”,也就是早期治疗的意思,所以正确答案为 B 项。

20.【答案】B

【详解】这篇文章主要是讲述什么呢?通过对全文的理解,概括,我们可推断出此文主要在介绍抑郁症这种病的症状、原因及治疗,所以正确答案是 B 项。A,C,D 都各强调了一个方面,并且不全面,所以不正确。

## Unit 7

### Section A

1.【答案】Individual motivation for work

【详解】要找一篇文章的主题,必须首先对全文有所了解。本题的答案是不可能从某一句话或某一个段落能够找得到的。这就需要有全局观念,要有概括能力,要从关键词中找答案。文章的第一段主要说明了有进取心的学生往往考分较高,这种进取心来自一种动力(motivation)。第二段主要说明那些有强烈愿望取得好成绩的人都有一种很强的 motivation,这种积极性或原动力是不需要有人监督的。第三段最后一句话高度概括了这一段落的中心意思。然后,把三段的主要内容融合在一起考虑,就不难得出这样的结论:Individual motivation for work。

2.【答案】By working well alone

【详解】要回答这个问题,需要从字行里去找答案,因为有些问题比较直观。个人如何获得好成绩?是靠个人的努力或是靠监管人员的监督?在第二段中有这样一句话:Their desire for accomplishment is a stronger motivation than any stimulation the supervisor can provide 由此可知,个人的成绩是靠自己的努力,靠动力,而不是靠别人的监管。

3.【答案】accept responsibilities for themselves or be responsible for themselves

【详解】根据短文意思,有争胜心的人往往会自己对自己负责。要回答这个问题是不能从某一句话或某一个段落中找答案的,而是必须对整篇文章进行概括和总结,并根据问题的提出进行推理。前面已经讲过,学习好要有动力,取得成功要有动力,要靠本人,而不是靠别人的监管和督促,从而也就得出结论:成功要靠自己(Be responsible for themselves)。

4.【答案】High achievement needs

【详解】招聘人员对大学高年级学生训练成管理人才的素质要求是:要有很强的事业心和进取精神。这可以从短文的第三段中看得出来,如:Thoughts concerning the achievement drive are often prominent in the evaluations made by the typical employment interviewer who interviews college seniors for excutive training. 这段文字的关键词语是:achievement drive(成就动力),也就是 achievement needs。

5.【答案】They are afraid of failing

【详解】What motivates some seniors to succeed? 要回答这个问题,需要仔细分析文章的最后一句话:Research indicates that some who do get ahead have an even stronger drive to avoid failure. 研究说明有些的确能够进取的人甚至有更强地避免失败的动力。这种避免失败的动力,实际上就是害怕失败。

### Section B

6.【答案】B

【详解】根据上下文来推断词义是遇到不熟悉词汇时的常用方法,另外结合平时接触的日常知识也是常用方法。一般对于很符合潮流,赶时髦的人被看做是 in the fashion,落伍或老土的人被形容成 out of fashion,另外从这句话中的 for their generation 也可猜出一点。这里的 their generation 当然是指年青的一代,而 in-thing 自然是年轻人崇尚的合乎潮流和时尚的事。所以,可以猜到答案是 B 项“时尚”,而 A 项“好奇”这在上文中已讨论完毕了。C 项“需要”显然说不通。D 项“压力”是下文将要讨论的问题,在此处还未出现。因此,通过排除法也可得出 B 项为最佳答案。

7.【答案】D

【详解】文章第一段很详细地解释了年轻人为什么会服食软性毒品,原因有三:好奇,压力和周围的环境。用排除法解答,我们可以在文中找到 A 项:他们认为软性毒品没什么伤害,B 项:他们对此感到好奇,C 项:他们碰到一些烦恼的事。而 D 项说由于他们的父母是吸毒者文中并未提及,由此得出结论。

8.【答案】D

【详解】文中第二段第一句就提到了社会工作者和心理学家们普遍认为解决年轻人吸毒成瘾的最好办法是校方、社会工作者,警方戒毒所一起做工作为年轻人提供所需的教育以抵制吸毒带来的影响和危险。因此最适合的答案应为 D 项。A 项过于强调戒毒所的单独作用。B 项过于强调了父母的作用,C 项更加与原文相反。

9.【答案】D

【详解】文中第一段末句提到属于一个团体的年轻人,若其他年轻人吸毒,他也会跟随一起吸毒以免被排斥。所以不难看出最佳答案为 D 项。而 A 项所认为的年轻人吸毒是适应趋势与文中怕被排斥的原因不符。

10.【答案】C

【详解】文中第二段花大篇幅强调了幸福的,充满爱的家庭对阻止年轻人吸毒的作用。文章最后一句还写道一个幸福的家庭就是一个无毒品的家庭这句话毫不夸张。所以最佳答案应为 C 项。

11.【答案】C

【详解】这一题要求我们抓住中心,不要混淆细节和中心的关系,第二段从头至尾都在说明各种食品的含水量,包括蔬菜和肉类,并未提及食物去水的方法或过程,所以很明显答案应为C项"食物的含水量"。

A项强调瘦肉和鱼的对比,但文中只叙述了瘦肉和鱼的含水量,并未就两者进行对比,所以A项不正确。B项强调食物水分和去除,从全文来看的确主要讲述的是水分的去除,但第二段却未提到去除食物水分,仅描述了各种食物的含水量,所以B项也不正确,D项强调的是水和食物的关系,此段只说明了含水量,也未就水和食物的关系展开论述,所以D是错误的。

12.【答案】C

【详解】这一题要求我们对细节部分有仔细的理解,由此题的题干我们无法推出考查点是什么,但仔细看完四个选项后,可以看出,这一题主要考查我们对第二段第一句话的理解,文中描述了鱼的含水量取决于它的脂肪含量,那么是脂肪量多的鱼含水量多还是脂肪量少的含水量多呢?细心的读者会发现文中描述的鱼的含水量视其脂肪量多少,含水量由80%向60%递减不等,即脂肪越多,含水量越靠近60%这一下限,因此,答案应是C项。

13.【答案】C

【详解】这一题考查对词汇的推测,在这里我们不仅可根据词根来推,还可由上下文来推,我们知道convention有"常规,惯例"的意思,那么其形容词conventional自然有一般,通常的意思,所以不难看出,此题的答案应是C项,另外,我们也可从下文中看出其意思,下一句话提到这是一种usual method,即conventional被usual所代替,据此,我们也可确定答案为C,即惯常且为人们所接受的。而A中的traditional是传统的意思,这与通常这一意思还有差距,B和D项的意思与原文相去更远。

14.【答案】C

【详解】这一题要求我们对全文有了整体理解后才能判断,且要注意选不正确的一项,A项认为"去除食物中的水有助于防止其腐坏"。由文中第二段的末句"若去除了水分,使食物变质的细菌的活动就被控制了"。就可确定A项与原文相符,所以不应选A。B项认为"用敞露的方法来干燥食物已有几百年的历史了"。文中第一段就写明了"几百年前人们就发现去除食物中的水分有助于储藏,并且最简单的方法就是将食物暴露于阳光或风中"。由此也可得出B项是与原文相符的,所以不应选B项,C项认为"在脱水过程中,热流的温度从进口到出口逐渐上升"。而文中写道热流的温度在进口处为110℃,而出口处为43℃,这说明温度是下降的,显然C项与原文不符,应选C项。

D项认为"干燥液体中的水分比干燥固体中的水分的过程复杂得多"。文中干燥固体食物如蔬菜,肉鱼等的过程很简单,只要将食物经由高温到低温的热流吹过即可,但液体食物却要经过好几步,很明显后者比前者复杂得多,所以D项也与全文相符,不应选D项。

15.【答案】B

【详解】抓住最后一段的中心,也就能做出最后一题,最后一段主要讨论干制食品的好处和方便之处。B项显然符合这一中心。A项只是文中略提到的一小点,而非中心。C项更与最后一段无紧要关系,文中只提到干制食品易于储存,并未提及储存方法。D项将听装和冷冻食品也包括在内,而文中只说明了干制食品较这两者的优点。所以应选B项。

## Unit 8

### Section A

1.【答案】They travel after being regulated in convoys

【详解】此题可在第一段中找到答案。即"regulating vehicles in convoys on motorways"。

2.【答案】Paying tolls, and coupling the car into an electronic convoy

【详解】此题可在第一段中找到答案。即"You would just pay a toll, couple your car into an electronic convoy and sit back to enjoy the journey."。

3.【答案】Motorways will gradually become more like railways

【详解】段落大意通常会出现在段首或段尾,有时也需要概括。第二段一开始就给出了该段大意,后面都是具体陈述高速公路如何更像铁路。

4.【答案】It is inflexible and doesn't go where you want

【详解】此题可在第三段中找到答案。即"Any rail system has in the end to be inflexible, it doesn't go where you want, especially in rural communities"。

5.【答案】Their flexibility, bringing people pride and convenient road network

【详解】此题可在第三段中找到答案。第三段中有"We're wedded to private cars, because of their flexibility and the pride people take in ownership—not to mention the huge sums we've spent on the road network."同时,第四段用了一个承上启下的句子"So cars aren't going to go away"。从这两处可以得出答案。

### Section B

6.【答案】D

【详解】见原文第一段说"如果(一个国家)人口太多"。"Thus, each person produces less and this means a lower average income than could be obtained with a small population."这样一来,每个人的生产率下降,就意味着平均收入要比人口少的国家要低。反之,也就是说,如果一个国家人口数量不大,那么个人的生产率就会很高,这样一来平均收入就会高。D项反映了这一观点,是正确答案。

7.【答案】B

【详解】见原文第一段“Other economists have argued that a large population gives more scope for specialization and the development of facilities such as ports, roads and railways, which are not likely to be built unless there is a big demand to justify them”人口多可以给许多发展提供机会,比如说港口,公路和铁路,这些设施的发展是建立在巨大的需求基础之上的。

8.【答案】B

【详解】见原文第二段中间“A decreasing birth rate may lend to a unemployment because it results in a declining market for manufactured goods.”人口出生率的下降会导致失业,因为人口减少,需求下降,会导致市场萎缩。因此B是正确答案。

9.【答案】B

【详解】见原文第二段末尾处“... the government of a developed country may well prefer to see a slowly increasing population...”。

10.【答案】C

【详解】见原文第二段开头“One of the difficulties in carrying out a world wide birth control program lies in the fact that official attitudes to population growth vary from country to country depending on the level of industrial development and the availability of food and raw materials”在全世界推行计划生育会遇到的一个困难是“国与国之间对人口增长的态度不一致,每个国家的态度取决于本国食品和原材料的供应能力”。C项准确说明了原因,是正确答案。

11.【答案】B

【详解】B. 内燃机作为农业主要动力的应用产生了巨大的影响。从历史知识可以知道:机械化运动始于对内燃机的应用。正是在内燃机的推动下,人们才发明了 threshing machine, hay rakes, hay loaders 等等这些农业机械,使它们成了农场生产的主要动力。说明了内燃机的应用对农业生产产生了巨大的影响。因此B项是正确答案。A. 农业生产引进机械在农民中间激起反对意见。不是农民反对,而是工人反对,见第一段。C. 1850年后农业生产的机械化使许多农民失去了工作,文中没有交代。D. 在19世纪60年代为了提高产量而对农业机械进行了精巧的改进。是作者在文中所举的一个实例(第一段),并非作者要表达的观点。

12.【答案】C

【详解】联系下文,作者列举事例说明了机器的使用迅速提高了农业生产的效率。因此,机器的应用不仅改变了农业生产的性质,而且极大促进了农业生产。据此我们可以推断 Skyrocketed 意思应是“迅速提高。”C项与该意思相同,因此是正确答案。

13.【答案】C

【详解】见原文第四段末尾“Some mechanization has reached the level of plantation agriculture in parts of the tropics, but even today much of that land is laborously worked by people leading draft animals pulling primitive plows.”在热带的某些地区也已使用机械生产,但是直到今天,热带的很多地方还是由人驱使动物拉犁进行耕种,也就是人类沿用了几千年的老办法。因此,C项是正确答案。

14.【答案】A

【详解】见原文第五段。“The problems of mechanization some areas are not only cultural in nature. For example, tropical soils and crops differ markedly from those in temperate areas that the machines are designed for...”机械化的问题在某些地区并不仅仅是文化上的差异。例如,热带的土壤和农作物和温带有明显的不同,而这些机器大都是为温带的土壤和农作物而进行设计的,说明它不仅仅是一个文化原因,还有一些经济方面的因素在内。A项说明了这一观点,是正确答案。

15.【答案】C

【详解】见原文最后一句“Introducing mechanization into such areas requires careful planing”此处的 such areas 指的是上一句话的“undeveloped countries”。

## Unit 9

**Section A**

1.【答案】By depositing charge slips in a bank

【详解】此题可在第一段中找到答案。即“by depositing charge slips in a bank or other financial institutions”。

2.【答案】Look for the card with the lowest interest rate

【详解】此题可在第二段中找到答案。即“If you will be one of the growing number of people who don't pay off their credit card transactions in full each month, look for the card with the lowest interest rate.”。

3.【答案】They can withhold payment

【详解】此题可在第三段中找到答案。即“if there is a problem with the products or service you purchase with your credit card, you have an opportunity to withhold payment by asking the credit card company to charge back to the retailer until the dispute is settled.”。

4.【答案】With finance charges and fees increasing, they go into debt

【详解】此题可在第四段中找到答案。即“In reality, your finance charges and fees only increase, and you go deeper into debt”。

5.【答案】What people should do when finding themselves in trouble

【详解】该段一开始说“If you do find yourself in trouble, do not ignore the bills”,那么,后面接下来就谈的是如何解决这些麻烦。

**Section B**

6.【答案】C

【详解】文章的题目应该是文章主题的反映。本文作者旨在告诉人们如何避免流沙的危险。C 项正确反映了这一主题，是正确答案。A. 流沙的方方面面，本文只重点讨论了流沙对人的危害。用此项做题目，涵义过宽。B. 怎样探测流沙，这只是如何躲避流沙危险的一种方法，用此做题目太片面。D. 在什么地方可以找到流沙，不是本文讨论的主题。

7.【答案】D

【详解】本文的主题已说明即如何躲避流沙的危险。A. 今天极少数人碰到过流沙；B. 流沙在很多地方还很普遍；C. 单凭肉眼无法识别流沙，均不能反映主题。故 D 是正确答案。

8.【答案】D

【详解】见原文第二段开头：Quicksand is usually found along the shores and in the beds of rivers. 流沙经常在河岸和河床地带发现，因此 D 答案正确。

9.【答案】C

【详解】见原文第三段"For test probing a pole or long stick should be used."应当用一根长杆或长木棍来进行探测，故 B 项是正确答案。其他各项都不是作者推荐的好办法。

10.【答案】A

【详解】从文中最后两段的描述中，可以看出，一个人遇到流沙时，不能惊慌要冷静地采取一系列措施，才能使自己脱险。不可激动，强壮和大胆都是无济于事的。因此 A 项是正确答案。

11.【答案】C

【详解】见原文第一段"... a King, who was so cruel and unjust towards his subjects that he was always called the Tyrant. So heartless was he that his people used to pray night and day that they might have a new king."可以看出，人们想换个新国王是因为现在国王对他们太残忍不公。因此 C 项是正确答案。

12.【答案】D

【详解】见原文第一段："My dear subjects, the days of my tyranny are over. Henceforth you shall live in peace and happiness, ..."从国王的话里看出，他想对自己的百姓好些。故 D 项是正确答案。

13.【答案】B

【详解】从文中第二段的描述，可以看出，国王履行了自己的诺言，而得到臣民的敬仰。因此 B 项是正确答案。

14.【答案】A

【详解】答案在原文第三段里。"As I was galloping through my forests one afternoon, I caught sight of a hound chasing fox."

15.【答案】C

【详解】见原文最后一句"I said to myself, he who does evil will sooner or later be overtaken by evil"。国王自己心里想，"作恶者迟早总会有恶报。"C 项正确反映了这一意思，是正确答案。

# Unit 10

## Section A

1.【答案】nothing important

【详解】文中对于伦理学家的态度有如下描述：But scientists assailed the moralists concerns as alarmist. 他们认为伦理学家的担忧和警告只是大惊小怪。因此可以得出结论，科学家认为伦理学家的警告是无关紧要。

2.【答案】be put into an embarrassing situation

【详解】试想一个人光着身子被人抓住是如何一种情景？当然是很尴尬的。再加上文中说明 because cloning human beings would serve no discernible scientific purpose. Now the cloning of human is within reach 更加可以确定这是一个很尴尬的场面。

3.【答案】US government highly concerns with cloning of humans

【详解】第二段阐述了克隆人类在伦理学上是否合理已经成为美国政府讨论的一个重要议题。克林顿总统也成立了一个由专家组成的咨询小组，专门讨论这个问题。因此，这也是本段的中心。

4.【答案】researchers

【详解】文中说到：The government could prohibit the cloning of human beings or issue regulations limiting what researchers can do. 很容易得出答案。

5.【答案】humans should not be cloned

【详解】根据文中 McCormick 的态度：any cloning of human is morally repugnant. 任何克隆人类的行为都是令人厌恶的，可以得出本题的答案。

## Section B

6.【答案】C

【详解】阅读本文，可以看出，文章的主题是探讨研究气象学的方法和途径。C. 气象科学的研究方法，能准确表达文章的主题，可以做本文题目，是正确答案。A. 气象预报的局限性，不是本文讨论的内容。B. 天气气象学的新进展；D. 动力气象学的基础。均不是文章重点探讨的内容。

7.【答案】B

【详解】见原文第三段开头：Synoptic meteorology is the scientific basis of the technique of weather forecasting by means of the preparation and analysis of weather maps...。天气气象学是天气预报技术的科学基础，它是靠对天气图……的绘制和分析来进行研究的。这说明天气气象学家的预测是建立在对天气图的分析的基础上的。B 答案与此意同，是正确答案。

8.【答案】B

【详解】B. 对人类生命的保护更大。从文中第二段的论述可知：精确的天气预报可以挽救人们的生命和财产。因此，其精确度越高,就可以更好地保护人的生命和财产安全。B 答案与文义相同,是正确答案。A.(可以给)气象学家研究带来更多的资金;C. 更大数量的专业预报人员;D. 更加专业化的天气气象学。此三项内容文中均未提及或论述。

9.【答案】A

【详解】第三段开头解释了天气气象学的定义,然后说明了气象学的研究目的和作用,强调了气象学的重要性。因此,本段的结构安排应是解释了一种科学理论并强调了其重要性。A 答案准确描述了这一结构,是正确答案。

10.【答案】C

【详解】从文章中第四段可直接找到本题答案。"The tools. . . are disciplines of mathematics and physics. . ."

11.【答案】A

【详解】见原文第二段前半部分:. . . ,people who rejected the railway produced well-known doctors who said that tunnels would be most dangerous to public health; they would produce colds and some other diseases. 反对铁路的人会搬出一些有名望的医生,这些医生说隧道对人体的健康危害最大,可以导致感冒和其他疾病。A 项符合文中意思表达,是正确答案。B. 隧道很冷和导致病并无直接关系,况且文中也并未交代。C. 找些在隧道里得过感冒的医生,与原文意思不符。D. 告诉人们他们自己在隧道里得过感冒。文中并无交代。

12.【答案】B

【详解】前面(文中第二段开头)说反对铁路的人找了些有名望的医生。(well-known doctors)。那么按照常理推测,那些支持铁路的人理所当然也会请些有名望的医生,因为这样才有说服力。因此,可以推测 eminent 的意思应是"著名的、有名气的"。B. outstanding 杰出的,与 eminent 意思最近,因此 B 是正确选项。A. optimistic 乐观的;C. desperate 绝望的;D. ordinary 普通的。均不能正确表达该词在此处的意思。

13.【答案】B

【详解】见原文第二段后半部分:More than a rapid and comfortable means of transport, they actually saw the railway as a factor in world peace. 他们(欢迎铁路的人)不仅仅把铁路看做一种迅捷、舒适的交通方式,而且认为它是世界和平的一个因素。这说明人们欢迎铁路,并不仅仅是因为它是一种更快的旅游方式。B 答案与这一观点相同,符合题意是正确答案。A. 是改变(地点)的便捷方式,仅是原因之一,以偏概全。C. 他们认识到很长时间里(火车)不会更快更舒服。与文中意思相反。D. 他们认为使军队的移动速度加快。这恰恰不是人们所期望的。

14.【答案】C

【详解】见原文末尾处: None of them expected that the more we are together, the more chances there are of war. Any boy or girl who is one of a large family knows that. 此处的 that 就是指前边的"the more we are together, the more chances there are of war"(人聚得越多,争斗的机会就越大)。家庭的孩子们对战争了解甚少,但对自己周围却了解很多,他们知道一个家庭里如果孩子太多,打架的机会就越大。因此 C 项是正确答案。

15.【答案】C

【详解】C. 战争是铁路发展的必然结果。无论从历史还是现实中看,这个观点都是不能成立的。文中作者也未表达。文中说:They did not expect that the railway would be just one movement for the rapid movement of hostile armies. None of them expected that the more we are together, the more chances there are of war. 他们(欢迎铁路的人)不希望铁路会成为敌对双方的军队迅速移动的又一种方式。不希望人聚得越多,打仗的机会就越大。C 项说法不正确,符合题目要求,是正确答案。A. 作者对铁路的态度是否定的,文中作者花了大量的笔墨来陈述铁路对人类带来的坏处,特别是在最后两句里,作者暗示铁路的发展使战争爆发的可能性更大,由此可以看出,作者对铁路的态度是否定的。B. 每一项新发明都会引发赞同和不同意见。这种说法符合文义(见原文第一句话)。D. 那些支持铁路的人说乘火车旅游可以睡得很好。这种说法也与文章内容相符。见原文第二段后半部分:They said that the speed and the swing of the train would. . . insure good sleep.(支持铁路的人说)火车的高速和摇摆会保证有一个好的睡眠(条件)。

# Unit 11

## Section A

1.【答案】A

【详解】这句话的意思是:虽然我们大多数的人不愿批评别人。apply to 意为"应用"。

2.【答案】B

【详解】这句话的意思是:我们无论如何也不愿赞美他人。reluctant 意为"不愿做,勉强的"。

3.【答案】C

【详解】这句话的意思是:我们很少有人懂得如何体面地接受赞美。gracefully 意为"体面地"。

4.【答案】D

【详解】这句话的意思是:相反,我们局促不安。embarrass 意为"使窘迫",是动词。

5.【答案】E

【详解】这句话的意思是:赞美尤其受那些做日常工作的人的欢迎。routine 意为"例行公事,日常事务"。

6.【答案】F

【详解】这句话的意思是:任何人都倾向重复做能带来快乐的事情。tend 意为"趋向,往往是"。

7.【答案】G

【详解】这句话的意思是：一组始终如一地受到赞扬。consistently 意为“始终如一地，一贯地”。

8.【答案】J

【详解】这句话的意思是：毫不吃惊的是，那些受到表扬的提高极大。surprisingly 意为“令人吃惊地”。

9.【答案】L

【详解】这句话的意思是：花五分钟时间写一封感谢信。花去时间用 spent。

10.【答案】O

【详解】这句话的意思是：让我们关注周围的平常之美。alert 意为“提防的，警惕的”。

**Section B**

11.【答案】C

【详解】有关本题的答案在原文第一段中有暗示：“As the total cost of study and living may be \$2,000 to \$3,000 a year these earnings are useful and often essential.”当每年的生活和学习费用达到 2000 至 3000 美元时，这些收入（学生通过打工挣来的钱）就是有益的而且经常是必要的。这说明学生在上学期间挣钱的原因是学校的费用较高。C 项说明了这一原因，是正确答案。A. 他们必须学会怎样挣钱；B. 他们被要求这么做；D. 他们的父母不答应他们付学费。这些在文中均无提及或暗示。

12.【答案】B

【详解】第一段结尾处明确地给出了本题答案“One popular occupation is that of porter at a supermarket, carrying housewives' groceries out to their cars.”

13.【答案】A

【详解】A. 美国政府非常重视教育。文章第二段详细陈述了美国联邦政府为了提高学生的经济地位，资助学生较好地完成学业而颁布的一些法令和给予学生贷款偿还一系列的优惠条件。充分说明了美国政府对教育的重视。因此 A 项是正确答案。B. 美国政府只重视高等教育；C. 获得高分的学生不交学费；均从文中找不到暗示。D. 确实很穷的学生不必还贷款。文中说：“Those who teach in depressed areas are specially favored and each year of depressed-area teaching wipes out fifteen per cent of the loan received.”在贫困地区教书的学生在偿还贷款方面享有特殊优惠。而不是说穷学生不用还贷款。因此 D 项内容与文中所述不符。

14.【答案】C

【详解】文中说（结尾处）：“... and each year of depressed-area teaching wipes out fifteen per cent of the loan received.”在贫困地区从事教学一年就可以免去学生接受贷款的 15%。如此推算，大约 7 年时间即可全部免去贷款额。因此 C 项是正确选项。

15.【答案】D

【详解】D. 在公立学校和贫困地区教书受到政府鼓励。在文中第二段后半部分，作者说明在公立学校教学只需偿还贷款的一部分。而在贫困地区教学的条件则更优惠。这说明，政府意在鼓励学生到此地方教学。因此 D 答案与文中内容相符，符合题意是正确答案。A. 美国大学里有些学生被迫去挣钱。从文中可以看出，学生打工多是自愿的，而不是受到强制去做的。B. 只有非技术性的工作才适合学生去做。文中说“Mostly students do rather unskilled work”。大多数情况下学生宁愿去做非技术性的工作。但并不是说学生只能做非技术性的工作。C. 政府贷款是学生接受教育的最后但是非常重要的途径。文中并没有说学生向政府贷款上学是在走投无路的情况下才做出的选择。

16.【答案】D

【详解】本文开头就指出了文章讨论的中心主题：愤怒是最难通过面部表情判断的情绪之一。文章的其余部分则详细介绍了由达拉斯·E·布兹比针对这个问题所做的实验的情况及其所得出的一些结果。可以看出，整段文章都是围绕“从表情判断愤怒”的话题展开的。D 项表达了这一观点，是正确答案。A. 愤怒和其他情绪之间的关系，不是文章讨论内容。B. 达拉斯·E·布兹比的发现。涵义过于广泛，因为本段只谈到他的一个实验，而并没有涉及其他的发现。C. 男女之间对于情绪（判断）的不同，仅仅是文章讨论中的一个细节，不是文章讨论的中心话题。

17.【答案】A

【详解】本文的主题是：愤怒是一种难于通过表情来判断的情绪。A 答案说明了文章主题，符合题意，是正确答案。

18.【答案】B

【详解】见原文后半部分：“Paradoxically, they found that psychological training does not sharpen one's ability to judge a man's emotions by his expressions but appears actually to hinder it.”近乎荒谬的是，他们发现心理学的训练并不能强化一个人通过表情判断情绪的能力，事实上，却对这种能力起到了阻碍作用。由此可见，接受心理学训练的人在这方面反而不如一般人。B 项反应了这一观点，符合题意，是正确答案。

19.【答案】B

【详解】文中说：“The investigation found further that women are better at detecting anger from facial expression than men are.”研究进一步发现在通过表情判断愤怒的情绪方面，女人比男人做得更好。而下文接着说一个接受心理学培训的人在情绪方面反而不如一个一般人。由此可以推断，要想获得最准确的判断，宁可去问一个目不识丁的女人，也不要去问一个心理学方面的专家。B 项与此意相同，符合题意，是正确答案。

20.【答案】C

【详解】见原文最后一句话：“In the university tests, the more courses the subject had taken in psychology, the poorer judgement scores he turned in.”大学测验表明：受实验者心理学的

知识越多,他在判断(通过表情判断情绪)方面的成绩就越差。根据上下文意思,可以推出,此处的 subject(受实验者)指的是一个接受通过表情判断情绪的实验的人。因此 C 项是正确的。

# Unit 12

## Section A

1.【答案】A little over four miles

【详解】短文第一句中"We' re more than halfway now"指他们已经走了一大半的路程;"It' s only two miles farther to the inn"指还有 2 里路就到了。由上可知从火车站到小客栈至少有 4 里,我们可用"over 4 miles"表达。

2.【答案】A girl

【详解】在通读全文后,我们可知,文中只有两个主人翁,一个是 girl,另一个是 boy,弄清文中多次用到"he","his"指代的是谁,就可知题目答案。从文中第三段第一句中"'You' ll have a cold drive going back,' he said anxiously"可知 You 指代 the driver,那么显然 the driver is a girl。

3.【答案】the East

【详解】本题问的是通过这篇文章,我们知道故事发生在什么地方。在查找文中细节后不难从原文中"He meant to say more but the east wind blew clear down..."推知他们现在是 in the east。

4.【答案】Dry and cold

【详解】通读整篇文章后,我们可发现有很多描写天气的词句,如"... the east wind blew clear down a man' s throat...","... at twenty below zero...",由此可总结那里的天气是 dry and cold。

5.【答案】At the inn

【详解】文中多次提到 the inn,我们可推知这个小客栈和文中两主人翁有一定联系,从第一句可知 the driver 对客栈比较熟悉,而最后一句中女孩说"I' m on my way home now."可推测 the driver 就住在 the inn。

## Section B

6.【答案】D

【详解】文章开头第一句话:One phase of the business cycle is the expansion phase. 扩展阶级是商业循环周期中的一个阶段。整段文章讨论了在这个阶段工商业的迅速扩展和膨胀。However, a time comes when this phase reaches a peak and stops spiraling upwards. This is the ending of the expansion phase. 当这阶段发展到一定的极点时,向上发展的势头就完全停止了,这就是扩展阶段的结束。根据逻辑推理,事物的发展在到达一个高潮之后就开始慢慢回落,商业也是如此。因此,可以推断,下一段内容就很自然地过渡到商业周期中的另一个阶段,衰退阶段。D 项与此意相同,是正确答案。

7.【答案】D

【详解】本短文讨论了商业循环周期中的一个阶段——扩展阶级。在这个阶段里,工商业迅速发展,社会经济和生活水平不断提高。因此,可以说,这个阶段是社会发展的大好时光。D. The Period of Good Times. 体现了这一意思,是正确答案。A. The Business Cycle 商业循环周期。用于此题目过大,因为本文只讨论了其中一个阶段。B. The Recovery Stage 恢复阶段,用于此题目过于片面。因为按文中所述,它只是扩展阶段中的一个分阶段。C. An Expanding Society 扩展的社会。不符合文章表达的内容,因为文章中只涉及社会生活的一个方面工商业。

8.【答案】A

【详解】见原文:As one part of the economy develops, other parts are affected. 当经济的一个环节发展时,其他环节就会受其影响。作者还以汽车工业的发展为例。汽车工业的发展导致了玻璃钢和橡胶工业的发展,同时带动了公路建设的发展等等。这说明了一种行业的繁荣发展,可以从其他许多行业中得到反映。A 项与此说法相同,符合题意,是正确答案。B. 会呈螺旋式的向上发展。与文义不符。文中已指出,当繁荣达到一定的极点时,它就会停止向上发展。C. 会对钢铁工业产生影响,用于此处不知所云。因为不是所有的行业繁荣都对会钢铁工业产生影响。D. will end abruptly 会突然停止。与文中意思不符。原文中说(末尾处):This prosperity period may continue to rise... 这种繁荣的时期还会继续发展。

9.【答案】B

【详解】本题题意是下列各项中哪种工业能较好地体现扩展阶段。A. 玩具;B. 机械工具;C. 粮食;D. 农业。不难看出玩具、粮食、农业的发展都离不开机械工具。实际上,几乎所有的生产企业都与机器工具分不开。因此机械工具行业的发展更能较好地显示扩展阶段。由此,B 项符合题意,是正确答案。

10.【答案】B

【详解】见原文第 3 ~ 4 行:There is an ever increasing optimism about the future of economic growth. 人们对未来的经济增长持一种将不断提高的乐观态度。这说明人们对未来充满信心。因此 B 项是正确选项。

11.【答案】C

【详解】见原文第一句话:The Ordinance of 1784 is most significant historically because it embodied the principle that new states should be formed from the western region and admitted to the Union on an equal basis with the original commonwealths. 1784 的法令具有最为重要的历史意义,因为它体现了这样一个原则:新的州应当在西部领土建立,而且应当在平等的基础上加入最初的联邦。也就是说新成立的州和最初成立的州享有同等的权力。C. 项答案体现了这一观点,符合题意,

是正确答案。A. 新加入联邦的州和最初的几个州的数量相同,文中并无提及。B. 联邦应把西部地区建成附属州。明显与文义不符。D. 广大的西部地区应当分成十二个州,原文并未提及。

12.【答案】B

【详解】见原文第2~3行:This principle,... was generally accepted by this time (1784)这个原则(即上文提到过的新州应当在平等的基础上加入最初建立的联邦)在这一时间(1784)得到了广泛的认可。B. 答案与此意相同,是正确答案。

13.【答案】D

【详解】原文最后一句话:What he dreamed of was an expanding union of self-governing commonwealths, joined as a group of peers. 他(杰斐逊)的梦想是建立一个由各州平等加盟的广泛的自治联盟。由此可见杰斐逊强调"自治"和各州的"平等地位"。是为了确保新成立的州的平等地位。因此D答案是符合题意的正确答案。

14.【答案】A

【详解】见原文结尾处:He had no desire to break from the British Empire simply to establish an American one—in which the newer region should be subsidiary and tributary to the old. 他(杰斐逊)不希望仅仅只是从大英帝国脱离出来后建立一个美洲自己的帝国——在这样的帝国里,新地区附属于老地区。这就暗示在英帝国内其他地方都是英国的附属地区。A项正确表达了这一意思,符合题意,是正确答案。

15.【答案】B

【详解】文章开篇指出1784的该法令具有重要的历史意义,因为它体现了州与州之间平等的原则,然后介绍了托马斯·杰斐逊对于该法的确立所做的杰出贡献。可以看出,文章的中心议题是1784年所颁布的法令所体现的原则。因此,可以推想,文章的下一段也应该围绕这个中心议题展开。B. 1784年法令的实施和文章的中心议题紧密相关,很有可能是下段将讨论的内容,符合题意。因此B项是正确答案。A. 杰斐逊在弗吉尼亚的老家;C. 英国殖民地向北美之外的扩张;D. 弗吉尼亚的经济发展。均偏离了本文的主题。

# Unit 13

## Section A

1.【答案】selling English wool to the colonists

【详解】从短文中可以知道,英国的炼铁者反对的是从美国进口生铁,以便使自己生产的生铁制成铁器运往美国,赚取利润。而铁器制造厂家则希望从美国免税进口生铁,其目的也是为了利润,虽然两者在生铁进口方面所持态度完全相反,但有一点是共同的——从生铁的进出口谋取利润,他们都不希望其他行业取代他们的地位和空间。如果把答案写成:high import taxes on American pig iron(反对美国生铁征收高额税金),这是冶铁者所反对的,而炼铁者则是赞成的。故不符题意要求。

2.【答案】low in cost and high in quality

【详解】有些问题在原文中可以直接找到,有些问题则需要从全文中总结和概括出来。本题的答案就比较直观,因为在文章的第一段中就有这样的词语:excellent quality。冶铁者为什么要买进口的生铁呢?主要是因它便宜。这样就不难发现:美国的生铁既便宜质量又好。

3.【答案】high import taxes had been placed on American pig iron

【详解】这个答案隐藏在文章的最后一句话里:The English wool industry supported the iron manufacturers, also, in the belief that the Americans would use the money received for shipments of crude iron to buy cloth made in England, thus discouraging the growth of wool manufacturing in America.(英国的毛纺工业也支持冶铁厂家,认为美国人会用从运输生铁所赚的钱来买英国的布料,从而遏制美国的毛纺业)。题目要求的是:如果对美国生铁征收高额关税,美国的毛纺业就会得到发展,这是个虚拟语气的句子,意思是实际上没有对美国的生铁进行高额征税。反过来讲,如果高额征了税,就不能够遏制住美国的毛纺业了,也就是说美国的毛纺业就会发展起来了。

4.【答案】growth of shipping between England and America

【详解】从文章内容上看,无论从美国进口生铁也好,还是出口到美国的铁器制品也好,都需要来来往往的运输,没有运输,就没有英美之关的交流和贸易。所以,这篇短文的主题不是关于英国的冶铁者的发展历史,也不是英国冶铁者同商人之间的矛盾,更不是关于是否征高额关税之争,而是英美之间的外贸往来的发展情况。

5.【答案】pay the cost of high import taxes

【详解】如果英国对从美国进口的生铁征收高税,那么美国人肯定会对从英国进口的货物,如铁器,毛料等征收高额关税,那么殖民地的居民就不得不从生铁中所获的利润去偿付进口货物的高额税金。

## Section B

6.【答案】B

【详解】从文中第三段可知:Then, in 1953, two scientists ... did a very simple experiment to find out what had happened on the Primitive Earth. 1953年,两位科学家……做了一次非常简单的试验试图找出原古地球上所发生的一切即人类和生物是怎样形成的。这说明从1953年起,人类才开始弄明白地球上的生物是怎样由原子和分子结合而产生的。因此B是正确选项。

7.【答案】A

【详解】答案从第二段开头可直接找到:..., scientists figured out that living things, including human bodies, are basically made of amino acids and nucleotide bases.

8.【答案】A

【详解】答案第三段中直接给出:..., Harold Urey and Stanley L. Miller did a very simple experiment to find out what had happened on the Primitive Earth. 紧接着作者又指出是在...4 billion years ago...。

9.【答案】D

【详解】见原文最后一句话:When Miller and Urey analyzed the liquid, they found that it contained....当米勒和尤里对液体进行分析时,他们发现它包含有……。很明显 it 指的就是从句里的 liquid。

10.【答案】C

【详解】文中第一段第二句话暗示了该题答案 Plants, fish, dinosaurs, and people are made of atoms and molecules...。

11.【答案】A

【详解】广告中的第一句话即说明了该广告的目的:Once again we require 10 excellent TOEFL teachers for our summer program. 要招聘 10 名 TOEFL 老师来实施夏季(教学)计划。接下去介绍了各种优惠待遇,这都是为吸引教师应聘的条件,所以说 A 项是正确答案。B. part -time 兼职的。广告中并未说明是要招聘兼职教师。C. permanent 永久的。文中说:Good possibility of longer term and permanent posts. 只是说有可能获得永久性的职位和待遇。因此不能说是招聘的主要目的。D. newly qualified 刚刚取得资格的。这与广告中的原意是不符的。原广告中需要的是"experienced teachers"(第二行)(富有教学经验的教师)。

12.【答案】D

【详解】答案从第 1—2 行中可以找到:... school offers a special package.... $ 1,500 and free accommodation for 200 hours teaching... 学校提供特殊的优惠条件:1500 美元,免费提供食宿,200 学时……。包含了 A,B,C 三项的内容,因此 D 答案是正确答案。

13.【答案】B

【详解】B. 夏季课程结束后可继续在学校任教,原文中第三行说:Overtime available 意在 200 个学时(从 7 月 2 日—8 月 4 日)完成之后继续上课。B 项符合文义和题目要求,是正确答案。A. 可以提前完成工作。与文中意思不符。原文中说:Shorter contracts available. 意思是:可以签为期更短的合同,并不含"可以提前结束工作。"C. 可以享受免费食宿更长一段时间,文中未交代。D. 可以随时辞掉工作。文中也未说明。

14.【答案】D

【详解】答案可以从广告的第二部分找到:Letters of application... to..., Churchill House school...。申请函……邮往 Churchill House 学校。

15.【答案】B

【详解】文中最后一行:Recognized by the British Council and a member of Arels-Felco 获 British Council 和 Arels-Felco 成员认可。从这里我们可以知道,Arels-Felco 是一个在教育界具有权威性的机构。它不大可能是一个学校,因为 Churchill House School 本身就是一个学校,它不可能被置于另一所学校的管理之下。因此 B. an educational organization(一个教育机构)是符合题意的答案。A,D 两项均与原文无关。

# Unit 14

### Section A

1.【答案】E

【详解】与 but 后所言周围有一大群人可知,前面应填一个人。

2.【答案】K

【详解】每个人都有过孤单的感觉。

3.【答案】A

【详解】通过加入俱乐部或社团等方式去解决他们孤单的问题。

4.【答案】M

【详解】没有简单的解决办法。

5.【答案】L

【详解】这里列举了几个觉得孤单的典型处境。

6.【答案】B

【详解】与后面知道该做什么相同的是充满自信。

7.【答案】O

【详解】与后面无助的同义的是 adrift(漂泊的)。

8.【答案】I

【详解】与怀疑同义的是不确定。

9.【答案】C

【详解】上句说在大城市中你很容易有除了你每个人的生活都充实,富足而忙碌的感觉。有人要去什么地方,你会认为他会去好玩又有趣的地方。

10.【答案】G

【详解】同别人要去的地方对应的是你的目的地。

### Section B

11.【答案】B

【详解】见原文第一段前半部分:A man cannot live in society without considering the interest of others as well as his own interests. 一个人在社会中生活不可能不考虑别人和他自己享有的利益。这说明一个人不能随心所欲地生活是因为别人和他享有同样的权力。B 项正确说明了这一原因,符合题意,是正确答案。A. 他对别人没兴趣,文中并无涉及;C. 他的决定总是不公平的;D. 他的决定可能会伤害别人。均与文义不符。

12.【答案】C

【详解】见原文第二段:Every man ought to behave with consideration for other men. He ought not to steal, cheat, or destroy the property of others. 每个人在行动时都应当为别人考虑,不可偷盗、欺骗或者破坏别人的财产。这说明在文明社会中,每个人都需要为自己的行为负责。C 与此意相同,符合题意,是正确答案。A. 相互诚实。只说明了其中的一个方面,不全面。B. 做一切事情都要小心谨慎,文中并无说明。D. 惩治犯罪行为,严格地说,只有司法机关才有权惩治犯罪行为。

13.【答案】A

【详解】文中说明了一个人在社会中生活为什么不能完全按照自己的意愿行事,每个人在行动时应当考虑别人的利益,应当为自己的行为负责。可以看出,作者写这篇文章的目的是为了告诉人们如何在社会中生活。A 项体现了文章的这一目的,符合题意,是正确答案。B. 举例说明法律的重要性,文中第二段谈到了法律的用处,但这不是文章重点讨论的内容,也不是文章写作的目的。C. 教人们如何阻止罪犯,不是文章讨论的主要内容。D. 劝告人们不要自做决定,文章的意思是人们在做决定时要考虑到别人的利益,而不是说人们不要做出自己的决定。

14.【答案】A

【详解】见原文第一段末尾:Too many road accidents happen through the thoughtlessness of selfish drivers. 太多的交通事故都是因自私的司机粗心大意,不为他人考虑而造成的。这说明,只要司机在开车时多为别人着想就可避免许多事故。A 项与此意相同,是正确答案。

15.【答案】C

【详解】阅读文章,可以发现,文章讨论的中心内容是一个人在行动时不要只考虑自己,而应该考虑其他人的利益,不应当做有损别人的利益的事情,从这里,我们可以得出一个结论,那就是:一个人在行动时应该严格约束自己,要对自己的行为负责。C 项与这一结论相同,符合题意,所以是正确答案。

16.【答案】C

【详解】原文第一句话直接点明了该题答案:In the past, the conception of marketing emphasized sales.

17.【答案】A

【详解】见原文:Marketing was the task of figuring out how to sell the products. Basically, selling the prodcut would be accomplished by sales promotion. In addition to sales promotion, marketing also involved the physical distribution of the product to the places where it was actually sold. 过去,销售学就是解决如何卖出产品的任务。基本上,产品的卖出,可以通过促销来完成……,另外,销售学还包括产品在实际销售地的实物分配。可以看出,传统的销售学实际上包括两个方面的主要内容,即产品的卖出和分配。因此,A 项是正确答案。

18.【答案】B

【详解】原文第一段结尾处可以找到答案:Distribution consisted of transportation, storage, and related services such as financing standardization and grading, and the related risks. 分配由运输、贮存,相关的服务如提供资金,标准化和分类以及相关风险四个方面组成,因此 B 是正确答案。

19.【答案】D

【详解】看原文第二段:It (the modern marketing) subscribes to the notion that production can be economically justified only by consumpion... This is very different from making a product and then thinking about how to sell it. 现代销售学赞同这一个观点,即产品的生产在经济上完全以消费为依据……,这与生产一种产品然后考虑把产品卖出是很不相同的(传统的销售学观念),也就是说现代销售学的目标是要达到一种生产和销售的平衡。D 项说明了这一观点,符合文义,所以是正确答案。

20.【答案】C

【详解】见原文第二段:..., the producer should consider who is going to buy the product—or what the market for the product is—before production begins. 产品制造者应当在生产某种产品之前考虑谁会买这种产品,或者这种产品的市场怎么样。也就是说,产品制造者应首先考虑消费的需求,然后再进行生产。C 答案与此意相同,是正确答案。

# Unit 15

## Section A

1.【答案】E

【详解】男人赚钱,所以他付账。

2.【答案】B

【详解】后一句话说男人是家里的主宰,显然他会作出大部分的决定。

3.【答案】F

【详解】后面说女人呆在家里照顾孩子和丈夫,那么她们很少外出工作,而不是 often“经常”。

4.【答案】J

【详解】女人在家做几乎所有的是家务事,可见她在家的工作非常重要。

5.【答案】O

【详解】本文一直讨论的是家庭婚姻问题,所以此处填 couples。

6.【答案】K

【详解】既然结婚,肯定要担负家庭责任。

7.【答案】C

【详解】后面列举了男女的一些选择,此处应填 choices。

8.【答案】H

【详解】他们会选择结婚或选择单身。

9.【答案】A

【详解】既然男女都有很多的选择,他们会选择让他们觉得惬意的。

10.【答案】I

【详解】后面说了男女在婚姻中很多决定由他们共同作出,责任共担。

**Section B**

11.【答案】B

【详解】见原文第二段:On 2nd December 1942 Enrice and Fermi and his colleagues produced the first controlled nuclear chain reaction. 这说明第一次控制核反应发生在 1942 年。那么 23 年前,应该是 1919 年。文中第二段开头又说:In 1919 Rutherford split the atom artificially. 1919 年实现人工分裂原子。B 项与此意同,是本题正确答案。

12.【答案】A

【详解】见原文第二段:The neutrons that are released in fission produce fission in other atoms. This is known as a chain reaction. 原子分裂所释放出来的中子又引起其他的分裂,这就是我们所熟知的链式反应。这说明链式反应是一个由一次反应的产物引发出更多的此类反应的过程。A. 答案与此意相同,因此是正确答案。B. 一系列连锁发生的事件,是反应而不是事件。C. 相互连接的一些东西,很明显与文义不符。D. 由分裂释放的一定数量的能量,只描述了一次反应的产物,而不是反应过程。

13.【答案】A

【详解】文中第二段:In 1939 Hahn and Strassman investigated the action of neutrons of urnium-235. They found that it was split into two equal pieces. This process is known as fission. 这说明发现原子分裂的时间是在 1939 年。见原文第三段:In 1951 electricity was first produced by using the heat from a nuclear reactor. 用原子反应堆开始发电的时间是 1951 年,因此,其间大约是 12 年,所以选项 A 为正确答案。

14.【答案】C

【详解】文章最后一段主要说明了原子能在现代社会中的广泛应用。C 项与此意相同,说明了本段的主要内容,因此是正确答案。

15.【答案】D

【详解】D. 原子能是由原子核释放的。与文中所述不符,原文第二段说:It (fission) released great amounts of energy. 能量是在分裂过程中释放出来的。D 项符合题目要求,是正确答案。A. 20 世纪以前,水,木头和化石燃料被用做能量来源。此说法在文中第一段可找到。B. 1942 年后,原子能被用于战争与和平,此说法符合文义,见原文第二段末尾和第三段开头。C. 研究原子的结构大约花了 1900 到 1925 年的 25 年的时间,符合文义,文中第一段末尾处可找到此意。

16.【答案】B

【详解】见原文第一段:A novel can be on the "best seller" list with a sale of fewer than 100,000 copies, but a popular TV show might have 70 million viewers. 一本小说的销售量不到 10 万册就可以上"最畅销书目"名单,而一个电视节目要有 7 千万电视观众才能称得上是受欢迎的节目,很明显作者在这里拿小说的销售量与看电视节目的人数进行比较。据此 A,C 两项被首先排除。audiences 指"现场观众",显然不符合文章意思,因此 D 项也被排除。viewers 看电视的人,符合文义。所以 C 项是正确答案。

17.【答案】B

【详解】见原文第二段:It turned out that the show producers who were pulling the strings, gave the answers to the most popular contestants beforehand. 结果证明是电视制作人在幕后操纵,他把答案给了那些最受观众喜欢的参赛者。因此制作人最应该受到谴责,所以 B 项正确。

18.【答案】C

【详解】见原文第二段:The result of this cheating was a huge scandal. 这种欺骗(电视观众)的行为结果成了一个巨大的丑闻。C 答案与此意同,是本题正确答案。

19.【答案】A

【详解】见原文第三段:Van Doren is no longer involved with TV. But game shows are still here, though they aren't taken as serioulу. 根据全文,此句意思应为:范多伦和电视已不再有联系,但是娱乐节目仍然存在,尽管它们已不再那么重要了。(因为娱乐节目对电视观众非常重要,人人都爱看),因此 A. importantly 重要地,符合文义,是正确答案。

20.【答案】D

【详解】作者在文中花了大量篇幅来叙述发生在电视娱乐节目中的丑闻,而且在文中最后一段作者这样评价娱乐节目:In fact, some of them try to be as ridiculous as possible. 事实上,有些娱乐节目非常荒谬可笑。可以看出作者对电视娱乐节目不赞成,持批判的态度。因此 D. critically 批评地,批判地,是正确答案。

四级考试核心考点
**精确打击**

# Part 5

# 完形填空

Cloze

# 一、完形填空

# 命题规律与应试技巧

## 1. 五大命题规律精确定位

完形填空是考查语言知识和语篇水平的综合测试方式，主要用来测试学生综合运用语言的能力。它要求考生具有一定的阅读理解能力，扎实的语法知识，利用已掌握的词汇的（lexical）、语义的（semantic）、句法的（syntactic）等知识，根据上下文的提示对空缺部分进行破译，填上最合适词。该部分一般共20题，考试时间15分钟。在一篇题材熟悉，难度适中的短文（300词左右）中留有20个空格，每个空格为一题，每题有四个选择项，要求考生在全面理解内容的基础上选择一个最佳答案，使短文的意思和结构恢复完整。填空的词项一般包括结构词和实义词。完形填空与单句填空在形式上相似，但考查的内容却包括词汇、语法、阅读和写作等各方面的应用能力。

完形填空的命题规律，可以大致总结为以下五种类型。

（1）固定搭配考查规律；

（2）关联词语考查规律；

（3）名词和代词一致性考查规律；

（4）文章主要内容考查规律；

（5）上下文线索考查规律。

## 2. 应试三大步骤精确突破

完形填空题的考查要点主要是词汇、语法、综合理解等方面，它要求考生在熟练掌握这方面的基础上，提高推理判断和分析归纳的能力，同学们在临场答题时，可以采用以下方法。

### 步骤1 快速浏览全文。

在浏览中，注意捕捉关键词，记忆相关信息；如文中的上下文关系，各层次之间的逻辑关系，所叙述的时间、地点、人物、事件、描述文中的空间关系等，以把握文章发展的基本线索。

### 步骤2 复读答题，先易后难。

首先选出那些比较直接而明显的，根据句子或上下文的意义就能确定的答案，如固定短语、常用句型、固定搭配等。如看到

depend，马上就能判断出与它搭配的介词“on”。

(1)利用词汇短语，如词语的固定搭配，惯用法，常见句型以及基本语法知识进行选择。

(2)利用已选出的答案帮助推断未确定的答案。

(3)根据相关内容及上下文的逻辑关系选择。

(4)根据全文内容或背景知识，从常识的角度去考虑选择。

**步骤 3　通读全文。**

将全文连同所选答案一起读一遍，检查前后是否连贯，内容是否清楚。此外，从词意、搭配、语法等方面仔细推敲，确保准确无误。

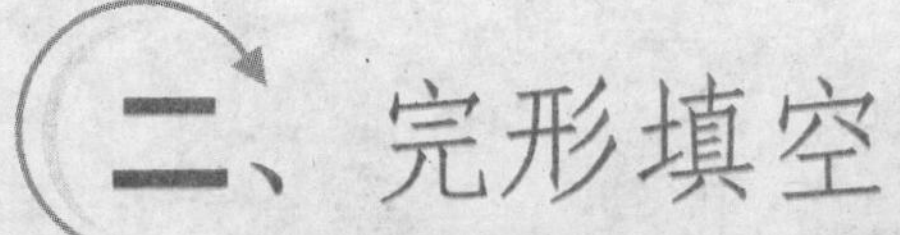

# 二、完形填空

# 核心考点精确打击

## Unit 1

**Directions:** *There are 20 blanks in the following passage. For each blank there are four choices marked A, B, C and D. You should choose the ONE that best fits into the passage. Then mark the corresponding letter on **Answer Sheet** 2 with a single line through the centre.*

Robert Edwards __1__ in an unusual accident many years __2__. He was also partially deaf __3__ old age. Last week he was walking near his home __4__ a thunderstorm __5__. He took shelter __6__ a tree and was struck by lightning. He was __7__ to the ground and woke up __8__ 20 minutes later, __9__ face down in water below a tree. He went into the house and lay down in bed. A short time later he awoke. His legs couldn't move __10__ he was trembling. __11__, when he opened his eyes he could see the clock __12__ the room in front of him. __13__ his wife entered he saw her for the first time in nine years. Doctors confirmed that he had __14__ his sight and hearing apparently __15__ the flash of lightning. But they were unable to explain the __16__. One possible explanation __17__ by one doctor was that Edwards lost his sight __18__ a hard blow in a terrible accident. Perhaps the only way it could __19__ was by __20__ blow.

1. A. blinded　　B. was blinded　　C. had been blind　　D. had been blinded
2. A. later　　B. before　　C. ago　　D. early
3. A. because of　　B. because　　C. at　　D. in
4. A. when　　B. while　　C. until　　D. where
5. A. fell　　B. blew　　C. formed　　D. approached

6. A. in B. on C. under D. near
7. A. thrown B. knocked C. fallen D. beaten
8. A. just B. some C. for D. within
9. A. to lie B. having lain C. lay D. lying
10. A. and B. when C. but D. while
11. A. Thus B. Therefore C. But D. Above all
12. A. across B. through C. into D. out of
13. A. While B. When C. Whenever D. As
14. A. gained B. gotten C. reminded D. regained
15. A. at B. in C. from D. on
16. A. result B. reason C. consequence D. content
17. A. offered B. contributed C. sought D. thought
18. A. because of B. owing to C. based on D. as a result of
19. A. restore B. be restored C. have restored D. have been restored
20. A. other B. the other C. another D. one

## Unit 2

**Directions:** *There are 20 blanks in the following passage. For each blank there are four choices marked A, B, C and D. Choose the best answer to fill in the blanks.*

The task of being accepted and enrolled in a university begins early for some students, long __1__ they graduate from high school. These students take special __2__ to prepare for advanced study. They may also take one of more examinations that test how __3__ prepared they are for the university.

In the final year of high school, they __4__ applications and send them, with their student records, to the universities which they hope to __5__.

Some high school students may be __6__ to have an interview with representatives of the university. Neatly __7__, and usually very frightened, they are __8__ to show that they have a good attitude and the __9__ to succeed.

When the new students are finally __10__, there may be one more step they have to __11__ before registering for classes and __12__ to work. Many colleges and universities __13__ an orientation program for new students. __14__ these programs, the young people get to know the __15__ for registration and student advising, university rules, the __16__ of the library and all the other __17__ services of the college or university.

Beginning a new life in a new place can be very __18__. The more knowledge students have __19__ the school, the easier it will be for them to __20__ to the new environment. However, it takes time to get used to college life.

1. A. as B. after C. since D. before
2. A. courses B. disciplines C. majors D. subjects
3. A. deeply B. widely C. well D. much
4. A. fulfill B. finish C. complete D. accomplish
5. A. attend B. participate C. study D. belong
6. A. acquired B. considered C. ordered D. required
7. A. decorated B. dressed C. coated D. worn
8. A. decided B. intended C. settled D. determiner
9. A. power B. ability C. possibility D. quality
10. A. adopted B. accepted C. received D. permitted

| | | | |
|---|---|---|---|
| 11. A. make | B. undergo | C. take | D. pass |
| 12. A. getting | B. putting | C. falling | D. sitting |
| 13. A. offer | B. afford | C. grant | D. supply |
| 14. A. For | B. Among | C. In | D. On |
| 15. A. processes | B. procedures | C. projects | D. provisions |
| 16. A. application | B. usage | C. use | D. utility |
| 17. A. major | B. prominent | C. key | D. great |
| 18. A. amusing | B. misleading | C. alarming | D. confusing |
| 19. A. before | B. about | C. on | D. at |
| 20. A. fit | B. suit | C. yield | D. adapt |

## Unit 3

**Directions:** *There are 20 blanks in the following passage. For each blank there are four choices marked A, B, C and D. Choose the best answer to fill in the blanks.*

The United States is well-known for its network of major high ways designed to help a driver get from one place to another in the shortest possible time. __1__ these wide modern roads are generally __2__ and well maintained, with __3__ sharp curves and many straight __4__, a direct route is not always the most __5__ one. Large highways often pass __6__ scenic areas and interesting small towns. Furthermore, these highways generally __7__ large urban centers which means that they become crowded with __8__ traffic during rush hours, __9__ the "fast, direct" way becomes a very slow route.

However, there is __10__ always another route to take __11__ you are not in a hurry. Not far from the __12__ new "superhighways", there are often older, __13__ heavily traveled roads which go through the countryside. __14__ of these are good two-lane roads; others are uneven roads __15__ through the country. These secondary routes may go up steep slopes, along high __16__, or down frightening hillsides to towns __17__ in deep valleys. Through these less direct mutes, longer and slower, they generally go to places __18__ the air is clean and the scenery is beautiful, and the driver may have a __19__ to get a fresh, clean __20__ of the world.

| | | | |
|---|---|---|---|
| 1. A. Although | B. Since | C. Because | D. Therefore |
| 2. A. stable | B. splendid | C. smooth | D. complicated |
| 3. A. little | B. few | C. much | D. many |
| 4. A. selections | B. separations | C. series | D. sections |
| 5. A. terrible | B. enjoyable | C. possible | D. profitable |
| 6. A. to | B. into | C. over | D. by |
| 7. A. lead | B. connect | C. collect | D. communicate |
| 8. A. large | B. fast | C. light | D. heavy |
| 9. A. when | B. for | C. but | D. that |
| 10. A. yet | B. still | C. almost | D. quite |
| 11. A. unless | B. if | C. as | D. since |
| 12. A. relatively | B. regularly | C. respectively | D. reasonably |
| 13. A. and | B. less | C. more | D. or |
| 14. A. All | B. Lots | C. Several | D. Some |
| 15. A. driving | B. crossing | C. curving | D. traveling |
| 16. A. rocks | B. cliffs | C. roads | D. paths |
| 17. A. lying | B. laying | C. laid | D. lied |
| 18. A. there | B. when | C. which | D. where |
| 19. A. space | B. period | C. chance | D. spot |

20. A. view B. variety C. visit D. virtue

# Unit 4

**Directions:** *There are 20 blanks in the following passage. For each blank there are four choices marked A, B, C and D. You should choose the ONE that best fits into the passage. Then mark the corresponding letter on* ***Answer Sheet*** *2 with a single line through the centre.*

Once upon a time a pair of crows were continually upset by a cobra. Every year the snake __1__ into their nest to eat the young crows __2__ they learned to fly. They asked their clever friend, the jackal, __3__ to do.

"Do not despair," he told them. "We cannot stop the cobra __4__, as we are not strong enough. We will have to use __5__ to destroy that __6__ beast. Just do __7__ I tell you and you will be safe."

The crow __8__ flew off to the river __9__ a princess was bathing, guarded by all her servants. He __10__ the most beautiful necklace __11__ on the shore and flew away just __12__ of the angry servants. Once they were running __13__ for the cobra's home he darted ahead and settled at the window __14__ to be lost.

The cobra attacked __15__. "Stupid crow." he thought. "He only just __16__ to escape, but dropped this __17__ necklace in his haste. What will his wife say? Now I will be the most magnificent cobra in the world." However, no sooner had he put it on __18__ the servants appeared and killed him to take the __19__ thing back.

That year the crow's family grew up healthy and __20__.

1. A. leaped B. crawled C. sauntered D. strolled
2. A. when B. while C. before D. after
3. A. how B. where C. when D. what
4. A. by chance B. by accident C. by force D. by scheme
5. A. hear B. craft C. enthusiasm D. courage
6. A. greedy B. sly C. clever D. mad
7. A. that B. how C. which D. what
8. A. happily B. obediently C. faithfully D. hopefully
9. A. that B. in that C. where D. which
10. A. snatched up B. took up C. deceived D. exchanged
11. A. lost B. fallen C. left D. hidden
12. A. within the reach B. out of reach C. between the reach D. to the reach
13. A. direct B. indirectly C. roundabout D. straight
14. A. to pretend B. pretend C. pretending D. pretends
15. A. at once B. slowly C. unhurriedly D. flurriedly
16. A. arranged B. managed C. planned D. designed
17. A. abundant B. wealthy C. valuable D. prosperous
18. A. when B. until C. as D. than
19. A. supercilious B. noble C. precise D. precious
20. A. safe B. safety C. safely D. safer

# Unit 5

**Directions:** *There are 20 blanks in the following passage. For each blank there are four choices marked A, B, C and D. Choose the best answer to fill in the blanks.*

A land free from destruction, plus wealth, natural resources, and labor supply—all these were important __1__ in helping England to become the center for the Industrial Revolution. __2__ they were not enough. Something __3__ was needed to start the industrial process. That "something special" was men—__4__ individuals who could invent machines, find new __5__ of power, and establish business organizations to reshape society.

The men who __6__ the machines of the Industrial Revolution __7__ from many backgrounds and many occupations. Many of them were __8__ inventors than scientists. A man who is a __9__ scientist is primarily interested in doing his research __10__. He is not necessarily working __11__ that his findings can be used.

An inventor or one interested in applied science is __12__ trying to make something that has a concrete __13__. He may try to solve a problem by using the theories __14__ science or by experimenting through trial and error. Regardless of his method, he is working to obtain a __15__ result: the construction of a harvesting machine, the burning of a light bulb, or one of __16__ other objectives.

Most of the people who __17__ the machines of the Industrial Revolution were inventors, not trained scientists. A few were both scientists and inventors. Even those who had __18__ or no training in science might not have made their inventions __19__ a groundwork had not been laid by scientists years __20__.

| | | | |
|---|---|---|---|
| 1. A. cases | B. reasons | C. factors | D. situations |
| 2. A. But | B. And | C. Besides | D. Even |
| 3. A. else | B. near | C. extra | D. similar |
| 4. A. generating | B. effective | C. motivating | D. creative |
| 5. A. origins | B. sources | C. bases | D. discoveries |
| 6. A. employed | B. created | C. operated | D. controlled |
| 7. A. came | B. arrived | C. stemmed | D. appeared |
| 8. A. less | B. better | C. more | D. worse |
| 9. A. genuine | B. practical | C. pure | D. clever |
| 10. A. happily | B. occasionally | C. reluctantly | D. accurately |
| 11. A. now | B. and | C. all | D. so |
| 12. A. seldom | B. sometimes | C. usually | D. never |
| 13. A. plan | B. use | C. idea | D. means |
| 14. A. of | B. with | C. to | D. as |
| 15. A. single | B. sole | C. specialized | D. specific |
| 16. A. few | B. those | C. many | D. all |
| 17. A. proposed | B. developed | C. supplied | D. offered |
| 18. A. little | B. much | C. some | D. any |
| 19. A. as | B. if | C. because | D. while |
| 20. A. ago | B. past | C. ahead | D. before |

# Unit 6

**Directions:** *There are 20 blanks in the following passage. For each blank there are four choices marked A, B, C and D. on the paper. You should choose the ONE that best fits into the passage. Then mark the corresponding letter on **Answer Sheet** 2 with a single line through the centre.*

At the docks you step into a shiny metal submarine that is shaped like a shark. The door is closed behind you, and __1__ engine *purrs to life*(发动起来) __2__ the submarine—like a shark __3__. Soon all sunlight __4__. The headlights of the diving ship are __5__. Then, you come to the sea bottom. From the diving submarine, you change to tiny jet boat. As you travel, you see strange fish and underwater mountains, cliffs, and valleys. But __6__ that—hotels and mining camps and farms and factories!

Let's go back to dry land and take a future journey __7__ way-up instead of down. Huge rocket liners take you into space to visit the

Moon Camp. You visit an observatory __8__ a giant telescope looks far into space—farther than anyone __9__ able to see from Earth.

Let's go back to Earth. Bicycles and perhaps skates may be __10__ by jet power, and a new thing __11__ may be a small flying saucer. What about the food of the future? Scientists think that much of it will be __12__ -made in factories __13__ such surprising things as coal, *limestone*(石灰石), air, and water. These things will be __14__ so skillfully by food chemists that the food of the future probably will be __15__. It probably will also be healthful because all the things that you need to live a long and healthy life will be put into it. Scientists of the future probably will find __16__ for most diseases. What about highways of the future? Electric __17__ will hold each car on the right road to get __18__ the "driver" wants to go. And it probably will be impossible for cars to __19__ together. Controls that won't even have to be __20__ will make all speeding cars miss each other or will put on the brakes.

The future should be a wonderful time in which to live.

1. A. strong B. powerful C. forceful D. efficient
2. A. slides down B. down slide C. down slides D. slide down
3. A. diving B. to dive C. being diving D. dives
4. A. goes B. has gone C. went D. is gone
5. A. turned on B. turned up C. turned down D. turned out
6. A. rather than B. more than C. other than D. less than
7. A. other B. another C. the other D. others
8. A. that B. which C. as D. where
9. A. has ever been B. has been ever C. has even been D. has been even
10. A. controlled B. run C. pulled D. directed
11. A. to be ridden B. for riding C. to ride D. of riding
12. A. artificial B. false C. artistic D. unnatural
13. A. of B. in C. from D. by
14. A. mixed B. blended C. confused D. breed
15. A. delicious B. delicate C. deliberate D. tender
16. A. cures B. heal C. remedy D. treat
17. A. symbols B. signals C. signs D. signatures
18. A. the place B. what C. wherever D. which
19. A. smash B. crash C. rush D. crush
20. A. removed B. seized C. held D. touched

**Directions:** *There are 20 blanks in the following passage. For each blank there are four choices marked A, B, C and D. You should choose the ONE that best fits into the passage. Then mark the corresponding letter on* ***Answer Sheet*** *2.*

Many of us may feel air-conditioners bring relief from hot, humid or polluted outside air, they gave rise to many __1__ health hazards. Much research has looked at __2__ the circulation of air inside a closed environment such as an office building can spread disease or __3__ occupants to harmful chemicals. One of the more widely publicized dangers is __4__ of *Legionnaire's disease*(军团病), __5__ was first recognized in the 1970s. This was found to __6__ people in buildings with air-conditioning systems in which warm air pumped out of the system's cooling towers was __7__ sucked back into the air *intake*(入口,进口), in most cases __8__ poor design. This warm air was, needless to say, the perfect environment for the rapid growth of disease-carrying bacteria __9__ from outside the building, where it existed in harmless quantities. The warm, bacteria-laden air __10__ cooled, conditioned air and was then circulated around various parts of the building. Studies showed that even people outside such buildings were at risk if they walked __11__ air *exhaust ducts*(废气管). Cases of Legionnaire's disease are becoming fewer with newer system designs and modifications to older systems, but many older buildings, particularly in developing countries, __12__ constant monitoring. The ways __13__ air-conditioners work to "clean" the air can *inadvertently*(无心地) cause health problems, too. One such way is with the use of an *electrostatic precipitator*(静电滤尘器), which __14__ dust and smoke particles from the air. What precipitators also do, __15__, is emit large quantities of *positive air ions*(正离子) into the ventilation system. A growing number of studies show that overexposure to positive air ions can __16__ headaches, fatigue and feelings of irritation. Finally, it should be pointed out that the artificial climatic environment created by air-conditioners can also adversely __17__ us. In a __18__ environment, whether indoor or outdoor, there are small variations in temperature and humidity. Indeed, the human body has long __19__ these normal changes. In an air-conditioned living or work environment, however, body temperature remain well under 37℃, our normal temperatures. This __20__ a weakened immune system and thus greater *susceptibility*(易感性) to diseases such as colds and flu.

1. A. possible B. potential C. available D. proficient
2. A. what B. when C. how D. which
3. A. expose B. explode C. export D. expand
4. A. those B. this C. these D. that
5. A. which B. it C. that D. and
6. A. affect B. have affected C. have been affected D. affecting
7. A. anyhow B. anyway C. somehow D. somewhat
8. A. due to B. according to C. as to D. because
9. A. originated B. originated C. to originated D. originating
10. A. was combined to B. was combined with C. combined to D. combined with
11. A. past B. passed C. through D. passing
12. A. acquire B. inquire C. require D. request
13. A. which B. that C. in which D. in that
14. A. removes B. takes C. brings D. dismisses
15. A. therefore B. however C. though D. accordingly
16. A. result from B. settle down C. lie in D. result in
17. A. effect B. effort C. afford D. affect
18. A. natural B. artificial C. normal D. unnatural
19. A. been accustomed to B. been accustomed with C. been familiar to D. been familiar with
20. A. leads B. leads to C. guides D. causes to

# Unit 8

**Directions:** *There are 20 blanks in the following passage. For each blank there are four choices marked A, B, C and D. You should choose the ONE that best fits into the passage. Then mark the corresponding letter on **Answer Sheet** 2.*

Japan's love affair with dance never seems to end. At community centers throughout the country, dance classes are always full, whether for ballet or *flamenco*(弗拉曼柯舞), which consistently __1__ young women, __2__ ballroom dancing, which is especially popular among middle-aged and older people. Recently, the dance scenes __3__ a colorful new infusion in the form of folk dances from around the world. Traditional Middle Eastern, African, and European dances, which __4__ recently were virtually unknown in Japan, are __5__ culture centers and community groups throughout the country. One example is belly-dancing, first __6__ by women in northern Africa and the middle east, was __7__ performed at celebratory occasions such as weddings. Japanese people have __8__ a passing familiarity with belly-dancing through movies. __9__, its popularity as a pastime began to spread about three years ago, __10__ more young Japanese women began taking trips to Turkey and other countries. Ever since then the popularity of belly-dancing classes at community centers and workshops run by individuals has skyrocketed. Most participants are initially __11__ by the lavish clothing and sexy movements, but they soon learn another of belly-dancing's merits; it __12__ a strenuous *workout*(训练). Fitness clubs are even starting to __13__ belly dancing in their programs of exercise classes. And a Turkish restaurant called Sofra, __14__ in Tokyo's Shinjuku district, invites customers to get up and dance with professional belly-dancers. A growing number of women are coming to the restaurant __15__ to dance. Of course, the number of people getting into folk dances is still small __16__ with the ranks of those involved in jazz dance and ballet-dance forms that __17__ popular in Japan for decades. But with more and more Japanese __18__ abroad to learn dances in their countries of origin and __19__ them when they come back to Japan, folk dance undoubtedly has a __20__ growing base of enthusiasts.

1. A. apply to　　B. appeal to　　C. suit to　　D. attach to
2. A. and　　B. that　　C. or　　D. with
3. A. has been getting　　B. has got　　C. got　　D. had got
4. A. /　　B. to　　C. until　　D. by
5. A. sweeping　　B. sweeting　　C. swallowing　　D. switching
6. A. was practised　　B. practised　　C. being practised　　D. practising
7. A. extremely　　B. violently　　C. intensively　　D. originally
8. A. gained　　B. won　　C. earned　　D. achieved
9. A. So　　B. However　　C. Though　　D. Therefore
10. A. while　　B. as　　C. since　　D. when
11. A. drawn on　　B. drawn of　　C. drawn in　　D. drawn up
12. A. provides　　B. supplies　　C. gives　　D. affords
13. A. contain　　B. include　　C. involve　　D. comprise
14. A. locating　　B. is located　　C. located　　D. locates
15. A. specifically　　B. specially　　C. particularly　　D. peculiarly
16. A. comparing　　B. compared　　C. to compare　　D. being compared
17. A. had been　　B. are　　C. were　　D. have been
18. A. head　　B. heading　　C. to head　　D. headed
19. A. teaching　　B. to teach　　C. taught　　D. teach
20. A. firmly　　B. steadily　　C. stably　　D. solidly

# Unit 9

**Directions:** *There are 20 blanks in the following passage. For each blank there are four choices marked A, B, C and D. You should choose the ONE that best fits into the passage. Then mark the corresponding letter on* ***Answer Sheet*** *2 with a single line through the centre.*

Some men spend much of their time under water. They are called __1__. If anything goes __2__ with a ship __3__ the waterline, a diver __4__ his suit and goes down with his tools __5__ the damage. He wears a special kind of suit __6__ rubber and canvas __7__ keeps __8__ the water. It covers his body from feet to neck __9__ leaves his hands __10__. His sleeves end in watertight cuffs __11__ the wrist. He puts on a heavy helmet __12__ with a tube which brings air from __13__ the surface of the water. The helmet has windows and __14__ to the neck of his suit. __15__ the diver may __16__, his suit is "padded" back and front with plates of lead and his shoes are weighted __17__ metal. His clothes __18__ about 150 pounds. Would you like to be a diver, and perhaps go down to the bottom of the ocean __19__ the lives of men in a __20__ submarine?

1. A. fishermen　　B. seamen　　C. sailors　　D. divers
2. A. wrong　　B. bad　　C. serious　　D. wrecked
3. A. under　　B. over　　C. below　　D. beneath
4. A. wears　　B. puts on　　C. dresses　　D. is dressed in
5. A. repairs　　B. repaired　　C. repairing　　D. to repair
6. A. made of　　B. made from　　C. consisted of　　D. composed of

7. A. in which　　B. on which　　C. which　　D. from which
8. A. away　　B. off　　C. far　　D. out
9. A. and　　B. but　　C. while　　D. however
10. A. empty　　B. useful　　C. free　　D. helpful
11. A. on　　B. round　　C. around　　D. at
12. A. linked　　B. connected　　C. combined　　D. attached
13. A. over　　B. above　　C. on　　D. onto
14. A. is fastened　　B. fastened　　C. fastens　　D. has been fastened
15. A. In order to　　B. In order　　C. In order which　　D. In order that
16. A. be sunk　　B. sink　　C. be sinking　　D. have sunk
17. A. with　　B. by　　C. to　　D. from
18. A. are weighted　　B. weight　　C. weigh　　D. are weighed
19. A. save　　B. to save　　C. saves　　D. saving
20. A. damaging　　B. dangerous　　C. damaged　　D. hopeful

# Unit 10

**Directions:** *There are 20 blanks in the following passage. For each blank there are four choices marked A, B, C and D. You should choose the ONE that best fits into the passage. Then mark the corresponding letter on* **Answer Sheet** *2 with a single line through the centre.*

It __1__ around nine o'clock when I drove __2__ home from work because it was already dark. As I approached the gates I switched off the head-lamps of the car __3__ prevent the beam from swinging in through the window and waking Jack, who shared the house with me. But I __4__. I noticed that his light was still on, so he was awake anyway—unless he'd __5__ asleep while reading. I put the car away and went up the steps. Then I opened the door quietly and went to Jack's room. He was in bed awake but he didn't __6__ turn towards me.

"What's up, Jack?" I said.

"For God's __7__ don't make a noise," he said.

The way he spoke reminded me __8__ someone __9__ who is afraid to talk in case he __10__ himself a serious injury. "Take your shoes off, Neville," Jack said.

I thought that he must be ill and that __11__ humor him to keep him happy. "There's a snake here," he explained. "It's asleep __12__ the sheets. I was __13__ on my back reading when I saw it. I knew that moving was out of __14__. I couldn't have moved even if I'd wanted to." I realized that he was __15__. "I was relying on you to call a doctor as soon as you __16__ home," Jack went on. "It hasn't bitten me yet but I dare not do anything to upset it. It __17__ wake up." "I'm sick of this," he said. "I __18__ that you'd be home an hour ago."

There was no time to argue or apologize __19__ late. I looked at him __20__ I could and went out to telephone the doctor.

1. A. had to be　　B. was to be　　C. must have been　　D. should have been
2. A. at　　B. back　　C. in　　D. to
3. A. so as to　　B. in order　　C. so that　　D. for
4. A. needn't bother　　B. didn't need to bother
C. needn't have bothered　　D. mustn't have bothered
5. A. become　　B. fallen　　C. gone　　D. grown
6. A. even　　B. just　　C. only　　D. rather
7. A. behalf　　B. love　　C. reason　　D. sake
8. A. from　　B. to　　C. of　　D. with

9. A. with pain B. in pain C. having pain D. having ache
10. A. would be B. does C. would make D. makes
11. A. I had rather B. I would rather C. I had better D. I would better
12. A. between B. beside C. below D. behind
13. A. lied B. laid C. laying D. lying
14. A. the bargain B. the question C. the chance D. the risk
15. A. in fact B. in serious C. in earnest D. in truth
16. A. would come B. have come C. were coming D. came
17. A. might B. can C. should D. shall
18. A. made it certain B. have been assured C. counted on D. took it for granted
19. A. for being B. to be C. on being D. to have been
20. A. as fiercely as B. as encouragingly as C. so bravely as D. so hopefully as

# Unit 11

**Directions:** *There are 20 blanks in the following passage. For each blank there are four choices marked A, B, C and D. You should choose the ONE that best fits into the passage. Then mark the corresponding letter on* ***Answer Sheet*** *2 with a single line through the centre.*

One hundred and thirteen million Americans have at least one bank-issued credit card. They give their __1__ automatic credit in stores, restaurants, and hotels, at home, across the country, and even __2__, and they make many banking services __3__ as well. More and more of these credit cards can be read automatically, making it possible to withdraw or __4__ money in scattered locations, whether or not the local branch bank is __5__. For many of us the "cashless society" is not on the horizon—it's already __6__.

While computers offer these conveniences to consumers, they have many __7__ for sellers, too. Electronic cash registers can do much more than simply ring up sales. They can keep a wide __8__ of records, including who sold what, when, and __9__ whom. This information allows businessmen to keep track of their list of goods by showing which items are being __10__ and how fast they are moving. Decisions to __11__ and return goods to suppliers can then be made. At the same time these computers record which hours are busiest and __12__ employees are the most efficient, allowing personnel and staffing assignments to be made __13__. Computers are __14__ on by manufacturers for similar reasons. Computer-analyzed marketing reports can help to decide which products to emphasize now, which __15__ develop for the future, and which to __16__. Computers keep track of goods in stock, of raw materials on __17__, and even of the production process itself.

Numerous __18__ commercial enterprises, from theaters to magazine publishers, from gas and electric __19__ to milk processors, bring better and more efficient services to __20__ through the use of computers.

1. A. owners B. banks C. shops D. cards
2. A. around B. abroad C. outside D. aboard
3. A. available B. sensible C. valuable D. practicable
4. A. put B. desert C. push D. deposit
5. A. large B. limited C. expanded D. open
6. A. here B. out C. inside D. now
7. A. inconveniences B. advantages C. conditions D. services
8. A. range B. deal C. scope D. extension
9. A. for B. to C. with D. at
10. A. taken B. given C. sold D. transported
11. A. make B. sell C. record D. arrange

12. A. what B. whose C. that D. which
13. A. actually B. accordingly C. similarly D. exactly
14. A. supported B. relied C. assisted D. centered
15. A. can B. rapidly C. to D. will
16. A. increase B. raise C. drop D. distribute
17. A. schedule B. business C. sale D. hand
18. A. others B. other C. another D. many
19. A. utilities B. necessities C. benefits D. unions
20. A. businessmen B. suppliers C. employees D. consumers

# Unit 12

**Directions:** *There are 20 blanks in the following passage. For each blank there are four choices marked A, B, C and D. Choose the best answer to fill in the blanks.*

Did you ever have someone's name on the tip of your tongue and yet you were unable to recall it? __1__ this happens again, do not __2__ to recall it. Do something __3__ for a couple of minutes, __4__ the name may come into your head. The name is there, since you have met __5__ person and learned his name. It __6__ has to be dug out. The initial effort to recall __7__ the mind for operation, but it is the subconscious __8__ that go to work to dig up a __9__ memory. Forcing yourself to recall __10__ never helps because it doesn't __11__ your memory; it only tightens it. Students find the preparatory method helpful __12__ examinations. They read over the questions __13__ trying to answer any of them. __14__ they answer first the ones __15__ which they are most confident. Meanwhile, deeper mental activities in the subconscious mind are taking __16__; work is being done on the __17__ difficult question. By the time the easier questions are answered, answers usually __18__ the more difficult ones will begin to __19__ into consciousness. It is often __20__ a question of waiting for recall to come to the memory.

1. A. As B. When C. While D. Whether
2. A. try B. what C. hesitate D. wait
3. A. simple B. apart C. else D. similar
4. A. unless B. and C. or D. until
5. A. some B. certain C. a D. this
6. A. then B. really C. only D. indeed
7. A. leads B. begins C. helps D. prepares
8. A. deeds B. activities C. movements D. procedures
9. A. light B. fresh C. dim D. dark
10. A. merely B. almost C. barely D. hardly
11. A. loosen B. weaken C. decrease D. reduce
12. A. into B. in C. about D. by
13. A. after B. besides C. before D. against
14. A. Thus B. But C. Therefore D. Then
15. A. of B. with C. for D. in
16. A. place B. shape C. charge D. action
17. A. too B. less C. not D. more
18. A. to B. of C. about D. for
19. A. appear B. grow C. extend D. come
20. A. nearly B. likely C. just D. even

# Unit 13

**Directions:** *There are 20 blanks in the following passage. For each blank there are four choices marked A, B, C and D. You should choose the ONE that best fits into the passage. Then mark the corresponding letter on* ***Answer Sheet*** *2 with a single line through the centre.*

Smoking is considered dangerous to health. Our tobacco-seller, Mr. Johnson, therefore, always asks his __1__, if they are very young, whom the cigarettes are bought __2__.

One day, a little girl whom he had never seen before walked __3__ into his shop and demanded twenty cigarettes. She had the __4__ amount of money in her hand and seemed very __5__ of herself. Mr. Johnson was so __6__ by her confident manner that he __7__ to ask his usual question. __8__, he asked her what kind of cigarettes she wanted. The girl replied __9__ and handed him the money. While he was giving her the __10__, Mr. Johnson said laughingly that __11__ she was so young she should __12__ the packet in her pocket in __13__ a policeman saw it. __14__, the little girl did not seem to find this very funny. Without even smiling she took the __15__ and walked towards the door. Suddenly she stopped, turned __16__, and looked steadily at Mr. Johnson. There was a moment of silence and the tobacco-seller __17__ what she was going to say. __18__ at once, in a clear, __19__ voice, the girl declared, "My dad is a policeman," and with __20__ she walked quickly out of the shop.

1. A. guests B. customers C. friends D. passengers
2. A. with B. for C. to D. by
3. A. nervously B. boldly C. hesitatingly D. heavily
4. A. some B. large C. exact D. enough
5. A. fond B. glad C. ashamed D. sure
6. A. worried B. surprised C. annoyed D. placid
7. A. forgot B. feared C. came D. remembered
8. A. Therefore B. So C. Instead D. Somehow
9. A. readily B. softly C. slowly D. patiently
10. A. change B. warning C. bill D. cigarettes
11. A. for B. as C. while D. though
12. A. cover B. lay C. hide D. take
13. A. that B. time C. fear D. case
14. A. Moreover B. Therefore C. Then D. Nevertheless
15. A. packet B. money C. advice D. blame
16. A. away B. over C. round D. aside
17. A. doubted B. wondered C. expected D. considered
18. A. And B. So C. But D. All
19. A. weak B. joking C. funny D. firm
20. A. which B. that C. him D. what

# Unit 14

**Directions:** *There are 20 blanks in the following passage. For each blank there are four choices marked A, B, C and D. You should choose the ONE that best fits into the passage. Then mark the corresponding letter on* ***Answer Sheet*** *2 with a single line through the centre.*

Columbus had no trouble getting men, money, and ships for his second voyage to the "East". He set __1__ again in September, 1493, this time __2__ 17 ships.

Columbus took with him men __3__ planned to settle on the new land, __4__ the pleasant life Columbus had described to them, and

__5__ rich. He took no women. The supplies on the second voyage __6__ seeds, tools, plants, horses, and __7__ domestic animals, and workmen—everything needed to __8__ a colony a success, he thought.

The king and queen told Columbus __9__ the Indians were to be treated __10__; any person who mistreated __11__ was to be punished. They were to be Christians, __12__ course.

Columbus's colony would carry __13__ trade. Spanish goods were to be be traded __14__ for gold. Columbus was to receive one-eighth of the __15__; Ferdinand and Isabella would get the __16__. When Columbus __17__ at Hispaniola, he found that the fort had been burned __18__. The men he had left __19__ had stolen gold and women, quarreled among themselves, and __20__ murdered one another.

| | | | |
|---|---|---|---|
| 1. A. down | B. out | C. up | D. in |
| 2. A. by | B. through | C. aboard | D. with |
| 3. A. whoever | B. which | C. who | D. whom |
| 4. A. see | B. begin | C. give | D. live |
| 5. A. lead | B. get | C. have | D. own |
| 6. A. excluded | B. contained | C. included | D. comprised |
| 7. A. other | B. another | C. rest | D. similar |
| 8. A. become | B. take | C. do | D. make |
| 9. A. which | B. that | C. where | D. thus |
| 10. A. badly | B. heavily | C. well | D. kind |
| 11. A. them | B. him | C. her | D. you |
| 12. A. in | B. of | C. on | D. for |
| 13. A. out | B. off | C. over | D. on |
| 14. A. hardly | B. only | C. scarcely | D. carefully |
| 15. A. property | B. profits | C. fortune | D. interests |
| 16. A. leftover | B. result | C. remains | D. rest |
| 17. A. came | B. reached | C. arrived | D. attained |
| 18. A. down | B. out | C. off | D. up |
| 19. A. as | B. there | C. that | D. who |
| 20. A. still | B. thus | C. hence | D. then |

# Unit 15

**Directions:** *There are 20 blanks in the following passage. For each blank there are four choices marked A, B, C and D. Choose the best answer to fill in the blanks.*

Today, most countries in the world have canals. Even in the twentieth century, goods can be moved more cheaply __1__ than any other __2__ of transportation. Some canals __3__ the Suez or the Panama, save ships weeks of time by making their voyage a thousand miles __4__. Other canals __5__ boats __6__ cities that __7__ on the coast. __8__ other canals drain lands where there is too much water, and help to irrigate fields where there is not enough water.

In places where __9__, irrigation canals drain water from rivers or lakes and __10__ the irrigation water. In places where there is too much water, canals can drain the water __11__ the land for use in farming. In Holland, acres and acres of land have been drained in this way. Since __12__ of this drained land is below sea level, the water in the canals has __13__ up to sea level. Dikes have been built in Holland to keep the sea __14__ covering the land, as it __15__ in the past.

Canals are also used to __16__ water to mills and factories. The water from a river is kept at a higher level than the river until it reaches the wheel of the mill. Then the water is poured over the mill wheel, __17__ it turn. The same __18__ is used in more modern factories and in hydroelectric generating plants. The force of the water, __19__ from a __20__ height, provides a cheap way of producing electricity.

1. A. by boat B. on boat C. on the sea D. on foot
2. A. method B. way C. approach D. means
3. A. as B. look like C. such like D. the same as
4. A. longer B. shorter C. wider D. narrower
5. A. let B. permit C. allow D. admit
6. A. reaching B. reached
C. to reach D. to have been reaching
7. A. is not located B. does not locate C. don't locate D. are not located
8. A. Yet B. Still C. But D. Any
9. A. does it not rain B. do not rain
C. it rains not very often D. it doesn't rain very often
10. A. providing B. provide C. provided D. to provide
11. A. of B. on C. off D. for
12. A. more B. much more C. most D. mostly
13. A. to pump B. been pumped C. pumped D. to be pumped
14. A. off B. out C. away from D. from
15. A. did B. does do C. will D. has done
16. A. bring B. take C. carry D. lift
17. A. made B. make C. making D. being made
18. A. procedure B. process C. principle D. principal
19. A. falling B. fallen C. fall D. fell
20. A. sure B. certain C. limited D. known

# 三、完形填空

# 答案详解

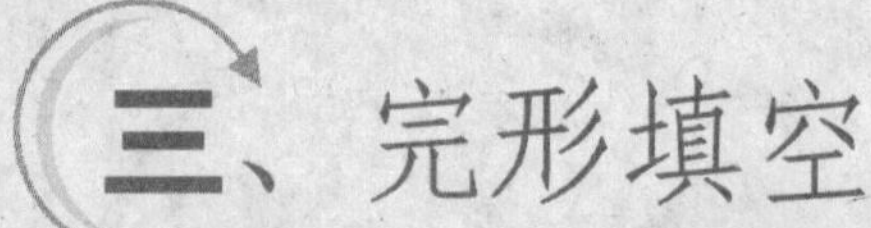

## Unit 1

1.【答案】B

【详解】运用排除法，从上下文中可以看出，Robert Edwards 是在很多年前的一场事故中双眼失明的，因此含有被动的意义。据此 A. 项被排除。C. had been blind 表示一种持续的状态，需要与表示一段时间的时间状语连用，用在此处不恰当，被排除。由于原文叙述的是"事情发生在过去"，而非强调"事情已经发生了，"因此 D. had been blinded 也被排除。B 项正确。

2.【答案】C

【详解】从上下文判断，事情是发生在"以前"，而非"以后"，故此 A. later 被排除。D. early 形容词无此用法,排除。B. before 以前,常用于间接引语里和动词的过去完成时一同使用。e. g. He said that he had built the swimming pool two years before. 他说他两年前修建了这个游泳池。用于此处不当,排除。C. ago 指从现在算起若干时间以前,和动词一般过去时一同使用。符合原文表达要求,为正确答案。

3.【答案】A

【详解】根据逻辑分析:耳聋和年老有因果关系。B. because 后只能跟从句,做原因状语,不符合原文结构要求。because of 后则跟名词或名词性短语,符合结构要求。因此,A 项为本题正确答案。C. at 和 D. in 均不能和 old age 连用。

4.【答案】A

【详解】本句意为:上周当暴雨来临时,他正在家周围散步。B. while 一边……一边,强调两个动作同时发生,表示主句动作发生在从句之中,即主句动作较短,发生在从句一段时间中。e. g. Please be quiet while I am talking to you. 我在对你们讲话时请不要作声。显然与原文意思不符,因为"散步"的动作在暴雨来临之前就开始了。C. until 直到……为止。主句的动作一直持续到从句的动作发生时为止。e. g. I will wait until he returns. 我要等到他回来。不能用于此处,因为原文主句用进行时态,不能表示动作的持续。D. where 在哪里。用于此处不当,因为这样给人的感觉是暴雨就只发生在他家周围,是不可思议的。A. when 当……的时候,引导状语从句,既可以表示主句动作与从句动作同时发生,也可以表示一定的先后性,用于此处恰当。故 A 项为正确选项。

5.【答案】D

【详解】原句意为:暴雨来临。A. fell 落下,降落,指物体从上到下的垂直运动,用于此处,搭配不当。blew(风)吹、刮,用于此处,也属搭配不当。C. formed 形成,指事物经历一个发展过程之后,最终形成,这与原文所表达的意义不相符,因为原文重点不是强调暴雨的"形成",而是"来临"。故此,approached 更合适,所以 D 项为正确答案。

6.【答案】C

【详解】此处应理解为他"躲在树下",故用"under a tree"。

7.【答案】B

【详解】C. fallen 摔倒,此处用被动式不当,排除。A. thrown 扔,强调动作实施者所做的"扔"的动作,与原文表达不符。D. beaten 打,敲,强调"打"的具体动作,一般用于具体意义。e. g. The bandits beat him badly. 匪徒们把他毒打了一顿。用于此处不当。B. knocked 敲、击、打,既可用于具体的,又可用于抽象的"打击,击倒",用于此处,符合原文表达需要,故 B. 项为正确答案。

8.【答案】B

【详解】原句意为:大约二十分钟后,他醒来了。A. just 刚好,表达过于精确,不符合上下文。C. for 后跟一段时间,表示动作的持续过程,如果用于此处,则 later 不能有,因此也不能用。D. within 在……内,表达过于精确。B. some 大约,同 about。符合原文表达,因此 B. 项为正确答案。

9.【答案】D

【详解】原句意为:(他)躺在树下,面部朝下浸入雨水中。主句的伴随状语,表示一种伴随的动作或状态,一般用分词或分词短语。A. to lie 是动词不定式,不能做伴随状语。C. lay 是动词 lie 的一般过去式,不能做伴随状语。B. having lain 是现在分词的完成体,不能做伴随状语。D. lying 是动词 lie 的现在分词形式,符合原文要求。故 D. 项为正确答案。

10.【答案】A

【详解】此处意为:他腿不能动,而且身子在颤抖。此处应为两个并列分句描述两个动作。因此 B. when, D. while 被排除。两个动作之间并无转折关系,所以 C. but 也被排除。A. and 为正确答案。

11.【答案】C

【详解】此处意为:但是,他睁开眼后,能看到……。联系上文,由于他以前双目失明,而现在突见光明,所以此处应用一个转折连词来完成上下文的过渡。A. Thus 这样;B. Therefore 因此;D. Above all 首先。均不是转折连词。C. But 但是,然而,转折连词,为正确答案。

12.【答案】A

【详解】此处意为:他看到在他对面的挂钟。D. out of 在……外部。钟表不可能挂在屋子外,故首先排除。C. into 到……里,着重表示动作的方向,用于此处不恰当。B. through 穿过,着重指物体从某一空间中间穿过。e. g. The train went through a tunnel. 火车穿过隧道。用于此处不当。A. across 在对面。符合原文表达要求,为正确答案。

13.【答案】B

【详解】此处意为:当他妻子进来(屋子)时,九年来他第一次看见了她。C. Whenever 无论何时,用于此处,与原表达意义不符。A. While, B. When, D. As 都有"当……的时候"的意思。as 着重表示从句动作与主句动作同时发生,持续时间不长。据此分析,as 用于此处不当。因为"她进来"应发生于"他看见"之前。while 用于此处不妥。when 可以表示主句动作发生在从句之后,符合原文表达需要。因此 B. 项为正确答案。

14.【答案】D

【详解】此处意为:医生证实他又恢复了视力……。联系上文,此处应为"恢复,再得到"。A. gained 得到;B. gotten 获得。均无此意。C. reminded 使回忆起……,提醒……,用于此处,意思不当。D. regained 重新得到,符合句意,为正确答案。

15.【答案】C

【详解】句意为:很明显,他恢复视力和听力得益于闪电的闪耀。A. at 在(某一时间、地点);B. in 在(地点、时间)里面;D. on 在……上面,在(某一天),均无此意。C. from 从……

(中得到,获益)。符合句意。

16.【答案】B

【详解】句意为:但是他们不能解释其原因。联系上下文,医生们试图解释"他为什么能恢复视力和听觉的原因",因此只有B. reason(原因)符合句意。A. result 结果;C. consequence 后果;D. content 内容。均无此意。

17.【答案】A

【详解】句意为:一位医生提供了一种可能的解释……。B. contributed 贡献,捐献。用于此处,意思不当。C. sought (seek 的过去式)寻找,同 look for,强调找的过程,不能表示结果。联系下文用于此处不当。D. thought 思考,想,用于此处,意思表达欠妥。A. offered 提供,提出,用于此处意思准确,结构正确,是正确答案。

18.【答案】D

【详解】句意为:爱德华兹在一次可怕的事故中受到了的猛烈的打击,从而导致失明。A. because of 因为、由于;B. owing to 由于。强调"原因和结果的关系",把"猛烈的打击"说成"失明的原因",似乎过于牵强。C. based on 以……为基础,与原文表达不符。D. as a result of 由于……的结果",强调"某事(行动等)引起的结果。"失明"是"受到猛烈打击"的结果。符合原文表达。

19.【答案】B

【详解】句意为:或许它(视力)能得以恢复的唯一途径就是……。A. restore, C. have restored 主动结构,不能用于此处。D. have been restored 被动语态完成体表示动作已完成。既已完成,就不存在"可能(perhaps)",因此 D. 项用于此处不当。B. be restored 被动结构,一般式,符合原表达要求,是正确答案。

20.【答案】C

【详解】句意为:(通过)另外一次打击。D. one 一次,用于此处意思不当。A. other 别的,其他的,其后常跟复数形式。e.g. There are other ways of doing this exercise. 做这个练习还有其他方法。用于此处,搭配不正确。B. the other 表示二者中的另一个。用于此处,意思表达有误。C. another 表示不定数目中的另一个。用于此处,意思准确,为正确答案。

## Unit 2

1.【答案】D

【详解】从本句结构看,四个选项中只有B、D能与 long 搭配,但从上文所说的"大学招生工作对某些学生来说很早就开始了"来看,不会是高中毕业后很久,只能是毕业前,故选 D。

2.【答案】A

【详解】四个选项都可表示"课程、专业"之意。其中 B. disciplines 和 D. subjects 意为"学科","科目",C. majors 意为"专业",A. courses 意为"课程",这里指学生选修的具体课程,且前面动词为 take,take courses 意为"选修课程",是固定搭配,故选 A。

3.【答案】C

【详解】副词除了修饰动词、形容词、副词之外,还可以修饰非谓语动词和介词短语。所以,从语法意义上来说,四个选项均可修饰 prepared,但从句子的意思来看,"准备好了"应该用 well prepared,不能用其它三个。

4.【答案】C

【详解】A. fulfill 和 D. accomplish 均指"完成(任务),达到(目的)";B. finish 意为"完成,结束",指某个具体动作的完毕;C. complete意为"使完整,完满",此处指学生填表格,故选 C。

5.【答案】A

【详解】participate 意为"参加,加入(某活动)",后边需用 in;C. study 后若与学校连用,也要用介词 in,表示"在某学校学习";D. belong 意为"属于",后面需连用 to;A. attend 意为"出席,参加",是及物动词,attend school 意为"上学"。从本句结构来看,which 引导出一个定语从句,在从句中没有出现 in 或 to,故不能选 B、C、D。

6.【答案】D

【详解】根据句意,学生进行面试应该是学校要求的,所以应该选 D. required,意为"要求";A. acquired 词形与 required 相近,但意思是"获得,取得",B. considered 意为"考虑,认为",C. ordered 意为"命令",均不符合句子的意义。

7.【答案】B

【详解】既是面试,就应注意由里及外的各个方面。从所给选项来看,这里指的是衣着整齐。"衣着,打扮"应该用 dress,A、C 在此不符合句意,worn 无此用法。

8.【答案】D

【详解】be determined to 是一个固定词组,意为"决心";decide to 意为"决定";intend to 意为"意欲,打算";settle to 意为"决定",但其前面均不用系动词 be,所以,此处只能选 D。

9.【答案】B

【详解】D. quality 意为"品质,素质",在此与后边的 to succeed 搭配不当,可首先排除。为了进入自己所希望的大学,在面试时,学生想展示的是良好的态度及取得成功的能力,而不是客观的可能性(possibility)。B. ability 与 D. power 均可表示能力,但 power 所指的能力常与权力有关,用在此处不符合句意。故只有 B 正确。

10.【答案】B

【详解】A. adopted 意为"采纳,采取",C. received 意为"收到",D. permitted 意为"允许",B. accepted 意为"接受";原句意为"最终,大学接受了新生",所以应选 B。另外,从文章第一句话中也可得到提示。

11.【答案】C

【详解】take a step 为固定搭配，意为"采取步骤（措施）"，此处 there may be one more step they have to take 意为"他们还需再做一件事情"。

12.【答案】A

【详解】根据句子的意义，学生注册分班，然后开始学习。四个选项中，只有 get to 意为"开始，着手"，其它三个选项均不符合句意。

13.【答案】A

【详解】offer 意为"（自愿）提供，供给"，符合句意。B. afford 意为"有时间做……，有经济能力做……，能负担起……"，C. grant 意为"授予，准许"；D. supply 也是"提供，供应"的意思，但它指的是提供具体的实物，所以不能选。

14.【答案】C

【详解】programs 指的是学校向学生提供的各种服务项目，此处应选介词 in，表示"在这些项目中"。

15.【答案】B

【详解】procedure 意为"（办事的）程序、方法"，通过上面所介绍的这些项目，学生可以了解如何报名、注册，如何咨询，学校提供这些项目的目的是让学生知道做这些事情的方法，所以此处应选 B。A. process 意为"过程，经过"或指法律上的诉讼程序，在此不符合句意。C. projects 意为"计划"，D. provision 意为"供应，准备"，均不符合上下文及句子的意义，不能选。

16.【答案】C

【详解】这里指的是学生对图书馆的使用，所以用 use。A. application 意为"应用"，B. usage 意为"用法"，D. utility 意为"实用，有用"，虽然都与"用"有关，但不符合本句的意义。

17.【答案】A

【详解】major 意为"主要的"，这里指除了列举出的之外，学校其它的一些主要服务，故选 A。而 B. prominent（显著的，突出的），C. key（关键的），D. great（伟大的）在此均不符合句意。

18.【答案】D

【详解】confusing 意为"令人迷惑的"，A. amusing 意为"滑稽的，逗人笑的"，B. misleading 意为"误导的"，C. alarming 意为"令人惊恐的"。根据下文：学生对学校了解得越多，就越容易适应新的环境。此处应选 confusing，如果对新的环境了解不多，那么，新的生活将会是"令人迷惑的"。

19.【答案】B

【详解】have knowledge about 意为"了解，知道"，其它三个介词用在此处不符合题意。

20.【答案】D

【详解】adapt to 是一固定词组，意为"适应"，符合上下文意义。fit 与 suit 意为"适合"，yield 意为"让步，服从"，均不合题意。

# Unit 3

1.【答案】A

【详解】从本句的意思来看，从句与主句之间应该是让步的关系，表转折，而非因果关系。四个选项中，只有 Although（虽然）可以用来引起表示让步的状语从句。

2.【答案】C

【详解】smooth 平坦的；stable 稳定的；splendid 壮丽的；complicated 复杂的。

3.【答案】B

【详解】A，C 修饰不可数名词。D 虽然修饰可数名词，但从句子的意思来看，这些宽阔的现代化公路应该没有什么急转弯，所以只能选 B。

4.【答案】D

【详解】sections 地段，路段；selections 选拔；separations 分离；series 一系列的事物。

5.【答案】B

【详解】enjoyable 使人愉快的；terrible 可怕的；possible 可能的；profitable 有益的。根据上下文的意思选 B。

6.【答案】D

【详解】pass by 经过；pass to 传递；pass into 逐渐变成；pass over 忽略。

7.【答案】B

【详解】connect 连接；lead 引导；collect 搜集；communicate 传达。

8.【答案】D

【详解】heavy 拥挤的。

9.【答案】A

【详解】when 在此处为关系副词，表示"在那时"。

10.【答案】C

【详解】almost 几乎；yet 尚，还；still 仍然；quite 相当。

11.【答案】B

【详解】从句子的意思来看，从句应是表示条件的状语从句。

12.【答案】A

【详解】relatively 相对地；respectively 各自地；regularly 定期地；reasonably 合理地。

13.【答案】B

【详解】less 较少，其它三项都会使句子意思前后矛盾，所以只能选 B。

14.【答案】D

【详解】此处是"some...，others..."句型，意思为"有些……又有些……"。

15.【答案】C

【详解】curving 成曲线形，此处指道路曲折蜿蜒。而 cross

through 的意思是“横穿”,与句子的意思不符,所以不选。

16.【答案】B

【详解】cliffs 悬崖,峭壁。而 rocks 岩石,用于此处意思不妥。

17.【答案】A

【详解】lying 位于。此处只能用现在分词形式,表示主动,所以不选 C,D。而 laying 的意思是“放置”,也不能选。

18.【答案】D

【详解】where 此处引导一定语从句修饰先行词 places。

19.【答案】C

【详解】chance 机会;space 空间;period 时期;spot 地点。

20.【答案】A

【详解】view 风景;variety 多样化;visit 参观;virtue 优点。

# Unit 4

1.【答案】B

【详解】蛇是爬行动物,故用“爬行”来描述蛇的行进是最恰当的。其他各项 A. leaped 跳跃;C. sauntered 闲逛、散步;D. strolled 散步、溜达。均无此意。

2.【答案】C

【详解】乌鸦一旦会飞,蛇就无法接近它们,因此,此处应填入 before 表示在它会飞以前。

3.【答案】D

【详解】乌鸦不知该怎么办,就去请教豺,应该做些什么,所以用 what。因为如果用 how,就表示知道做什么,但不知道以何种方式做。显然与原文义不符。when, where 用于此处没有道理。

4.【答案】C

【详解】从下一句“... as we are not strong enough”是因为力气不够。所以是不能用“武力”来阻止眼镜蛇。故此用 by force。

5.【答案】B

【详解】既然不能用武力阻止,就得用“计谋、诡计”,所以用 craft 最合适。A. hear 听,动词,用于此处搭配不正确。C. enthusiasm 热情;D. courage 勇气。仅有“热情”和“勇气”是不够的,要付诸实施才能有效果。而从文中可以看出,乌鸦对此已经有“热情”“且有”勇气。

6.【答案】A

【详解】此处用“贪婪”来形容眼镜蛇很恰当,因为上文说它们每年都要吃。B. sly 狡猾的;C. clever 聪明的;D. mad 疯狂的。在文中并无暗示。

7.【答案】D

【详解】A. that 不能用于此。因为 that 引导宾语从句在从句中不做句子成分。这样宾语从句结构不完整,tell 缺直接宾语。B. how 表示方式,用于此处不当。因为此处是交代要“做什么”而不是“如何做”。C. which 哪一个,不符合题意。D. what 什么,表示要做的内容最为恰当。故 D 项为正确答案。

8.【答案】B

【详解】乌鸦听了豺的话,顺从地按照它的话去做,因此此处填入 obediently(顺从地)最合题意。A. happily 高兴地,眼镜蛇一日不除,乌鸦就不可能高兴。C. faithfully 忠诚地、忠实地,不符原文意思。D. hopefully 充满希望地。这时候说满怀希望为时过早,因为计谋刚刚开始。

9.【答案】C

【详解】此处为定语从句,修饰“river”,引导词在句中做地点状语,故用 where。

10.【答案】A

【详解】描述乌鸦的动作用“抓起来”最合适。A. snatch up 突然抓起来,符合文义。B. took up 开始做,从事某项活动,与文义相去太远。C. deceived 欺骗;D. exchanged 交换、交流。意思不当。

11.【答案】C

【详解】项链应该是放在岸上的。故用 left。其他项均不能正确表达文义。

12.【答案】B

【详解】联系上下文,乌鸦的用意在于引这些人去眼镜蛇的地方,故应该是“刚好抓不着”。因此用 out of reach。

13.【答案】D

【详解】联系上下文,此处应为“径直向眼镜蛇的家跑去”。因此 C. roundabout 绕圈子的,不符合题意;另外,roundabout 是形容词,不可修饰动词。B. indirectly 间接地、迂回地,不合文义。A. direct 直接的,形容词,不能修饰动词。D. straight 径直的(地),直截了当的(地)。既可做形容词,又做副词可修饰动词,因此 D 项是正确答案。

14.【答案】C

【详解】此处为伴随状语,用现在分词表示伴随发生的动作,符合文义。

15.【答案】A

【详解】眼镜蛇的进攻应该是迅速的,立即的,故用 at once。

16.【答案】B

【详解】原文此处应理解为“它(乌鸦)只顾着设法逃跑”。眼镜蛇的想法。A. arranged 安排;C. planned 计划;D designed 设计;均含有事先周密安排之意,而眼镜蛇是不可能想到这一点的。故 B. managed 设法为正确答案。

17.【答案】C

【详解】A. abundant 丰富的;B. wealthy 富有的;C. valuable 值钱的、贵重的;D. prosperous 旺盛的、繁荣的。很显然用 valuable 修饰 necklace 最合适。

18.【答案】D

【详解】固定搭配:no sooner... than 刚……就。

19.【答案】D

【详解】此处 thing 指上文的 necklace。A. supercilious 目空一切的、傲慢的;B. noble 贵族的、高尚的;C. precise 精确的;D. precious 宝贵的、珍贵的。可以看出只有 D 答案可以修饰 necklace (thing)。

20.【答案】A

【详解】不及动物词后跟形容词做补语表示一种伴随状态。healthy 是形容词,并列结构词性相同。故 safe 应用形容词形式。B. safety,名词;C. safely,副词。均不合要求。并列成分语法结构相似,故 safe 也不能用比较级,所以 A 项正确。

# Unit 5

1.【答案】C

【详解】factors 意为"因素",符合句子的意义要求。A. cases (情况),D. situations(形势,局面)用在此处意思明显不合适。B. reasons 可以表示"原因,理由",但后面的介词常用 for,而不用 in,所以此处 C. 为最佳选择。

2.【答案】A

【详解】这里需根据上下文的逻辑关系选择一连接词。上文作者提到了促使英国成为工业革命中心的几个重要因素,然后,作者接着说这些是不够的,语气上发生了明显的转折。四个选项中只有 A. But 能表达语气的转折,因而是答案。

3.【答案】A

【详解】从上文所说的"仅有前面提到的那些因素是不够的"可知,工业革命的开始还需要其它的东西。something,anything 可以与 else 连用,意思是"其它的东西",something else 用于肯定句,anything else 用于否定句。B. near(近的)与 D. similar (相同的,相似的)用在此处意思不恰当,C. extra 意为"额外的,多余的",它强调的是正常数额之外的,而不是其它的,也不能入选。

4.【答案】D

【详解】从下半句的定语中可以看出,这些人能够发明机器、发现新能源、建立新的商业机构来改造社会,因而应该选 creative(有创造力的)。B. effective 意为"有效的",C. motivating 意为"激发的,促动的",A. generating 意为"产生的,生成的",均不用于修饰人。

5.【答案】B

【详解】sources 除了"来源"之外,还可指"资源",sources of power 意为"能源"。A. origins(根源),C. bases(基础),D. discoveries(发现)均不符合句意。

6.【答案】B

【详解】上文提到在工业革命中还需要能发明机器的有创造性的人。下面又提到这些人大多数是发明家,而不是科学家,所以此处应选 B. created,意为"发明,创造"。C. operated 意为"操纵",D. controlled 意为"控制",A. employed 后与物连用,意为"使用",单从本句来看,这三个词都可以选用,但从上下文来看,意思就不合适了,所以,不能入选。

7.【答案】A

【详解】四个选项均可与 from 搭配。come from 意为"是……地方人,来自……家庭背景";stem from 意为"源自,来自",不能用来描述人,只能用于描述事物的来源;arrive from 意为"自某处抵达",appear from 意为"从……出现"。本句意为"在工业革命中发明机器的人们来自于不同的家庭背景,不同的行业"。所以,只能选 A。

8.【答案】C

【详解】本题测试固定结构 more... than 的用法。除了构成形容词与副词的比较级之外,more... than 结构还可以表示"与其说……,不如说……"的意思。如:He is more mad than stupid. 与其说他蠢,倒不如说他疯。这个结构中的 more 不能用其它任何比较级代替。所以,A、B、D 三项均不能入选。

9.【答案】C

【详解】pure 意为"纯粹的,纯理论的",从句意来看,一个只对科学研究感兴趣,而不是为了使自己的发现得到应用,而进行研究的人是一个从事纯理论研究的科学家。C. pure 符合此意,为正确答案。A. genuine 意为"真正的",用在此处,显然意义不全面,因为科学家还包括从事实用研究的。B. practical 意为"实用的,注重实际的",正好与句子表达的意义相反,D. clever 意为"聪明的",也与上下文意义不符。

10.【答案】D

【详解】此处应根据句意选择一个恰当的副词。科学家主要的工作是进行研究,从四个选项的意义来看,科学研究不能根据人的情绪愉快地(happily)或不情愿地(reluctantly)进行,也不会只是偶尔地(occasionally)进行,应该是准确地(accurately)进行,所以 D. accurately 最符合句意,为正确选择。

11.【答案】D

【详解】这里需要注意上下文的逻辑关系。前面说过从事纯理论研究的科学家只对研究感兴趣,而不是为了使他的发现得到应用进行研究,显然这里表示的是目的,所以选 D. so,so that 可以引导目的状语从句。now that 意思是"既然",相当于 since,在这里不符合上下文的逻辑意义。all 后的 that 从句是它的定语,它不能与 that 构成连接词。and 在句中连接两个并列成分,而在此句中,前面没有与 that 从句并列的成分,所以也不能入选。

12.【答案】C

【详解】本句讲的是发明家的工作性质与特点，而这种特点是一贯的、固有的，所以应该选用 C. usually(通常)，不能用 A. seldom(很少)，B. sometimes (有时)或 D. never(从不)。

13.【答案】B

【详解】在本句中，作者告诉我们发明家也就是对实用科学感兴趣的人，那么他制造的东西应该是能实际使用的。本段的最后一句也说：不管采用什么方法，发明家都是为了得到一个具体的结果，如制造一台收割机，使电灯泡发光等。而这些都是具有实用价值的，所以应选 B. use. 另外，从搭配意义上来说，something has a concrete plan/idea/means 也不正确。

14.【答案】A

【详解】这里需要搞清楚 theories 与 science 的关系，非常明显，它们之间的关系是所有性质的，所以应选用 of，意为"科学的理论"。

15.【答案】D

【详解】本文说发明家通常都努力想制造出某种有实际用途的东西，而此句所举的例子如制造收割机，使灯泡发亮还有其它的目标都是具有实用价值的，那么他的研究要得到的是一种具体的、特定的结果。四个选项中，惟有 D. specific 意为"具体的，特定的"，是答案。

16.【答案】C

【详解】根据句意及所给选项，A. few 在此意义明显不对；B. those 用来指前面提到过的人或事物，但在此句中没有它所指代的对象；D. all 与 C. many 都可以用来表示数量，但是对于一个发明家来说，让他达到所有目标，解决所有问题是不可能的，相比之下，C. many(许多)更符合实际，所以是答案。

17.【答案】B

【详解】上文说到在工业革命中发明机器的人是发明家而不是科学家，那么，此句中的发明家指的就是那些发明机器的人。四个选项中，只有 develop 有"研制、开发"的意思，所以是答案。

18.【答案】A

【详解】从本句的结构及意义看，要求选用一个与 no 意义相近的词，四个选项中，只有 A. little 意为"几乎没有"，表示否定意义，其余三词都不表示否定意义，不符合本句要求。

19.【答案】B

【详解】本句的结构很明显是一个虚拟条件句。表示与过去事实相反，但是由于主语带有修饰成分，比较长，因而可能会影响理解。本句意为："如果没有科学家早年打下的基础，那些在科学上接受过很少或根本没有受过训练的人就不可能有发明创造。"因此这里需要选用表示条件的连接词 if. C. because 和 A. as 用来引导表示原因的状语从句，D. while 可以用来引导时间状语或并列句表示语气的转折，用在此句中均不符合上下文的意思。

**20.**【答案】D

【详解】这里只要注意到了谓语动词的时态 had not been laid 就能确定答案为 D。before 意为"以前"，表示在过去某个时间以前，常与过去完成时连用。A. ago 意思也是"以前"，但它指的是现在某个时间以前，通常与一般过去时连用。B. past 意思是"过去的"，表示时间时一般不用作副词。C. ahead 意为"前面的"，它通常指将来的时间，而不是过去的。

# Unit 6

1.【答案】B

【详解】这句话的意思是：强大的发动机噗噗地发动起来。strong 意为"强壮的"；powerful 意为"强大的"；forceful 意为"坚强的，有说服力的"；efficient 意为"有效率的"。

2.【答案】C

【详解】这句话的意思是：潜水艇开始下潜。此句中，介词在句首，因此，要倒装。

3.【答案】A

【详解】这句话的意思是：像鲨鱼潜水一样。此处用一个动名词来形容正在下潜。

4.【答案】D

【详解】这句话的意思是：很快光线消失了。gone 在此处相当于一个形容词，意为"已去的，过去的，消失"。

5.【答案】A

【详解】这句话的意思是：正在下潜的潜艇前灯亮了。turn on 意为"打开"；turn down 意为"拒绝"；turn up 意为"出现"；turn out 意为"结果是，被证明是"。

6.【答案】B

【详解】这句话的意思是：但是不止这些。rather than 意为"宁愿"；more than 意为"不仅仅"；other than 意为"除了"；less than 意为"比……少"。

7.【答案】C

【详解】这句话的意思是：以另一种方式做一次未来旅行。从破折号后看，有两个方向，即 up 和 down，所以，要用 one 和 the other。

8.【答案】D

【详解】这句话的意思是：在天文台那里有一架巨大的望远镜可观测到太空深处。该语法为疑问副词引导定语从句。

9.【答案】A

【详解】这句话的意思是：比地球上任何人观测到的都远。ever 意为"曾经"；even 意为"甚至"。副词一般放于助动词之后，实义动词之前。

10.【答案】B

【详解】这句话的意思是：自行车或溜冰滑板可能由喷气动力推动。control 意为"控制"；run 意为"操纵，开动"；pull 意为"拖，拉"；direct 意为"指引，指挥"。

11.【答案】C

【详解】这句话的意思是:一种新型的可开的东西。此处用不定式表示目的,ride 意为"适于乘的,适于骑的"。

12.【答案】A

【详解】这句话的意思是:多数食品将是人工制造的。artificial 意为"人工的";false 意为"假的,不正确的";artistic 意为"艺术的";unnatural 意为"矫揉造作的,不自然的"。

13.【答案】C

【详解】这句话的意思是:在工厂里由令人吃惊的原料制成。be made of 意为"用……制造",产品看得出原材料;be made from 意为"用……制造",产品看不出原材料;be made in 意为"在……制造"。

14.【答案】B

【详解】这句话的意思是:这些东西被精心调和。mix 意为把几种东西保持原状的混合,意味着很乱;blend 意为"混合",但保持着整体的统一性,很和谐;confuse 意为"使混乱,混淆";breed 意为"使繁殖,饲养"。

15.【答案】A

【详解】这句话的意思是:未来的食物味道鲜美。delicious 意为"美味可口的";delicate 意为"娇嫩的,精致的";deliberate 意为"故意的";tender 意为"柔软的,温柔的"。

16.【答案】A

【详解】这句话的意思是:科学家们有可能找到大多数疾病的治疗途径。cure 意为"(疾病)治愈,治疗方法";heal 意为"(外伤,伤口等)治愈";remedy 意为"药物";treat 意为"治疗",是动词。

17.【答案】B

【详解】这句话的意思是:电子信号将控制每辆车。symbol 意为"象征";signal 意为"信号":sign 意为"标志";signature 意为"签名"。

18.【答案】C

【详解】这句话的意思是:到司机想去的任何地方。在主句中,get 缺少一个宾语,在从句中,go 缺少一个宾语,而 wherever 等于 the place where,因此,选 C。

19.【答案】A

【详解】这句话的意思是:撞车是不可能的。smash 意为"猛撞,猛冲";crash 意为"坠落,坠毁";rush 意为"冲";crush 意为"压碎,碾碎"。

20.【答案】D

【详解】这句话的意思是:控制系统不需触摸。remove 意为"移动,搬开";seize 意为"抓住";hold 意为"拿着,握住";touch 意为"触摸,碰"。

# Unit 7

1.【答案】B

【详解】这句话的意思是:他们会给健康带来潜在危害。possible 意为"可能的";potential 意为"潜在的";available 意为"可用到的,可利用的";proficient 意为"精通"。

2.【答案】C

【详解】这句话的意思是:封闭环境内的空气循环如何传播疾病。如何传播疾病只能用 how。

3.【答案】A

【详解】这句话的意思是:使拥有者暴露在有害的化学物质下。expose 意为"使暴露";explode 意为"爆炸,爆发";export 意为"出口,输出";expand 意为"扩张,使膨胀"。

4.【答案】D

【详解】这句话的意思是:最为人所知的威胁之一是军团病的危害。that 代指 danger.

5.【答案】A

【详解】这句话的意思是:人们最初是在二十世纪七十年代认识军团病的。which 在此引导非限制性定语从句。

6.【答案】B

【详解】这句话的意思是:军团病已经影响了居住在有空调的房子里的人。不定式的完成时态表示动作发生在过去。

7.【答案】C

【详解】这句话的意思是:热空气又以某种方式被吸进气孔。anyhow 意为"无论如何,不管怎样";anyway 意为"无论如何,不管怎样";somehow 意为"以某种方式";somewhat 意为"有点,稍微"。

8.【答案】A

【详解】这句话的意思是:在很多情况下,是由于设计太差。according to 意为"根据";due to 意为"由于,因为";as to 意为"关于,至于";because 意为"因为"。

9.【答案】D

【详解】这句话的意思是:携带疾病的细菌来自建筑物外。细菌来自建筑物外,因此,要用现在分词表主动。

10.【答案】B

【详解】这句话的意思是:携带细菌的空气与冷空气结合起来。be combined with 意为"与……结合"。

11.【答案】A

【详解】这句话的意思是:如果他们经过废气管。"经过"用 past;through 意为"通过";passed 和 passing 为 pass 的分词形式。

12.【答案】C

【详解】这句话的意思是:许多老式建筑物需要经常的监督。acquire 意为"获得";inquire 意为"询问,查究";require 意为

“需要,要求”;request 意为“请求,要求”。

13.【答案】C

【详解】这句话的意思是:空调滤净空气的方式。以……方式为“in the way”,因此,关系代词 which 前必须有一介词 in。

14.【答案】A

【详解】这句话的意思是:静电滤尘器将空气中的灰尘和烟尘粒子排走。remove 意为“移动,移走”;take 意为“带走”;bring 意为“带来”;dismiss 意为“解散,开除”。

15.【答案】B

【详解】这句话的意思是:然而,滤尘器也将大量的正离子带入了循环系统。therefore 意为“因此”;however 意为“然而”;though 意为“虽然”;accordingly 意为“因此”。

16.【答案】D

【详解】这句话的意思是:过分暴露在正离子下会导致头疼。result from 意为“归因于,来源于”;settle down 意为“定居”;lie in 意为“存在于”;result in 意为“导致”。

17.【答案】D

【详解】这句话的意思是:人为的空气环境也会影响我们。effect 意为“影响”,是名词;effort 意为“努力”;afford 意为“供得起”;affect 意为“影响”,是动词。

18.【答案】A

【详解】这句话的意思是:在自然的环境里。natural 意为“自然的”;artificial 意为“人工的,人造的”;normal 意为“正常的”;unnatural 意为“矫揉造作的”。

19.【答案】A

【详解】这句话的意思是:人的身体长期以来习惯了正常的变化。be accustomed to 意为“习惯于”;be familiar to 意为“(东西)对于(人)很熟悉”;be familiar with 意为“(人)熟悉(东西)”。

20.【答案】B

【详解】这句话的意思是:这导致免疫系统减弱。lead to 意为“导致”;guide 意为“引导”;cause 意为“导致”,其后不能跟 to。

## Unit 8

1.【答案】B

【详解】这句话的意思是:他们不断地吸引年轻的女子。apply to 意为“应用”;appeal to 意为“吸引”;suit to 意为“适合”;attach to 意为“附属于”。

2.【答案】C

【详解】这句话的意思是:不管是芭蕾舞,弗拉曼柯舞,还是舞厅的舞蹈。此处连词的用法是 whether. . . or. . .

3.【答案】A

【详解】这句话的意思是:最近,这种舞蹈场景又融入了一种新的元素。由于句中出现了 recently,且这种元素的融入是从过去到现在而且仍旧在进行,所以要用现在完成进行时。

4.【答案】C

【详解】这句话的意思是:这些舞蹈直到近来才为日本人所知。此句中,unknown 隐含 not. not. . . until 意为“直到……才”。

5.【答案】A

【详解】这句话的意思是:这些舞蹈正席卷文化中心。sweep 意为“席卷,扫过”;sweet 意为“甜的,可爱的”;swallow 意为“吞下,咽下”;switch 意为“转向”。

6.【答案】B

【详解】这句话的意思是:肚皮舞首先是北非和中东的妇女跳的。舞蹈是被跳,因此用过去分词来修饰。

7.【答案】D

【详解】这句话的意思是:肚皮舞起初是在庆祝场合跳。extremely 意为“极端地,非常地”;violently 意为“猛烈地,激烈地”;intensively 意为“强烈地,集中地”;originally 意为“最初,原先”。

8.【答案】A

【详解】这句话的意思是:日本人过去是通过电影熟悉肚皮舞。gain 意为“获得,得到,赢得”;win 意为“赢”;earn 意为“赚得”;achieve 意为“完成,达到”。

9.【答案】B

【详解】这句话的意思是:然而,它的流行起于三年前。两句间是转折关系。

10.【答案】D

【详解】这句话的意思是:那时,越来越多的日本年轻女子到土耳其和其他一些国家旅行。由于先行词是 three years ago,且从句中缺少状语,因此用 when。

11.【答案】C

【详解】这句话的意思是:大多数的参与者最初是被它华丽的服饰和性感的动作所吸引。draw on 意为“提款”;draw in 意为“吸引”;draw up 意为“起草”。

12.【答案】A

【详解】这句话的意思是:它提供了强有力的训练。provide 意为“提供,供给”;supply 意为“供给(缺乏和不足之物)”;give 意为“给”;arrange 意为“安排”。

13.【答案】B

【详解】这句话的意思是:健身俱乐部甚至开始将肚皮舞作为训练课程。contain 意为“包含”;include 意为“包括”;involve 意为“涉及,使陷于”;comprise 意为“由……组成”。

14.【答案】C

【详解】这句话的意思是:这家餐馆位于东京的 Shinjuku 地区。位于……“be located in”,因此用 located。

15.【答案】A

【详解】这句话的意思是:越来越多的妇女专门来这家餐馆是

为了跳舞。specifically 意为“特定地，明确地”；specially 意为“特别地”；particularly 意为“独特地，显著地”；peculiarly 意为“特有地，特别地”。

16.【答案】B

【详解】这句话的意思是：跳民族舞的人相比之下很少。与……相比为 compare... with...，因此要使用过去分词。

17.【答案】D

【详解】这句话的意思是：爵士和芭蕾在日本流行已有几十年了。由于出现了 for decades，因此要使用现在完成时。

18.【答案】B

【详解】这句话的意思是：随着越来越多的日本人到国外学习舞蹈。介词 with 后固定用名词或动名词。

19.【答案】A

【详解】这句话的意思是：他们回日本后教授舞蹈。介词 with 后固定用名词或动名词。且 teaching 和 heading 构成并列结构。

20.【答案】B

【详解】这句话的意思是：热心民族舞的人无疑会稳定地变多。firmly 意为“坚定地，稳固地（形容不易动摇）”；steadily 意为“稳定地（稳定、经久不变）”；stably 意为“稳定地，坚固地（强调事物在进行过程中不偏离，不波动，有规则地稳步前进）”；solidly 意为“坚硬地”。

## Unit 9

1.【答案】D

【详解】大部分时间都在水下的人应称作潜水员。D. divers 潜水员是正确答案。实际上，答案可以从原文第四行中找到“a...，a diver...”A. fisherman 渔民；B. seamen 水手；C. sailors 海员。均不合适。

2.【答案】A

【详解】原文此处的意思是“如果船出了毛病”something/anything goes wrong with... 意为……出了毛病，习惯搭配。其他各项无此习惯用法。

3.【答案】C

【详解】原文此处意为“在（船的吃水线）下。”B. over 在……之上，与文义不相符，首先排除。A. under，C. below，D. beneath 都有“在……下边”的意思。under 含有“在……之中”，或“在……之内”的意思。e. g. swim under the water 在水下游泳。由于 waterline 只是一条“线”，所以用 under 意思不当。beneath 含有“在……正下方”或“在（紧靠着）……底下”。below 可以用来表示位置、职位等“在……下面。”e. g. 100 metres below sea level 海平面下 100 米；通过以上辨析，可以看出 C 项符合文义，是正确答案。

4.【答案】B

【详解】原文此处意为“潜水员穿上潜水服，带上工具下去……”本题四个选项均含有“穿”的意思。A. wears 穿着、戴着，强调穿的结果或状态。e. g. He wears a military uniform. 他穿着军装。B. puts on 穿上……，强调穿的动作或过程。C. dresses 给……穿衣，及物动词。e. g. Dress yourself more neatly. 穿得整齐一点。D. be dressed in...，穿着，和 wears 当“穿着”讲时用法相同。根据文义分析，此处的“穿”不是强调结果，而是强调穿的动作。因此 B 项是正确答案。

5.【答案】D

【详解】原文此处意为“……下去修补破损之处。”跟在动词短语 go down 之后做目的状语。只能用带 to 的动词不定式。其他各项形式均不正确。

6.【答案】A

【详解】原文此处意为“他穿着一种由橡胶和粗帆布制成的特殊潜水服……”首先运用排除法：C. consisted of D. composed of 由……组成，指组成某一完整事物的若干部分。用于此处不当，首先排除。A. made of B. made from 两组短语均含有“由……制成”的意思。但是 be made of 指所有的材料在成品中保持原来的样子，或构成混合物及化合物的成分。be made from 指用这种原料造出来的产品的样子与原来的原料已有所不同。据以上辨析，A 项符合原文意思，是正确选项。

7.【答案】C

【详解】分析原文句子结构。可以看出引导词在句子中做主语，因此 C 项是正确选项。

8.【答案】D

【详解】原文此处意为“把水隔在外边”。keep out 使……在外，使……不入内。因此 D 项是正确答案。A. away，keep away 站开，使离开；B. off，keep off 让开，不接近；C. far，不能单独与 keep 连用。

9.【答案】B

【详解】原文此处意为“（衣服）把（潜水员）从头到脚都包了起来，只有手可以自由活动”。A. and 用于此前后关系不正确，排除。B. but C. while D. however 均可表示转折关系。while 含有“对比”的意思 e. g. She was dressed in red while her sister was dressed in blue. 她穿红色衣服，而她妹妹却穿蓝色的衣服。用于此处不当，因为文中并无对比含义。however 可是、但是，多插在句中，前后都要用逗号隔开，有时也放在句首或句尾。e. g. She was，however，aware of the circumstances. 不过她还不知道这些情况，文中并无标点符号隔开，故不用。B. but 表示一般的转折关系，用于此结构合理，是正确答案。

10.【答案】C

【详解】联系上下文“从头到脚都被衣服裹了起来”，只有手露在外边，是因为要用手进行修理工作，也就是说，手是可以自由活动的，不受衣服的限制。因此 C 项与文义相符，是正确答案。A. empty 空的；B. useful 有用的；D. helpful 有帮助的。均不正确。

11.【答案】D

【详解】原文此处意为"(潜水员衣服)的衣袖在手腕处形成一个不漏水的袖口。"D. at 在……,可用于表示在某一地方,某一点。是本题正确答案。A. on 在……上面,用于此处不当。B. round 在……周围、围绕。手腕是在衣袖口内,而不是被(很多)袖口围着。C. around 在周围,一般用做副词。

12.【答案】B

【详解】原文此处意为:他(潜水员)戴上一顶连接有软管的重钢盔……D. attached 贴上、系上,不合文义,首先排除。A. linked; B. connected; C. combined 都含有"连接"的意思。link 强调不同事物之间的联系。e. g. The interest of individuals is indissolubly linked with that of the country. 个人利益和国家利益不可分割地联系在一起。combine 含有"联合、合并"的意思,强调把不同的事物结合在一起。常指抽象事物。connected"衔接、连接",表示把一种事物与其他事物连接在一起。据以上辨析,联系原文,可知 B 项最合文义,是正确答案。

13.【答案】B

【详解】原文此处意为"(软管)从水面以上吸入空气"。D. onto 到……上,表示方向,与文义不符,排除。A. over; B. above; C. on 均含有"在……上"的意思。on 表示一事物在另一事物的表面之上,两事物之间表面接触。over 指垂直的上方。above 则是指笼统的上方。可以看出 B. over 最符合文义,故 B 项是正确答案。

14.【答案】C

【详解】原文此处意为:钢盔(前面)有窗状开口,而且和潜水服的脖子紧紧系在一起。C. fastens 紧扣、紧系。fasten to 和……紧系在一起。fasten 做不及物动词。符合题意,结构正确,是正确答案。A. is fastened 被动语态,强调主语与谓语动词是被动关系。强调被动动作的发生。用于此处,不符合文章意思,因为文章并不是要强调一种被动的动作。B. fastened 时态与文中不符。D. has been fastened 语态不正确。

15.【答案】D

【详解】原文此处意为"为了便于潜水员(下沉……)"D. in order that 为了,以便,后跟从句。符合文中结构要求,是正确答案。B. In order; C. In order which 结构不正确。A. In order to 后跟动词不定式,不能跟从句。

16.【答案】B

【详解】B. sink 下沉,不及物动词。动词不定式一般式表示动作未发生,符合文义(见 85 题)是正确答案。A. be sunk, sunk 作"下沉"讲是不及物动词,没有被动语态。C. be sinking 进行体表示正在进行的动作,与文中意思不符。D. have sunk 完成体表示动作已发生,不合文义。

17.【答案】A

【详解】原文此处意为"鞋由金属加重了重量"。be weighted with…由……加重重量。其他各项与 weight 搭配无此意。

18.【答案】C

【详解】原文此处意为"他的衣服重 150 磅"。C. weigh 重量为及物动词,符合文义,结构合理,是正确答案。A. are weighted, weight 作动词意为"增加重量",用于此处不当,故 B. weight 也不正确。D. are weighed, clothes 和 weigh 之间是主动关系,不用被动语态。

19.【答案】B

【详解】跟在动词短语 go down 之后做目的状语,用带 to 的动词不定式。

20.【答案】C

【详解】原文此处意为"……在一艘被毁的潜水艇"。C. damaged 过去分词做定语,表示被动结构,符合文义。A. damaging 语态不准确。B. dangerous 危险的。submarine 本身是没有危险性的。D. hopeful 满怀希望的,与文义不符。

## Unit 10

1.【答案】C

【详解】根据下文"... because it was already dark"作者从外边漆黑一片推断当时的时间。must + have done 表示对过去的推测"一定是……",符合原文义表达。A,B 项没有推测之意,因此不能用。D. should + have done 表示应该做而没有做,明显不符合文义。

2.【答案】B

【详解】开车回家,故用 back。不用 to 的原因是 home 是副词,不做介词宾语。

3.【答案】A

【详解】从上下文分析,可知"我"关上车灯就为了不让车灯的亮光打搅杰克睡觉。so as to 为了……表示目的,后跟动词原形,符合文章结构要求,为正确答案。in order 后加 to 则为正确答案。so that 后面不能直接跟动词,它引起目的(结果)状语从句。for 后要跟动名词,而不能跟原形动词。

4.【答案】C

【详解】从下文分析。因为"我"看到杰克的灯还亮着,说明他没睡,因此我关车灯的做法就是不必要的。此处表达应为"本不必做而做了"。用 needn't have done 可以表达这一意思。因此 C 项正确。A,B 两项不涉及"过去"而被排除。must + have done 只能用于肯定式表示对过去的推测,不能用于否定形式。

5.【答案】B

【详解】fall asleep 熟睡、睡着了,固定搭配,其他各项无此搭配。

6.【答案】A

【详解】按照常理“我”和杰克共住一套房子。如果他没有睡着,他应该向“我”打招呼,而现在他连头都不转过来看我一眼。因此用 even 更能体现原文涵意。

7.【答案】D

【详解】for God' s sake 看在上帝份上,表示加强语气的请求,固定用法。其他各项无此用法。

8.【答案】C

【详解】remind sb. of……使某人回忆起,习惯搭配。其他各项无此搭配。

9.【答案】B

【详解】in pain 疼痛,在痛苦中,固定搭配。其他各项均不符合英语习惯的表达。

10.【答案】B

【详解】do sb. an injury 伤害某人,属习惯搭配。in case 万一,引导的从句不用虚拟语气。

11.【答案】C

【详解】我想他一定是生病了,因此我最好给他开个玩笑让他高兴。因此用 I' d better(最好)而不用 I' d rather(宁愿)更能确切传达原文意思。I' d better = I had better 而不是 I would better,因此 C 答案正确。注意 I' d rather = I would rather,而不是 I had rather。

12.【答案】A

【详解】根据下文判断,这条蛇应是在床单的夹层之间睡着了。A 答案符合文义。B. beside 在……旁边。如果在床单旁边,杰克不至于不敢动。C. below 在……以下。指位置位于某物体或在某物下方,所指范围较宽。床单下面是没有空间的。因为床单是紧贴床面的,所以用 below the sheets 意思不通。D. behind 在……后边。“在床单后边”无法理解。

13.【答案】D

【详解】此处原文意思为“我正仰面躺着看书……”应用过去进行时态。lay 放置、搁,及物动词用于此处意思不当。另外,lay 可作动词 lie 的过去式(lie“躺、平躺、位于”讲时)D. lying,是动词 lie 的现在分词形式。用于此处符合文义。注意:lie 做“躺、平躺”讲时,过去式和过去分词为 lay, lain; lie 做说谎讲,过去式和过去分词是 lied, lied。动词 lay(放置、搁)的过去式和过去分词为 laid, laid。

14.【答案】B

【详解】out of the question 不可能,固定用法。注意:out of question 毫无疑问。

15.【答案】C

【详解】此处应理解为“我意识到他是认真的”。A. in fact 事实上,做状语,不可以做表语,故不能用于此。B. in serious 搭配不正确。可以说,in all seriousness 非常严肃。C. in earnest 认真、当真,既可做状语又可做表语,符合文义,为正确答案。D. in truth 事实上,同 in fact,不可做表语。

16.【答案】D

【详解】主句用过去进行时,根据主句与从句时态一致原则,从句也应用过去时态,用一般过去式,不能用过去将来时。as soon as 引导的从句不能用过去进行时态。故 D 项是唯一正确答案。

17.【答案】A

【详解】此处为杰克的推测“它(蛇)也许会醒来”。故用 might 表示推测可能。can 表示“可能性”,指可能性极大而且会发生,显然不符合文义。C,D 两项均无“推测”之意。

18.【答案】D

【详解】此处原文义为“我认为你一个小时前就该回来了。”杰克根据以往的经验知道“我”通常回来的时间。因此他就想当然地认为“我”今晚也会像往常一样准时到家,而没想到我会回来晚。A. made it certain that 确实弄清楚……,实际上是不清楚(“我”要回来的时间),所以 A 项不能选。B. have been assured 确信……,含“得到别人的保证,一定会……”的意思。显然不符合文义;另外,前后时态不一致。因此 B 项也不能选。C. counted on 指望、期待。后不能跟 that 引导的宾语从句。也被排除。D. took it for granted 认为……当然……。后跟 that 从句,做真正的宾语(it 是形式宾语)符合文义,结构正确,故 D 项是正确答案。

19.【答案】A

【详解】apologize for (sth. or doing sth.)为……向某人道歉,习惯搭配。不能直接跟 on,也不能接动词不定式。

20.【答案】B

【详解】原文此处意思应为“我尽量用目光来鼓励他,然后出去给医生打电话。”因为杰克已经等了很久,而且蛇随时都可能醒来,他内心一定充满了恐惧,所以我应当用目光来激励他要挺住。B 项符合文义为正确答案。A. as fiercely as 严厉地,按情理讲,是不该这么做的。C. so bravely 勇敢地,用于此处是指“我”自己勇敢地看着他,与原文意思不符。D. so hopefully as 满怀希望地。“我”不用“满怀期待”地看着他,因为危险不在“我”身上。

## Unit 11

1.【答案】A

【详解】原文此处意为:它们(信用卡)给它们的持有者(拥有信用卡的人)在商店、旅馆……提供自动赊货。填入什么词,关键是要弄清楚句子的主语 they 的指代关系。从上下文看出,they 指的是 credit cards。因此 owners(拥有者)符合文章内容表达,是正确答案。

2.【答案】B

【详解】看一看此空前边的短语,... in stores, restaurants, and hotels, at home, across the country and even 在商店里,饭馆里,旅馆里,在家里,在全国各地,沿着此种思维方式再进一步,那就是"在国外"。因此,B. abroad 在国外,符合文章表达,是正确答案。其他各项均无此意。

3.【答案】A

【详解】原文此处意为:"它们(信用卡)还使得(人们)能够利用银行的许多服务项目(使银行的很多服务项目可利用)。A. available 可利用的,符合文义,是正确答案。D. practicable 可行的、行得通的,服务项目一般来说不存在行不行得通的问题,只存在人们到底能不能享受服务的问题。B. sensible 可感觉的,明智的,与文义不符。C. valuable 有价值的。用于此处,文章意思不顺。因为文章讨论的不是服务的价值,而是人们可享受的服务。

4.【答案】D

【详解】原文此处意为:……(信用卡)使取钱和存钱在人口分散的地区成为可能……"此处用 deposit(存款)是和 withdraw(取款)相对而言的,符合文章意思。因此 D 项是正确答案。

5.【答案】D

【详解】从上下文判断,原文此处意为:……不管当地的支行是否营业,(人们都可以存款或取款)。因此 D. open(营业着的)是正确答案。其他各项均不能正确表达文章的意思。

6.【答案】A

【详解】根据上文判断,此处意思应为:对于我们很多人来说"无货币社会"已不是遥远的事情了——它已经在我们面前了。A. here 在这里,符合原文意思表达是正确答案。B. out 外边;C. inside 在……里边,均不合文义。D. now 现在,副词,描述动作发生的时间,不作表语。

7.【答案】B

【详解】联系上下文,原文此处意为:计算机给消费者提供方便的同时,也给卖货人带来了好处。B. advantages 优点、好处。符合题意,是正确答案。A. inconveniences 不便,不符合上下文意思,因为下文讨论的内容是计算机给卖主带来的好处。C. conditions 条件;D. services 服务。均不能正确表达文章意思。

8.【答案】A

【详解】此处原文义为:它们(计算机)保留了广泛的记录,包括谁卖了什么东西,什么时间(卖给)某人。A. range 范围,符合题意,是正确答案。B. deal 可以说 a great deal of 大量的,但不可与 wide 搭配。C. scope 范围,但与 range 不同的是,scope 指的某一活动,某一影响所能达到的程度及影响,还包括一个人的理解能力的范围。用于此处意思表达不当。D. extension 伸长、延长,不符文义。

9.【答案】B

【详解】根据上文,此处意为"卖给谁",和前边的词语从逻辑上构成恰当的关系:"... including who sold what, when, and to whom."因此 B 项是正确答案。

10.【答案】C

【详解】原文此处意为:这种信息使商人们能通过显示哪些商品正在销售以及货物的转运速度来掌握货物的线索。C. sold 卖、销售。动词 sell 的过去分词。与 being 一起构成被动语态进行体,表示货物正在销售中,符合文章意思,是正确答案。A. taken 拿;B. given 给……,不符合原题。D. transported 运输、运送。与"moving"意思重复不能用于此。

11.【答案】C

【详解】原文此处意为:然后就可以做出登记货物和把货物返回给供应商的决定了。C. record 记录、登录,符合题意,是正确答案。事实上,答案可以从下面的"these computer record..."找到暗示的答案。其他各项用于此,使句子意思与下文连接不紧密。

12.【答案】D

【详解】原文此处意为:同时这些计算机记录下来哪个小时最忙,哪个雇员办事更有效率。表示哪一个应当用 which。其实,从上边的"which hours are business"找到暗示的答案。

13.【答案】B

【详解】此处原文义为:……可以据此制定相应的人才和工作人员计划安排。从上下文推断,此处含顺承关系。B. accordingly 因此,相应地,照着办。能正确表达原文句之间的逻辑联系,因而是正确答案。其他 A. actually 事实上;C. similarly 相似的;D. exactly 确切地。均不能正确表达原文逻辑关系。

14.【答案】B

【详解】此处意为:为着同样的原因,生产者也要依靠计算机。B. relied 依赖、依靠。rely on 依赖……。符合搭配要求,是正确答案。A. supported 本身是及物动词,不与 on 连接。C. assisted 帮助、援助,不与 on 连接。D. centered 集中。center on 集中……于……。与文章意思不符。

15.【答案】C

【详解】根据并列部分结构一致的原则。此处 develop 前应加上 to,变为动词不定式和上文的"which products to emphasize"结构对应。

16.【答案】C

【详解】根据上文意思,由计算机制定的市场分析报告有助于(生产者)决定哪些商品应现在重点生产,哪些产品将来要发展,那么此处的意思应该为"哪些产品的产量应下降",才使文章意思表达完整正确。因此 C. drop 下降、降低是正确答案。

17.【答案】D

【详解】原文此处意为"计算机记录贮备中的货物,现有的原材料……"。D. hand, on hand 手头,在跟前,现有。符合文义,是正确答案。A. schedule 时间表 on schedule 按时,用于此与原句语意不通。B. business 事务,on business 因公、有事,与文章无关。C. sale 出售。on sale 出售的、上市的。根据原文意思判断,此处的原材料是用来进行生产的,而不是用于出售的。

18.【答案】B

【详解】根据上下文意思判断,此处原文的意思应是:大量的其他商业企业。因为下文提到商业企业名称是上文未提到的。所以 D 项首先被排除。C. another 表示另外任何一个,用于此不当。A. others 相当于"other + 名词"的复数形式,用于此处结构不正确。因此,只有 B. other 是正确答案。

19.【答案】A

【详解】此处原文的意思是……,从煤气和电力公用事业到牛奶加工厂…… A. utilities 公用事业(常用复数)。此处用 utilities是为了和其后的 processors 形成结构上的对应,都是表示"企事业单位"。其他各项都不能满足这一结构要求。

20.【答案】D

【详解】根据上文,原文此处意为:(商业企业)通过利用计算机向消费者提供更多更好的高效服务。因为企业生产的最终目的,就是为消费者服务的。所以此处用 D. consumers(消费者)比其他各项都合适。

## Unit 12

1.【答案】B

【详解】此处是一个表示时间的从句,四个选项中,A. As B. When C. While 都可用来引导时间状语。A. As 多用于口语,强调"同一时间"或"一先一后",有时还可指"随着"。B. When 则强调特定时间,C. While 也表示同一时间,它所表示的时间不是一点,而是一段。根据此句所表达的意义,应选用 B。

2.【答案】A

【详解】根据作者在下边所说的可以看出,当发生这种情况时,做点其它事情,放松了,那么,你就会自然而然地想起忘记了的事情。因而作者在此提议不要努力去回想。所以 A. try(试图,努力)符合句意,是答案。C. hesitate(犹豫)与 D. wait(等候)用在此处意为"赶快做……",与上下文意思相反,不能选用。

3.【答案】C

【详解】作者在这里建议不要努力去回想,而是假设做点其它事情。C. else 可与不定代词 something, anything 等连用,意为"其它的"。符合句意,是答案。A. simple 意为"简单的",做点别的事情,目的是为了使大脑放松,任何事情都可以,不论简单与否。B. apart 意为"分开的",与 A. 同样,不是句子所要求的。D. similar 意为"相同的,相似的。"用在此处,与上下文意思矛盾。

4.【答案】B

【详解】在英语中,有一种固定的表达结构,即前边是一个祈使句,后边跟一个由 and 连接的句子,前边的祈使句相当于一个条件句,后边的句子表示结果,所以选项 B. and 在此符合句子的意义及结构,是正确选择。A. unless 意为"除非",表示一种否定的条件,用在此处,前后两句意思矛盾,C. or(或者)用来连接两个并列句,此处也不能选用,D. until(直到……为止)表示动作或状态的继续,用在此处与上一句中的"for a couple of minutes"语义上发生矛盾,也不能选。

5.【答案】D

【详解】从上下文意义看,此处所说的 person 含有特指意义,指你见过又忘记了他的名字的人,所以不能选用表示泛指的不定冠词 a。B. certain 意思是"某个,某些",表达的也是不定的意义,A. some 作形容词,除了表示"一些,若干"之外,也可指未知而不是特定的事物,意为"某个,某些";D. this 意为"这,这个",具有特指意义,因而是正确选择。

6.【答案】C

【详解】上句说既然你见过这个人,知道了他的名字,它就会在你的记忆里,如果想不起来时,惟一要做的事情只是从记忆中把它找出来。所以 C. only(只,仅仅)符合上下文及句意,是答案。A. then(那么)表达的是总结性的语气,与上下文逻辑不符。B. really(确实,实际上)与 D. indeed(确实)意义大致相同,与上下文表达的语气不符,都不能入选。

7.【答案】D

【详解】从句子结构可以看出,此处需选择一个能与介词 for 搭配的动词,四个选项中,只有 D. prepares 符合要求,prepare... for... 意思是"为……准备……",是答案。其余三个选项均不能与 for 搭配使用。

8.【答案】B

【详解】activities 意思是"活动",根据上下文得知,作者在这里讲述的是如果忘记了一个人的名字,怎样再把它从模糊的记忆中回想起来,而记忆与回想无论是清醒状态中还是下意识的,都是大脑的活动,而且在下文中,也可以找到"deeper mental activities in the subconscious mind ..."(潜意识中较深层的智力活动)的表达。因此选 B. activities, A. deeds 指人的"行为,功业",C. movements 指的是"运动,移动",D. procedures 指的是办事情的"程序,方法",用在句中,均不符合句意。

9.【答案】C

【详解】dim 意思是"模糊不清的";A. light 意为"明亮的",B. fresh 意为"新鲜的,清新的",D. dark 意为"黑暗的"。从意思上来说 A. light 和 D. dark 不能修饰 memory, B. fresh 还有"(印象)鲜明"的意思,它可以用在"sth. is fresh in one's memory"结构中,意思是"某人对某事记忆犹新"。但它也不能直接作定语修饰 memory,只有 C. dim 可以与 memory 搭配,意思是"模糊不清的记忆",符合上下文及句意,所以是正确选择。

10.【答案】B

【详解】从句子结构来看，此处要求填入一个修饰 never（从不）的副词。never 是一个表示绝对概念的词，但是为了留有余地，使口气缓和一些，常用 almost（几乎）来修饰，因此 B. almost 是答案。A. merely 意思是“仅，只不过”；C. barely 意为“仅仅，几乎没有”；D. hardly 意为“几乎不，几乎没有”，都是表示否定意义的，用在此处，与 never 构成双重否定，与句子意义相反，因此都不能选用。

11.【答案】A

【详解】根据分号后边句子中的动词及上下文的语义，此处显然需要填入一个与 tighten（使绷紧）意思相对的动词，选项 A. loosen（使松弛）用在此处符合上下文的意义，因此是答案。C. decrease 与 D. reduce 都是“降低，减少”的意思，B. weaken 意为“削弱”，均不符合句子要求。

12.【答案】B

【详解】从下文所讲的可以看出，学生在考试中采用此种方法尤为有用，而英语“在考试中”介词用 in，不能用其它三个。

13.【答案】C

【详解】根据上句所说，这种方法是“preparatory method”（准备方法），那么，学生应该在答题之前先看问题，然后才能知道哪些容易哪些难，所以，此处 before（在……之前）符合上下文逻辑，是答案。A. after（在……之后）与句子要求正好相反，B. besides（除……之外，还……），D. against（与……相反）在此不符合句意。

14.【答案】D

【详解】此处这几句讲的是学生在考试中答题的步骤顺序，所描述的动作是一个接一个顺次进行的。四个选项中，A. Thus 意为“于是，因而”，C. Therefore 意为“因此，所以”，它们表示的是结果。B. But（可是，但是）表示语气的转折。只有 D. Then 是表示序列的，意为“然后，接着”，最适合句意，是答案。

15.【答案】A

【详解】根据句子结构，这里需要选一个能与 confident 搭配的介词，confident 意为“有信心的，确信的”，与介词连用，构成短语 be confident of，意思是“对……有信心（把握）的”，用在此处，符合句子的意思。B. with 和 C. for 不能与 confident 搭配，D. in 可以与名词 confidence 连用，构成短语 have confidence in，意思是“对……有信心”，但不能与形容词 confident 连用。

16.【答案】A

【详解】从句子结构及所给选择项看，这里需要选一个名词与 take 构成短语来说明主语 activities. 所给的四个选项都能与动词 take 连用。take place 意为“发生，进行”，take shape 意为“（思想）定形，成形”，take charge 意为“负责，掌管”，take action 意为“采取行动”。根据句子上下文的意义，这里应选 A，指学生在回答容易回答的同时，潜意识中的深层智力活动一直在进行。其余短语用来说明 activities，意思不通。

17.【答案】D

【详解】这里上下文的线索非常清楚，学生在答那些比较容易的问题时，潜意识中的智力活动一直在考虑另一些问题，根据下一句所说的，较容易的问题答完时，那些较难的问题的答案将开始出现在脑子里，可以得知，潜意识中一直考虑的是那些较难的问题，而且后边一句中也出现了“more difficult”的表达。所以应选 D。

18.【答案】A

【详解】answer 意为“答案”，其后常跟介词 to 引出短语作它的定语，意为“……的答案”，属于固定搭配，所以选 A。

19.【答案】D

【详解】根据句子的意思，这里指的是答案开始进入意识中，也即出现在大脑中。A. appear 意为“出现，显露”，多指具体事物的出现，不用于抽象意义。B. grow 本意为“生长，成长”，短语 grow into 意为“成长为……，习惯于……”。C. extend 指的是空间或时间的“伸展，延长”，come into 意为“进入”，四个选项中，只有 D. come 从结构及意义上都符合句子要求，所以是正确选择。

20.【答案】C

【详解】just 意为“只，仅仅”，相当于 only，根据文章所说的你见过的人，学过的知识，都会保存在你的记忆中，虽然你一时可能想不起来，但这只是一个等待的问题，它们早晚会出现在你的记忆中，所以 just 用在此句中意思正确，总结了全文。A. nearly 意思是“接近地，大约地”，不符合全文所表达的意义，B. 意为“可能，或许”，常与 very、most、quit 连用，表示可能性，在此也不符合句子的要求。D. even 意思是“甚至，即使”，用在句中意思不合适。

## Unit 13

1.【答案】B

【详解】根据上下文判断，此处应为“顾客（买烟的人）”。因此 B. customers（顾客），是正确答案。

2.【答案】B

【详解】原文此处意为：烟是给谁买的”。根据句子结果判断：whom 是宾格形式，做 brought 的间接宾语，应加入一个介词 for，bring（sth.）for sb. 给（某人）买（某物）。其他介词均不能用。

3.【答案】B

【详解】根据下文判断，此处原文义为：一个他从未见过的小姑娘大胆地走进他的商店……。因此 B. boldly（大胆地）符合文义，是正确答案。其他 A. nervously（紧张不安地），C. hesitantly（犹豫地）D. heavily（沉重地）均与文义不符合。

4.【答案】C

【详解】根据下文判断。此处原文义为:她手里有刚好(买一包烟的)钱……。因此,C. exact(确切的)是正确答案。

5.【答案】D

【详解】根据上下文判断。原文此处意为:她看起来非常自信。D. sure 确信。be sure of oneself 有自信心。因此 D 项是正确答案。其他各项 A. fond(喜欢),B. glad(高兴),C. ashamed(羞耻)均不符合文义。

6.【答案】B

【详解】根据上下文判断,约翰逊先生对她的这种非常自信的做法应感到非常惊讶。因此 B. surprised(惊奇的)是正确答案。其他各项 A. worried(忧虑的、担心的)C. annoyed(生气的)D. placid(平静的)均无此意。

7.【答案】A

【详解】根据下文判断。此处意为:他忘了问她一般要问的问题。因此 B. forgot(忘记)是正确答案。

8.【答案】C

【详解】根据上下文判断,此处应填入一个表示转折的词,C. Instead(不是、而是)表示转折关系,是正确答案。其他 A. Therefore(因此),B. So (因此), D. Somehow(有点儿)均无转折含义。

9.【答案】A

【详解】根据上下文,原文此处意为:姑娘很快做了回答……A. readily 很快地,符合文义,是正确答案。B. softly 柔和地,文中并无暗示。C. slowly 缓慢地,与文义不符。D. patiently 耐心地。文中并无暗示。

10.【答案】D

【详解】根据上下文,原文此处意为:他把烟递给她……。因此 D 项是正确答案。

11.【答案】B

【详解】根据下文意思判断。此处应填入一个词引导原因状语从句,据此 C. while, D. though 两项被排除。A. for, B. as 都是表示原因的连词。for 引导的句子放在(结果)后面,表示附带说明的理由或推断的理由。用于此处结构不当。而 as 引导的原因状语从句则可以置于(结果)前面,因此 B 项是正确答案。

12.【答案】C

【详解】根据上下文,原文此意为:她应该把烟盒藏在口袋里。因此 C 项是正确答案。

13.【答案】D

【详解】根据上文意思,原文此处意为"……以防警察看见"。D. case, in case 免得、以防,符合题意,是正确答案。其他各项无此用法。

14.【答案】D

【详解】根据上下文意思判断,此处应填入表示转折的连词。A. Moreover 而且;B. Therefore 因而;C. Then 那时;D. Nevertheless 然而,可是。可以看出,只有 D 项是正确答案。

15.【答案】A

【详解】原文此处意思应为:……她拿起那包香烟向门口走去。因此 A. packet(小包烟盒)是符合文义的是正确答案。

16.【答案】C

【详解】根据上下文,此处原文的意思为:她突然停住,转回身镇静地看着约翰逊先生。C. round, turn round 转回身。符合题意是正确答案。A. away, turn away 转身不看,不理睬;B. over, turn over 移交、交给;D. aside 与 turn 搭配少见。

17.【答案】B

【详解】根据上下文,原文此处意为:沉默了一会儿,这个卖香烟的人不知道她要说什么。B. wonder 想知道(而不知道)符合文义,是正确答案。A. doubted 怀疑,指不相信或对某人的言行产生怀疑。与文义不符。C. expected 期待、期望;D. considered 考虑、思考。均与文义不符。

18.【答案】D

【详解】根据上下文,原文此处意为:突然…… D. all, all at once 突然,是正确答案。其他答案与 at once 连接均无此意。

19.【答案】D

【详解】根据上下文,原文此意为:姑娘清楚而坚定的说……其前的形容词修饰语是 clear,其后的动词是 declare(宣告、断言)。因此,此处用 D. firm(坚定的)最合文义。

20.【答案】B

【详解】原文此处意为:她说着那些话"(我父亲是个警察)",迅速走出了商店。C. him; D. what 首先排除,因为这两项均使文义无法理解。从句子结构判断,此处是由 and 连接的两个分句,因此 and 后不是定语从句,用 which 不当。可以用 that 代指前边的内容。所以 B 项是正确答案。如果把 and 去掉,则 A. which 是正确答案。

## Unit 14

1.【答案】B

【详解】根据上下文,原文此处意为,他又一次失去……。set out 出发,因此用 B 项,其他各项搭配均无此意。

2.【答案】D

【详解】原文此处意为……这次,随行的有 17 条船。表示"随行,随身带有……",用介词 with,因此,B 项为答案正确。

3.【答案】C

【详解】原文此处意为:哥伦布随行带着那些想在新的陆地上安家的人。根据意思及原文结构判断,此处填入一个定语从的引导词,表示人,在从句中做主语,因此用 who。

4.【答案】D

【详解】原文此处意为:……过上哥伦布向他们描绘过的那种快乐的生活。live a . . . life 过……样的生活,因此 D 项是正确答案。

5.【答案】B

【详解】原文此处意为:"……而且变得富有"。根据句子结构判断,rich 是形容词,可以跟在连系动词后做表语。B. get 变得,成为,连系动词。符合文义,结构合理,是本题正确答案。其他各项 A. lead, C. have, D. own 都不能做连系动词,后边不能直接跟形容词,因此不能用于此处。

6.【答案】C

【详解】原文此处意为:第二次航行随船携带的给养足包括种子,工具,植物,马……。A. excluded 排除,与文义不符,首先排除。B. contained 包含,容纳,强调内部"包含,含有",e. g. Sea water contains salt. 海水里含有盐分,与文义不符,被排除。C. included;D. comprised 两词都含有"组成"或"包括"的意思。comprised 强调组成整体的各个部分有共同的特征。e. g. The course comprises two lessons 这一教程包括十篇课。不适合用于此处,因为组成"给养"各部分并不一定具有共同的特征。C. included 是一般用词,说指一般意义的包括,符合文义,是本题正确答案。

7.【答案】A

【详解】原文此处意思是:……,马和其他驯养动物,……。马本身就是一种驯养动物,因此,后边的应是"除了马之外的,其他的驯养动物"。A. other 其他的,表示"除了……以外的",符合文义,是正确答案。B. another 另一个,后边接可数名词单数形式,不能用于此处;C. rest 其余的人(或物),名词,不符原文结构要求。D. similar 相似的,用于此,文章意思不通顺。

8.【答案】A

【详解】原文此处意为:成功建立殖民地所需的一切东西……。D. make 使变成为……,make 后可以跟名词作宾语补足语,make a colony a success 使殖民地成为一个成功的殖民地……,符合文义及结构要求,是正确答案。其他各项均不符合原文结构搭配要求。

9.【答案】B

【详解】原文此处意为:"国王和王后告诉哥伦布对待印第安人要……"。从文义和句子结构判断,此处应填入一个宾语从句引导词,在从句中不担任何句子成分,只起引导作用。据此判断,B 项正确。

10.【答案】C

【详解】根据下文判断,原文此处意为"印第安人应被很好地对待"。故此 C. well 好地,副词,符合原文意思,结构搭配合理,是正确答案。D. kind 友爱的,形容词,不能用来修饰动词。A,B 两项意思与文义不符。

11.【答案】A

【详解】根据句子结构判断,此处应填入一个人称代词,代指上文已经提到的"the Indians",由于是复数形式,因此 A. them 符合要求,是正确答案。

12.【答案】B

【详解】原文此处意为:"当然,他们要成为基督徒"。of course 当然,习惯用法。其他各项均无此习惯用语。

13.【答案】D

【详解】根据下文,原文此处意为:"哥伦布的殖民地将继续开展贸易"。D. on, carry on 继续开展,进行下去;符合文义,是正确答案。A. out, carry out 开展,实施,指开始实施某项活动,措施等,与文义不符。B. off, carry off 夺走,抢走;C. over, carry over 转,留到……。均不符合文章意思。

14.【答案】B

【详解】根据下文,原文此处意为"西班牙货物只与黄金交换"。B. only 仅,符合文义,是正确答案。D. hardly 几乎不;C. scarcely 极不;D. carefully 认真地。均不符合题意。

15.【答案】A

【详解】原文此处意为:哥伦布接受(交易)利润的八分之一。贸易的所得应是利润,因此 B. profits(利润)是正确答案,其他各项均无此意。

16.【答案】D

【详解】根据上下文,原文此处意为:费迪南德和伊萨贝拉获取剩余的(利润),因此,此处用 D. rest 剩余的物(人等)。

17.【答案】C

【详解】原文此处意为:"当哥伦布到达……",C. arrived, arrived at 到达,符合文义,符合结构搭配要求,是正确答案。A. came 来,came to 到达,不与 at 连接。B. reached 到达,及物动词,直接跟目的地,不与 at 连接;D. attained 达到(理想的状态),一般与 to 连接,不与 at 连接。

18.【答案】A

【详解】原文此处意为:……他发现已经被烧毁……。A. down, burn down 烧毁,符合文义,是正确答案。B. out, burn out 烧光,烧完,用以强调燃料,燃烧的东西烧光,燃完。不符合文义。C. off, burn off 烧掉,用于指有目的地把某物放于火中烧掉,一般用于某一具体的东西,不符合文义。D. up, burned up 烧掉,做此意时与 burn off 含意相同。

19.【答案】B

【详解】根据上下文,原文此处意为"他留在那儿的人……"。据句子结构判断,此处应填入一个表示地点的词,做状语,代指上文出现的,"(in) the fort"因此,只有 B. there(在那里)符合结构要求,是正确答案。

20.【答案】D

【详解】原文此处意为"……(他们)偷了黄金和女人,彼此争吵不休,进而相互谋害。"从文中可以看出,此处应填入一个表示时间上有顺承关系的副词:然后,进而……。D. then 然后,符合文义,是正确答案。A. still 仍然,不符合文义。B. thus 这样一来,没有表示出时间上的先后关系。C. hence 因此,不符合题意。

# Unit 15

1.【答案】A

【详解】by boat 用小船运货;如:by ship or by plane;on boat 没有这种形式;on the sea 在海上,但不表示海上运输;on foot 步行,不符合题义。

2.【答案】D

【详解】means 的意思是“方法”、“办法”,单复数同形,by means of 表示“……的方式”或“……的方法”;method 和 way 用作“方法”时区别不大,都可指做某事的具体方法或方式,但 method 是可数名词,而 way 常用单数形式,且多有一个描绘性的词语来修饰。如:I adopted their method of making machines. 我采用了他们制造机器的办法。What is the correct way to address the Queen? 什么是称呼女皇的正确方法? 另外 way 可用于一些固定词组,这时不能用 method 替换。by the way 顺便说一句;out of the way 别挡路;get/in the way 挡路,阻碍。approach 是方法、手段的意思,但经常和 to 搭配。This is the approach to the problem.

3.【答案】A

【详解】as 用作介词,表示状态、特征、情况、工作等,常译为“作为”如:He was famous as a soldier. 作为工兵,他很出名。另外,as 可以做 for instance 解释,故选 A。C. such like 英语中没有这种形式。D. the same + n./pron. + as + 从句与……一样,如:This is the same bag as I lost yesterday.

4.【答案】B

【详解】shorter 形容词作 make 的宾补,符合题意。

5.【答案】B

【详解】permit 允许,虽然 let, allow 和 permit 都表示“允许”之意,但 let 之后需带不加“to”的动词不定式作宾补,也可加动名词作宾语,与 allow 相比,permit 在较为正式场合使用。根据题意所以选 B。

6.【答案】C

【详解】permit 一词后需加带“to”的动词不定式作宾补,故选 reach。

7.【答案】D

【详解】主谓一致,定语从句中谓语动词的单复数应与先行词保持一致。

8.【答案】B

【详解】这里要表示“还有一些运河”,应该选 still;“Still other...”是“还有一些……”的意思。

9.【答案】D

【详解】“不常下雨”是 It doesn't rain very often. 而不是 It rains not very often. 英语中如果要否定谓语后面的状语,习惯用法是将 not 提前移到谓语中去。如:He didn't come here by bike.

10.【答案】B

【详解】provide 与动词 drain 是并列关系,同时做谓语。providing 与 provided 均可引导条件状语从句。如:Providing(Provided) that it rains tomorrow, the meeting will be cancelled.

11.【答案】C

【详解】drain(sth.) off/away 为固定搭配,意为“(使液体)流走”。

12.【答案】C

【详解】most 用作形容词时,意思是“大多数,大部分”,直接用在名词或是带一个形容词的名词前,不能用 the, these 或物主代词。作代词用时,常用 most of..., 其后要跟带有 the、these、those 物主代词的名词或人称代词宾格。如:Most boys like outdoor games. Most of us feel the same about the way. most 用作副词可以是 much 的最高级,mostly 只能用作副词,是 mainly 的意思,还可译为“大部分地”“几乎全部”:The medicine was mostly sugar and water.

13.【答案】D

【详解】have to do... 意为“不得不”,且这里应用被动态语。

14.【答案】D

【详解】from 与有些动词,如 prevent、stop、keep 等搭配,意思是“阻止某人做某事”。

15.【答案】A

【详解】根据句中时态“in the post”表过去,所以应该用一般过去时。

16.【答案】C

【详解】bring 带来;take 带走,life 抬起。

17.【答案】C

【详解】making 现在分词短语充当结果状语。

18.【答案】C

【详解】principle 原理;principal 校长,重要的;process 过程;procedure 程序。

19.【答案】A

【详解】这里是独立主格结构,falling 现在分词短语表主动意义。

20.【答案】B

【详解】certain 有“某一,某些”(修饰可数名词)及“一些、一点儿”(修饰不可数名词)的意思,而 sure 不能这样用。

四级考试核心考点
**精确打击**

# Part 6

# 翻译（汉译英）

Translation

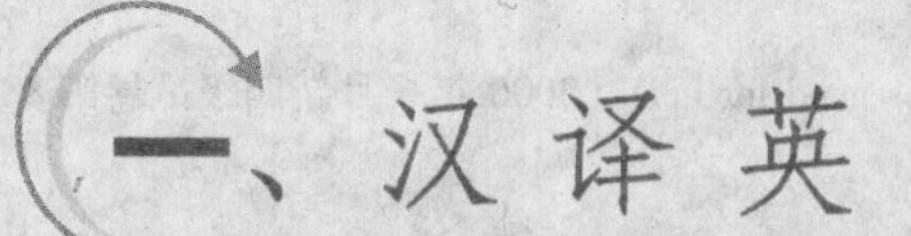

# 一、汉译英 三大命题规律精确定位

在汉译英这一部分，考生要求在5分钟内完成对5个不同句子汉语部分的翻译。每个句子长度在20词左右。所需翻译的部分在10个词以内。纵观新四级的样题、2006年6月的试题(新题型)和往年出现的少数翻译题，我们可以发现，这一题型考查的并不是真正意义上的翻译能力，而是测试应试者对英语词汇、短语、固定搭配、固定表达方式以及句型(尤其是特殊句型)的掌握程度。考虑到该题型给出的应试时间只有5分钟，这更是对考生上述基础知识熟练程度的一次检验。

对样题和2006年6月的试题(新题型)中出现的翻译题加以分析，我们可以把它们归为三类：

## 1. 对基本词汇、短语的考查

例1：Though a skilled worker, ________(他被公司解雇了).(2006年6月新四级试卷第91题)

【答案】he was fired/discharged/dismissed by the company

【解析】本题与其说是翻译了一个句子，更不如说就是考查对解雇这一词的英文表达。笔者认为这一题是2006年6月新四级试题中最简单的一个，因为用英语表达解雇这个意思的方法有很多，比如fire，discharge，dismiss等。

例2：By contrast, American mothers were more likely ________ (把孩子的成功归因于)natural talent.(新四级样卷第91题)

【答案】to attribute their children's success to

【解析】本题是典型的对短语的考查。应试者要求能正确熟练的使用短语attribute...to...除此之外，其实本题还间接考查了be likely to这个短语。往往有人只注意归因于该怎么翻，而忽视了加上to使be likely to这个短语完整。

## 2. 对固定表达方式的考查

例1：The substance does not dissolve in water ________ (不管是否加热).(新四级样卷第87题)

【答案】whether (it is) heated or not

【解析】这是典型的对固定表达方式的考查。学生如果知道whether...or not这一表示选择的固定表达方式，本题很容易做出来。

例2：Your losses in trade this year are nothing ________(与我的相比).(新四级试卷第89题)

【答案】compared with mine

【解析】本题是对表示比较的固定表达方式的考查。在用compare一词进行比较时，一种是用其过去分词形式，即如答案所示。还有一种就是使用它的名词形式：in comparison with mine。

例3:Having spent some time in the city, he had no trouble ________(找到去历史博物馆的路).(2006年6月新四级试卷第87题)

【答案】finding the way to the history museum

【解析】本题考查的是 have no trouble doing 这一固定结构的构成,同样类型的还有 no use/good doing 等等,要求学生知道这种结构中必须要用现在分词形式。

例4:________(为了挣钱供我上学),mother often takes on more work than is good for her.(2006年6月新四级试卷第88题)

【答案】In order to earn money/finance for my schooling/education

【解析】本题考查的就是目的状语 in order to,当然在"供我上学"这一部分有一些考生出现问题,翻的冗长,如 for me to go to school 或 for me to study 等等。其实只用说 for my schooling 或 education 就行了。

## 3.对特殊句型的考查

例1:Not only ________(他向我收费太高), but he didn't do a good repair job either.(新四级样卷第88题)

【答案】did he charge me too much

【解析】本题是考查倒装句的掌握情况。在句首出现否定词,要求我们助动词前置,形成部分倒装。如不小心注意,这种问题很容易造成失分。

例2:On average, it is said, visitors spend only ________(一半的钱)in a day in Leeds as in London.(新四级样卷第90题)

【答案】half as much

【解析】本题考查的是比较句型 as...as 结构,注意修饰钱的数量使用 much,而不是别的什么词。

例3:The professor required that ________(我们交研究报告).(2006年6月新四级试卷第89题)

【答案】we hand in our research report

【解析】本题是对虚拟语气的考查。要求考生正确使用从句中动词的形式,要么为 should hand in 要么省略 should。

例4:The more you explain, ________(我愈糊涂).(2006年6月新四级试卷第90题)

【答案】the more confused I am

【解析】本题考查的是 the more..., the more... 这一特殊句型。本题要注意的是形容词 confused 要紧跟 the more 后。

总的来说,翻译题型可以划分为上述的三大类。但这三大类型不是绝对就分开来考,往往这三者是交织在一起。比如上述2006年6月新四级试题第89题既考了虚拟语气的特殊形式(即宾语从句中的(should)do/be 结构),又考了短语 hand in。从这一点看,又要求应试者对基础知识把握全面。

# 二、汉译英

# 应试三大步骤精确突破

## 步骤1 分析清楚整个句子的结构。

拿到题目，要做的第一件事不是看中文部分的意思，而是首先分析清楚整个句子的结构。明白自己要翻译的处于句子的哪一部分；是主语、谓语、宾语还是仅仅作为整个句子的定语成分或状语成分，亦或是句子中的从句？只有了解了这些，你才能做到翻译出来的答案能够和句子的其他部分正确的放在一起。比如：判断翻译部分中是否应该用动词，用动词的话应该用什么时态和语态，是名词的话应该用单数还是复数，这些问题，必须在分析了整个句子结构和意思的情况下才能作出回答。例如下面一题：

The traditional approach ________ (处理复杂问题) is to break them down into smaller, more easily managed problems.

【答案】to dealing with complex problems

【解析】这是四级考试中出现过的一道题。其中文部分“处理复杂问题”为动宾短语，但在句子中它并不是谓语成分，真正的谓语是后面的系动词。“处理复杂问题”实际上是对前面 approach 的一个限定。而又由于 approach to 后面要求跟名词性短语，所以翻译时要用 deal 的分词形式。

我们再看另一个例题：

The sports meet oringinally due to be held last Friday ________ (最终因天气不好而取消了).

【答案】was canceled because of the bad weather

【解析】这是四级考试中出现过的一道题。不少考生将 due 当作动词而错误分析整个句子结构，从而导致失分。如果知道 due to 是形容词短语，则很好分析出本题已给出部分实际就是主语部分，要求考生补出谓语部分。另外，做好对全句的理解，不难发现应使用过去时的被动态。

## 步骤2 把中文部分转换成英文。

在了解了整个句子结构和意思后，我们要做的是把中文部分转换成英文。这个时候就需要用到我们在题型分析中谈到的基本技能了。即

**(1)词汇的掌握**

对于翻译题，一般不会考单个的词汇，更多的是把词汇放在短语中去考查，就词汇本身而言，大家应该注意以下几点：

对于名词，判断其是否可数，从而判断用不用冠词，用什么冠词，同时解决单复数的问题。

对于动词，判断其是否及物，时态及语态。

对于形容词和副词，注意比较级和最高级的使用问题。同时请注意如下几组词：

①close 与 closely

close 意思是“近”；closely 意思是“仔细地”。

②late 与 lately

late 意思是“晚”；lately 意思是“最近”。

③deep 与 deeply

deep 意思是“深”，表示空间深度；deeply 时常表示感情上的深度，“深深地”。

④high 与 highly

high 表示空间高度；highly 表示程度，相当于 much。

⑤wide 与 widely

wide 表示空间宽度；widely 意思是“广泛地”，“在许多地方”。

⑥free 与 freely

free 的意思是“免费”；freely 的意思是“无限制地”。

对于虚词，如连词、介词相对变化较少，而且多出现在短语或句子连接部分，故在此不多做讲述。

**(2)短语及搭配的掌握**

我们看到的最多的便是对形容词短语和动词短语的考查，在 2006 年样题和历年的真题中我们都看到这样的题目，又比如下面这两题：

例 1：Young people ________(不满足于) stand and look at works of art, they want art they can participate in.

【答案】are not content to

【解析】考查形容词短语 be content to (do)

例 2：Not only the professionals but also the amateurs ________ (将受益于) the new training facilities.

【答案】will benefit from

【解析】考查动词短语 benefit from

除了动词短语和形容词短语，比较难以让人掌握的便是介词短语。介词短语可以充当定语，有时也可以做表语，所以很容易和形容词短语甚至动词短语发生混淆。比如下面的一道四级题：

Finding a job in such a big company has always been ________ (做梦也想不到的).

【答案】beyond his wildest dreams

【解析】一看到做梦，许多考生脑海里第一个反应便是用动词，但在本题中用动词是很不好表达的，如果掌握 beyond 的用法，本题就迎刃而解。

另外还有一类搭配，属于并不是很固定的搭配，也需要引起注意，如：

attach/lay/put importance/emphasis/stress to

commit mistake/suicide/crime/sin

**(3)句型的掌握**

定语从句中需要注意的是用 which 或 as 指代全句这样的类型，典型的如 as is known。

状语从句中需要注意的是几个时间状语从句，比如：

hardly/scarcely... when/before, no sooner... than 和 as soon as

都可以表示“一……就……”的意思。例如：

I had hardly / scarcely got home when it began to rain.

I had no sooner got home than it began to rain.

As soon as I got home, it began to rain.

我刚到回家，就下起雨来了。

注意：如果 hardly, scarcely 或 no sooner 置于句首，句子必须用倒装结构。例如：

Hardly / Scarcely had I got home when it began to rain.

No sooner had I got home than it began to rain.

但我们应把更多的精力放在一些特殊句式上：

①强调句型

It's very ________(你很体谅人) not to talk aloud while the baby is asleep.

【答案】considerate of you

【解析】本题实际上就是一个以“it”作形式主语开头的强调句型，形容词修饰本身时后用 of 否则用 for，知道这一点，则本题没有难度。

②比较句型

On average, it is said, visitors spend only ________(一半的钱)in a day in Leeds as in London.(新四级样题)

【答案】half as much

【解析】本题考查的是比较句型 as...as 结构,注意修饰钱的数量使用 much,而不是别的什么词。

属于比较句型的还有

表示倍数的:

twice/three (four...) times as ... as ...

使用比较级的:

比较级 + than

no more/less than ... 不超过/不少于

否定词 + 比较级的,例如:

I can't find a better vase than this! 我再也找不到比这更好的花瓶了!

the more ... the more...,例如:

The more you explain, ________(我愈糊涂).(2006 年 6 月新四级试题)

【答案】the more confused I am

【解析】本题考查的是 the more... the more...这一特殊句型。本题要注意的是形容词 confused 要紧跟 the more 后。

③倒装句型

因否定词在句首而进行的倒装,如 Not only...but also, Hardly/Scarcely...when, No sooner... than 等:

Not only ________(他向我收费太高), but he didn't do a good repair job either.(新四级样题)

【答案】did he charge me too much

【解析】本题是考查倒装句的掌握情况。当句首出现否定词,要求我们助动词前置,形成部分倒装。如不小心注意,这种问题很容易造成失分。

as, though 引导的倒装句,如:

Young as he is, he can recite 100 poems. 尽管他这么小,却能背 100 首诗。

虚拟语气中的倒装,如:

Were I you, I would try it again. 我要是你的话,就再试一次。

④虚拟语气

虚拟语气的常规用法,我们除了要注意上述的虚拟语气中的倒装外,还要注意一些特殊的虚拟语气形式,如:

The professor required that ________(我们交研究报告).(2006 年 6 月新四级试题)

【答案】we hand in our research report

【解析】本题是对虚拟语气的考查。要求考生正确使用从句中动词的形式,要么为 should hand in 要么省略 should。

又如:

________(如果没有帮助)of their group, we would not have succeeded in the investigation.

【答案】But for the help

【解析】but for 是一种变形的虚拟语气,是对过去事实的一种相反假设。

再如:as if/though 后的从句也可能用虚拟语气。

He looked as if he had been drunk. 他看上去像喝醉了似的。

## 步骤 3 检查。

第三步也就是最后一步,便是检查了,首先检查拼写,大小写是否有误,检查词汇的使用是否有误(采用步骤 2 中介绍的方法);接着再看看句子结构。有一种错误很不好检查出来,便是中式英语,看下例:

—I'd like to inform you that your membership in the club has expired.

—________(我知道了,谢谢).

【答案】I see, thank you

【解析】很多学生想当然的翻译成了 I know,甚至 I have known,这便是不了解英语惯用法产生的错误。要避免这种错误,只有平时多看,多背,增强语感。

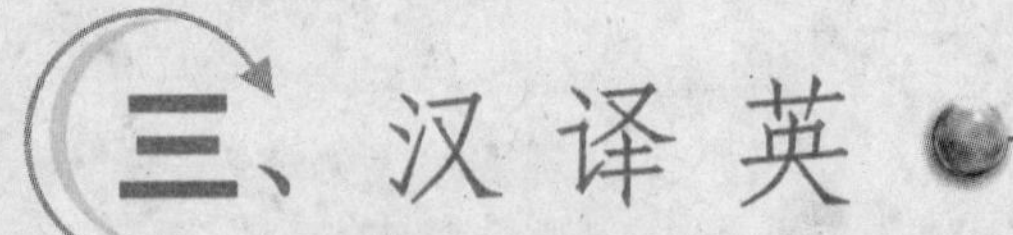

# 三、汉译英 核心考点精确打击

**Directions:** *Complete the sentences on **Answer Sheet** 2 by translating into English the Chinese given in brackets.*

## Unit 1

1. ____________________(作为一个年轻人他进行了几次航行) in the Mediterranean, where the greatest mariners of antiquity were bred.
2. America was ____________________(由哥伦布意外发现的) and was named for another man.
3. Warnings about the dangers of smoking seem to ____________________(对这个年龄组没有丝毫影响).
4. ____________________(出乎我意外的是), I found I had a lot in common with this stranger.
5. ____________________(要不是他们的帮助), we should not have succeeded.

## Unit 2

1. What's more, ____________________(生物有能量并能再生), while the chemicals on the earth 4 billion years ago were lifeless.
2. ____________________(人比动物高级) in that they can use language as a tool to communicate.
3. I don't doubt ____________________(他是一个智慧的科学家), but can he teach?
4. The government has decreed that ____________________(废除燃气税).
5. Foreign bankers are cautiously optimistic ____________________(国家经济的未来).

## Unit 3

1. I'd rather you ______________________(礼貌地对她讲话).
2. As regards your problem, I am sorry to say that ______________________(我爱莫能助).
3. Dick didn't want to walk home because ______________________(他习惯于每天都有车来学校接他).
4. She is expecting some visitors today and she likes to ______________________(让自己的房间看起来干净整洁).
5. Tommy was ______________________(这个班上最幸福的孩子之一).

## Unit 4

1. ______________________(根据发言人所说), there are some supernatural phenomena for which no scientists can offer any reasonable explanation.
2. The information in last year's tourist guide is ______________________(不再有用).
3. We're waiting to see what our competitors will do ______________________(在我们行动之前).
4. ______________________(抓住这次机会), and I promise you won't regret it.
5. Her publisher knew they were ______________________(冒险) when they agreed to publish such an unusual novel.

## Unit 5

1. It is difficult to imagine ______________________(没有人必须工作的社会).
2. He issued an order to them ______________________(所有人 5 分钟内进去).
3. I don't think she'll pardon me ______________________(告诉史密斯她的秘密).
4. The runner succeeded in ______________________(保持领先位置) in the last round of the race.
5. ______________________(因为地震), the roof of the house caved in, keeping inside at least 7 people.

## Unit 6

1. He was motivated by ______________________(希望达成某种妥协).
2. A well-dressed man, ______________________(他的外表和谈吐都像个美国人), got in the car.
3. They ______________________(尽最大努力帮助病号和伤员).
4. ______________________(你肯帮我的话) will make my success certain.
5. The sentense is passive in form ______________________(而意思是主动的).

## Unit 7

1. ______________________(早知如此), I would not have gone back.
2. The old man was ______________________(把手插在口袋里在田野里散步).
3. That dress was so expensive ______________________(我买不起).

4. ________________________(在那个国家失业现象) is very serious.
5. The children are ________________________(期待着圣诞节的到来).

# Unit 8

1. In 1944 a 22 – year – old Army medic was answering a battle field cry for help when ________________________(一个爆炸的德国弹片击中了他).
2. One of the difficulties in carrying out a world – wide birth central program lies in the fact that official attitudes to population growth vary from ________________________(各个国家工业发展水平) and the availability of food and raw materials.
3. In time, the number of certain farm machines that came into use skyrocketed ________________________(这改变了农耕的性质).
4. But ________________________(实现机械化最大的障碍) is the fear in underdeveloped countries that the workers who are displaced by machines would not find work elsewhere.
5. But the mass production of goods resulting from the Industrial Revolution in the 19th century made ________________________(面对面的推销方式) less efficient than it previously was for most products.

# Unit 9

1. Today we are especially pleased and happy to receive in our capital Beijing the delegation ________________________(以琼森先生为首的).
2. China holds ________________________(对[南沙群岛]无可置疑的主权) the Nansha Islands.
3. This is ________________________(决不)targeted at our Taiwan com – patriots, but at foreign forces attempting to interfere in China's reunification and seek the independence of Taiwan.
4. All the students should ________________________(德智体发展).
5. ________________________(看太多电视节目) will do great harm to the eyesight of children.

# Unit 10

1. When they advise your kids to "get an education" if you want to raise your income, they ________________________(没有全部说对).
2. ________________________(他们的真正意思是) to get just enough education to provide manpower for your society, but not too much that you prove an embarrassment to your society.
3. ________________________(你想获取硕士学位的话), make sure it is an M. B. A., and only from a first – rate university.
4. A Ph. D. ________________________(是你能得到的最高学位), but except in a few specialized fields such as physics or chemistry where the degree can quickly be turned to industrial or commercial purposes, you are facing a dim future.
5. Not for our needs, mind you, ________________________(而是出于我们的需求).

# Unit 11

1. I had gone to Belgrade on a ten – month scholarship to learn Serbo – Croat, ________________________(一门我只有些最基本的认识的语言).

2. On my first day at the Language Institute in Belgrade I was graded and put into a class of twelve people, ______________ (共有约九个民族), including myself.
3. The lab bit was useful, ______________ (但相当重复而烦人) after a while. The class sessions were extremely useful, with a variety of very competent teachers.
4. Each text ______________ (后面都有一个单词表). In class the texts were thoroughly explored and a grammar element was fed in by the teacher.
5. The teachers differed ______________ (在方式上).

## Unit 12

1. It takes ______________ (不只是漂亮的外表) to be an air stewardess.
2. This adds up to a pretty tall order, ______________ (因此回电话的那么多人中很少有人能被选中).
3. ______________ (如果女孩们工作做得好) it is because of proper training.
4. They must know ______________ (如何用三国的货币算出一包香烟的价格); know how to pour coffee for four and not spill a drop.
5. They must also know something about the theory of flying and the parts of an aircraft ______________ (详细的).

## Unit 13

1. I knocked at the door, but ______________ (没有人回应).
2. The invention of plane is a great ______________ (贡献) traffic.
3. If we ______________ (比较一下这两本小说), we'll find there are many differences in plot.
4. We must save all the children from the big fire ______________ (不惜一切代价).
5. The November nights in the valley grew very cold, and the lost campers ______________ (轮流睡觉), to keep a fire going.

## Unit 14

1. I can't afford a new pair of shoes, ______________ (更不用说买冬天穿的大衣).
2. She treated the little boy very kindly, ______________ (好像她是他妈妈).
3. Let me ______________ (借此机会) to thank you for all your help.
4. I wear glasses because ______________ (我的视力很差).
5. The two pictures are similar, at least ______________ (就背景颜色而言).

## Unit 15

1. During the famine a lot of people have ______________ (没有选择只能吃草和树叶).
2. ______________ (陆教授献身教学) earned him the respect of both his colleagues and students.
3. Some people view life as an unending conflict between the force of ______________ (正与邪).
4. ______________ (考虑到境况已改变), he urged that the meeting be postponed to the next month.
5. This is ______________ (我们梦想的完美的房子). Let's take it.

# 四、汉译英

# 答案详解

## Unit 1

1. As a youth he made several voyages
2. discovered by Columbus purely by accident
3. have little impact on this age group
4. To my surprise
5. But for their help

## Unit 2

1. living things have energy and can reproduce
2. Human beings are superior to animals
3. that he's a brilliant scientist
4. the gasoline tax be abolished
5. about the country's economic future

## Unit 3

1. spoke to her politely
2. there is nothing I can do
3. he was used to being picked up at school every day
4. have her room look clean and tidy
5. one of the happiest children in the class

## Unit 4

1. According to the speaker
2. no longer useful
3. before we take action
4. Take this opportunity
5. taking a risk

## Unit 5

1. a society in which nobody has to work
2. that every one should go in in 5 minutes
3. for telling Smith her secret
4. keeping his lead
5. Because of the earthquake

## Unit 6

1. a desire to reach a compromise
2. looked and talked like an American
3. did their best to help the sick and the wounded
4. Your assistance
5. but active in sense

## Unit 7

1. If I had known this beforehand
2. walking in the field with his hand in his pockets
3. I couldn't afford it
4. Unemployment in that country
5. looking forward to Christmas Day

## Unit 8

1. fragments of an exploding German shell tore into him
2. country to country depending on the level of industrial development
3. and changed the nature of farming
4. the greatest obstacle to mechanization
5. person to person selling

## Unit 9

1. headed by Mr. Johnson
2. undisputable sovereignty over
3. by no means
4. develop morally, intellectually and physically
5. Too much exposure of TV programs

## Unit 10

1. tell you only half the truth
2. What they really mean is
3. If you go for a master's degree
4. is the highest degree you can get
5. but for our demands

## Unit 11

1. a language of which I had only a very basic knowledge
2. containing some nine different nationalities
3. but rather repetitive and boring
4. was followed by a vocabulary list
5. in their approach

## Unit 12

1. more than good looks
2. hence so few are chosen from the many who answer the call
3. If the girls do their job well
4. how to work out the price of a packet of cigarettes in three currencies
5. in detail

## Unit 13

1. nobody answered
2. contribution to
3. compare this novel with that novel
4. at all costs
5. slept by turns

## Unit 14

1. let alone a winter coat
2. as if she were his mother
3. take this opportunity
4. my vision is weak
5. as far as background colour is concerned

## Unit 15

1. no choice but to eat grass and leaves
2. Professor Lu's devotion to teaching
3. good and evil
4. Considering the changed circumstances
5. the perfect house we've dreamed of

# Part 7

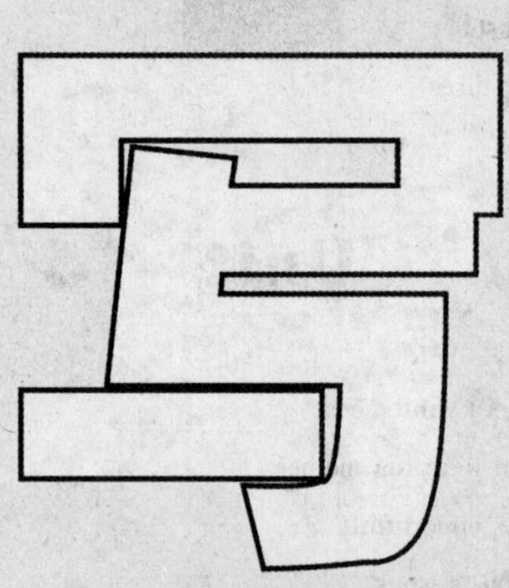

# 写作

## Writing

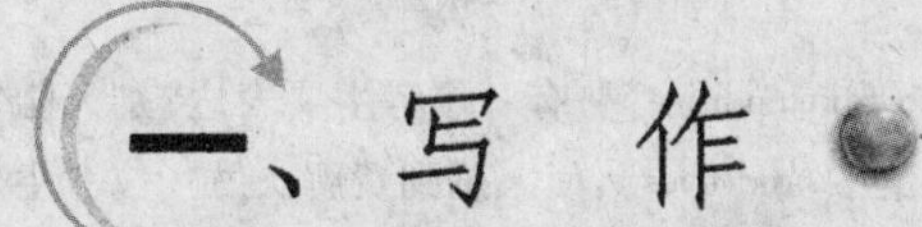


# 题型聚焦

大学英语新四级考试中作文只有一题，考试时间为30分钟。要求考生写出一篇至少120词的短文。试卷上可能给出题目，或规定情景，或写出看图表作文，或给出段首句要求续写，或根据所给文章（英语或汉语）写出摘要或大意，或给出关键词写短文等等。写作的内容包括日常生活、科技、社会和文化等方面的一般常识，不涉及知识面过广、专业性太强的内容。新近又增加了应用文的写作，如书信、简历、通知、便条、总结报告等。对作文的要求是：切题，文理通顺，表达正确，意思连贯，无重大语言错误。

短文写作部分的目的是测试学生的英语书面表达能力。

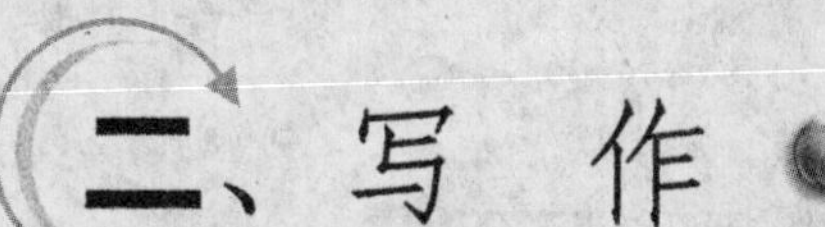


# 三大症结精确描述

大学英语四级考试作文题采用总体评分方法，这种评分方法更注重内容和篇章结构，便于考查考生表达思想的能力。

评分是从内容和语言两个方面对作文进行综合评判。因为内容和语言是一个统一体，作文应表达题目所规定的内容，而内容要通过语言来表达。所以四级作文评分时既要考虑作文是否切题，是否充分表达思想，也要考虑是否用英语清楚而确切地表达思想，也就是要考虑语言上的错误是否造成理解上的障碍。

纵观历年四级考试作文阅卷的情况，普遍存在着以下几方面的问题，而这些问题正是考生作文拿不到高分的关键所在。

1. 平时缺乏训练、没有掌握写作能力。许多考生的写作仍停留在句子水平，还不能从语篇的角度组织语言，因此写出的文章不流畅很生硬。还有一些学生的语言基础薄弱，中学所学的语言基础知识在大学里没能够得到巩固。

2. 对四级作文的答题方法缺乏了解。四级考试的题型以写 Argument 类型的文章为主，偶尔会出现图表作文、书信或描述性、说明性的题目。

3. 母语干扰的痕迹非常明显。例如，有一次作文题目为“Harmfulness of Fake Commodities”。有一位考生要表达“别人赚钱有人眼红”这样的意思，他不会使用“envy”或“envious”等词语，而是直接把眼红翻成了“red eyes”，使人感到费解。另一位考生想要讲述自己买了一双假冒皮鞋，不久鞋上开了一个洞这样的意思，就写出了“There was a cave in the shoe.”这样的句子。这些虽然都是极端的例子，但也反映了母语干扰的一个方面。

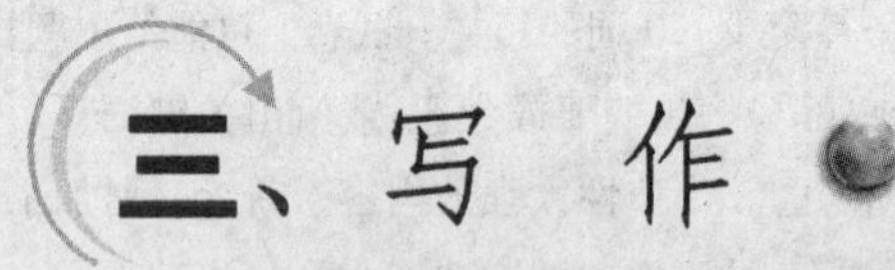

# 三、写作

# 应试技巧精确突破

## 技巧1 审题透彻。

切题是评分的基础。如果试题让你说东，你却偏偏说西，作文写得再好也是枉然，不可能得高分。

## 技巧2 言简意赅，逻辑性强。

由于受时间、篇幅的限制，考生在写作时必须选取最能说明问题的语言，以最合乎逻辑的方式安排层次，使之环环相扣，说服力强。这就是说，考生应对自己要从几个方面阐明主题，每个方面需用怎样的方法来说明等心中有数，不可天马行空，想起一句写一句地凑字数。

## 技巧3 语言规范，避免语法错误。

无论你对主题的认识怎样，都需用正确的语言来表达。语言错误太多最直接的恶果有两个：一是妨碍考生顺利表达思想，使阅卷老师不知所云；二是给阅卷老师留下不良印象，认为该生英语水平很差，进而影响其总分。所以，考生除了平时要加紧语言基本功的训练外，考试时尽量选用自己熟悉的句型和表达方式也很重要。也许你的用词或表达法不是最好的，但总比犯错误好。

技巧4 字迹清晰，标点正确。

由于四级考试集中阅卷，时间紧、压力大。清晰的字迹、整洁的卷面既能为阅卷老师减轻一定的工作强度，又可为自己争取到良好的印象，何乐而不为呢？至于标点，看似简单，其实不然。许多考生有一字一点的坏习惯，还有些考生中、英文标点不分，这些毛病一定要注意改掉。

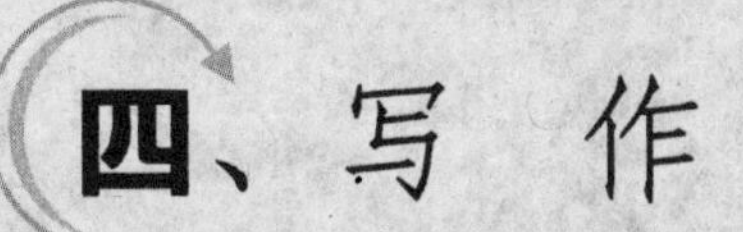

# 四、写作

# 核心考点精确打击

## Unit 1

**Directions**: *For this part, you are allowed* 30 *minutes to write a composition on the topic* "***Should Private Cars Be Encouraged in China***". *You should write at least* 120 *words and you should base your composition on the key words below. Remember to use them properly.*

fantastic spur/ private cars/ on the rise/ different views/ in favor of/ convenience and mobility/ automobile industry develop/ trigger other industries/ against/ problems/ pollution/ accidents/ energy consumption/ traffic congestion/ threat to the existence

## Unit 2

**Directions**: *For this part, you are allowed* 30 *minutes to write a composition on the topic* "***Wealth and Happiness***" *in three paragraphs. You are given the first sentence or part of the first sentence of each paragraph. Your part of the composition should be at least* 120 *words, not including the words given. Remember to write neatly.*

1. Wealth has always been what some people long for.
2. There is no doubt that wealth brings happiness especially in the modern society.
3. But there are exceptions.

## Unit 3

**Directions**: *For this part, you are allowed 30 minutes to write a composition on* "**Can Road Accidents Be Avoided**?" *in three paragraphs. You are given the first sentence of each paragraph. Your part of the composition should be at least 120 words, not including the words given. Remember to write neatly.*

1. There are more and more road accidents in our cities.
2. Some people say that traffic accidents can hardly be cut down.
3. In fact, most road accidents can be avoided.

## Unit 4

**Directions**: *For this part, you are allowed 30 minutes to write a composition about* "**Television**". *You should write at least 120 words following the outline given below in Chinese*:

1. 电视给我们展现了一个生动的世界;
2. 电视的教育作用;
3. 电视的弊端。

## Unit 5

**Directions**: *For this part, you are allowed 30 minutes to write a composition on the topic* "**Recreations**". *You should write at least 120 words and you should base your composition on the outline (given in Chinese) below*:

1. 娱乐的必要性。
2. 两种娱乐的好处(physical activities and intellectual activities)。
3. 将这两种形式进行比较,谈谈自己的看法。

You can write the composition in one or more paragraphs. Remember to write it neatly.

## Unit 6

**Directions**: *For this part, you are allowed 30 minutes to write a composition on* "**Treasure Trees**". *Your composition should be at least 120 words. Remember to write neatly.*

## Unit 7

**Directions**: *For this part, you are allowed 30 minutes to write a composition on the topic* "**Are Prizes Good Things**?" *according to the following outline (given in Chinese). Your composition should be at least 120 words. Remember to write your composition clearly and neatly.*

1. 有奖竞争的作用
2. 奖励的弊端
3. 我的观点

## Unit 8

**Directions**: *For this part, you are allowed 30 minutes to write a composition on "**Competition**" in three paragraphs. You are given the first sentence of each paragraph. Your part of the composition should be at least 120 words. Remember to write neatly.*

1. Competition makes people creative.
2. Competition produces better things.
3. Society, through competition, has developed a lot.

## Unit 9

**Directions**: *For this part, you are allowed 30 minutes to write a composition on the topic "**A Letter to Alice Brown**". Your composition should be based on the situation (given in Chinese) below. You can write the composition in one or more paragraphs. Remember to write it neatly.*

*Situation*：你是一家美国公司的雇员。一位名叫 *Alice Brown* 的女士向你的公司申请一个秘书的职位。由于她缺乏足够的知识和经验,公司决定不录用她。人事科长（*Personnel Director*）*John Smith* 让你写一封回信,将公司的决定通知她并表示同情。

## Unit 10

**Directions**: *For this part, you are allowed 30 minutes to write a composition on "**Humor**" in three paragraphs. You are given the first sentence of each paragraph. Your part of the composition should be at least 120 words, not including the words given. Remember to write neatly.*

1. A sense of humor is regarded as the most valuable character asset.
2. Humor is beneficial for people's mind and body.
3. Humor helps people to get along well with each other.

## Unit 11

**Directions**: *For this part, you are allowed 30 minutes to write a composition on the title "**On Advertisement**". Your composition should be at least 120 words. You should also base your composition on the outline (given in Chinese) below*:

1. 广告的用途；
2. 目前的问题；
3. 解决办法。

## Unit 12

**Directions**: *For this part, you are allowed 30 minutes to write a composition on the title "**Is Studying Really of No Use?**". Your composition should be at least 120 words. You should also base your composition on the outline (given in Chinese) below*:

1. 我们生活的社会需要我们掌握越来越多的知识；

2. 然而,有相当多的人却认为读书无用;
3. 作为大学生,你的看法是……

## Unit 13

**Directions**: *For this part, you are allowed* 30 *minutes to write a composition on* "***Movies of Violence***", *based on the following outline*:

1. Why are there so many movies of violence?
2. The harm of movies of violence;
3. What should we do?

Your composition should be at least 120 words. Remember to write neatly.

## Unit 14

**Directions**: *For this part, you are allowed* 30 *minutes to write a composition on* "***On Exams***" *in three paragraphs. You are given the first sentence of each paragraph. Your part of the composition should be at least* 120 *words, not including the words given. Remember to write neatly.*

1. A student has to take various exams in order to eventually graduate.
2. However, exams may be unfair or fail to justify the work of a student.
3. Therefore, we must have a right attitude towards exams.

## Unit 15

**Directions**: *For this part, you are allowed* 30 *minutes to write a composition with the title* "***On Water***". *Your composition should be at least* 120 *words. Remember to write your composition neatly. You should also base your composition on the outline (given in Chinese) below*:

1. 水的重要性;
2. 水在我国的现状;
3. 怎么办?

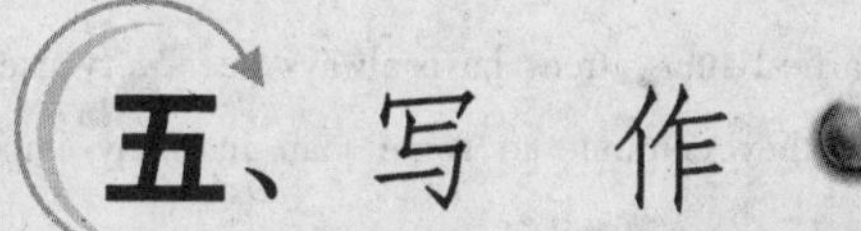

# 五、写作

# 答案详解

## Unit 1

### Should Private Cars Be Encouraged in China

With the fantastic spur both in industry and in economy in China, the number of people who own private cars is on the rise. And there have been two quite different views on this phenomenon.

Some people argue that private cars can bring convenience and mobility to the owners. If more people buy cars, automobile industry will develop dramatically. What's more, the growth of automobile industry can trigger the boom of other important industries such as iron and steel production, energy and technological application. In brief, they are in favor of developing private cars.

However, others hold that automobiles will give rise to a series of problems such as more serious environmental pollution, more traffic accidents and more energy consumption. Besides, there are 1.2 billion people in China. If every family owned a car, too many cars would run on this land, which will certainly lead to traffic congestion and pose great threat to the existence of the people. Therefore, they are against developing private cars in China.

## Unit 2

### Wealth and Happiness

Wealth has always been what some people long for. It is true that most of them try to acquire wealth by means of honest labor. Their efforts contribute to the welfare of the society and at the same time to the accumulation of their wealth, and hence their happiness.

There is no doubt that wealth brings happiness especially in the modern society, where various kinds of modern conveniences, new fashions and entertainments make their appearances with each passing day to make life more comfortable and colorful. Only money can turn admiration into reality.

But there are exceptions when wealth does not go hand in hand with happiness. Wealth may encourage those weak-willed persons to be addicted to some harmful habits, such as drug-taking or gambling, and bring about their own ruin. Also, a person may lose his reason and go astray if he is passionately devoted to seeking wealth. Therefore, one can never rely only on wealth to achieve happiness.

## Unit 3

### Can Road Accidents Be Avoided?

There are more and more road accidents in our cities. For example, every time you are traveling in the street, you are likely to see one or more accidents involving cars or bikes. For another example, news of road accidents has become common topics for newspaper reports. In a word, traffic accidents have increased at such a rate that people are showing a growing concern for this problem.

Some people say that traffic accidents can hardly be cut down. For one thing, the rapidly increasing population has made the city a very crowded place. Besides, each day a lot of new cars are produced and put onto the road. Finally, the road condition is poor for lack of repairs. All this makes it difficult for people to travel safely on the street.

In fact, most road accidents can be avoided. A large number of traffic accidents are related to carelessness. Therefore, they can be greatly reduced if people take care while traveling on the street. Of course, traffic condition also needs improving so as to meet the demand of the increasing population and vehicles.

## Unit 4

### Television

Television presents a vivid world in front of us. We can see the news, plays, films, and sports, etc. Television in our time serves as a very enjoyable entertainment for every person.

Television can also play an educational role in our daily life. Besides regular courses for TVU students, we can also watch various educational programs. Among them, the English on Sunday is my favorite.

However, television can also be harmful. It will hurt one's eyes if one watches it for a long time without a break. Young children often watch too much television to pay proper attention to their homework. Yet, television is still very popular for it has more advantages than disadvantages.

## Unit 5

### Recreations

Recreations are an important part of people's life. For example, after hours of attentive study, students feel like having a football game to relax their nerves. Workers, too, find it very satisfying to sit in front of a TV set for an hour or two when they come back from a day's tiring work. Besides, recreations serve as a pleasing way for the retired people to pass their excessive time. Everywhere you go, you will find that during their spare time, people are engaged in recreational activities of one kind or another.

Generally speaking, there are two kinds of recreations: physical activities and intellectual activities. Physical activities, on the one hand, keep one fit and develop team spirit. Basketball is an example. On the other hand, intellectual activities such as playing chess and reading novels can train one's brains and provide temporary escape from one's troubles.

In my opinion, there should be a balance between the two forms of activities. This is because physical activities are necessary for good health while intellectual activities are beneficial to one's mind. Therefore, in order to make his life enjoyable, one should go in for both kinds of pastimes.

## Unit 6

### Treasure Trees

Since the earliest time, trees have always been very useful to man. Even today they continue to serve man in many important ways.

Trees provide man with food, fuel for burning and building material in the form of wood. Without them it would be impossible for man to build houses, boats and even bridges. Furniture such as tables, chairs and cupboards is also made of wood.

Trees give man shade. In some hot regions or countries, they protect him from the fierce heat of the sun. They are also useful in preventing good and fertile top soil from being washed away by heavy rain.

Trees help to prevent drought and floods. Unfortunately in many parts of the world, people cut them down in large numbers. They never bothered to plant new trees. Strong winds gradually blew away the rich surface soil and eventually the land was turned into a worthless desert.

So, for the benefit of man, trees and forests must be taken good care of.

## Unit 7

### Are Prizes Good Things?

Nobody denies that competitions with prizes can bring people's initiative into full play. Encouraged by a strong desire to win a prize, one will go all out to seek his greatest success. Consequently, the best achievements will be obtained in a game with awards.

However, the prompting of a prize does not always lead to desirable results. For example, a competitor who cannot overcome the temptation of a prize is likely to take a stimulant. And the hopeless competitor sees nothing but the prize, which might cause improper behavior.

In my opinion, the remedy lies in moral education of competitors. This will help them understand the slogan: "Friendship first and competition second." Then, prizes will play a better part in any contest. Generally speaking, I believe that prizes are good things.

## Unit 8

**Competition**

Competition makes people creative. Competition is a product of the development of society and it gives people a sense of the pursuit of excellence. This is people's inborn nature. Nothing can stop them from bringing their intelligence into full playing competition.

Competition produces better things. This is the guarantee of enterprises' existence. They have to produce better things. Otherwise, they will have no customers. So, in fact, they produce better things for their own benefit.

Society, through competition, has developed a lot. Without competition, people would create nothing. They would feel satisfied with their present condition. Therefore, with no competition, there would be no progress.

## Unit 9

Dear Miss Brown,

We received your application last week. We regret to inform you that our company has hired another applicant for the job of secretary. Having carefully studied your qualification, we have decided that we need someone with more office experience for this particular position.

You have impressed us, however, with your desire to learn and your ability to work hard. We therefore hope that you will apply for a position with my company at some time in the future. We would like to talk to you again after you have gained on-the-job experience.

We would like to thank you for your interest in our company. Please contact us if there are further questions.

Sincerely yours,

John Smith

Director of Personnel

## Unit 10

**Humor**

A sense of humor is regarded as the most valuable character asset. It lies in each person's mind. However, it has to be cultivated. A person without humor is just like a river without water. Therefore, in a way, humor decides a person's character.

Humor is beneficial for people's mind and body. When people meet with tension, burden and frustration, humor can help them get over these problems. And also, humor can make sorrows and tiredness vanish like smog.

Humor helps people to get along well with each other. A case in point is that in a family the couple had a quarrel. The husband said something that hurt his wife. At that time, if he said something humorous, his wife would burst into smile. And the crisis would be passed. So there is no doubt that humor can bring happiness either to us or to others.

## Unit 11

**On Advertisement**

China is now thick with commercial and advertisements which do many important things for society. First of all, they convey business information. Another use of advertisements is to make the public interested in what manufacturers want to sell.

It is because advertisements have powerful influence that the profiteer tricks the public by them. His bread, in truth, is made with soy flour, while he advertises it as white bread. Moreover, surplus advertisements have interfered in people's normal life. Every ten minutes a television program will be interrupted by commercials for a couple of minutes which ruins good movies and exciting television shows. Worse still, some profiteers advertise their goods by means of dirty pictures to poison people's minds.

These problems can't be ignored.

The best way to solve the problems is to make laws in consumers' interests. According to the laws, advertisements must be completely truthful and healthy. Let good advertisements facilitate communication between the businessmen and the public, and help keep the business world moving.

## Unit 12

**Is Studying Really of No Use?**

The society we live in requires us to master more and more knowledge. And knowledge is accumulated through constant studying. No matter who we are, no matter what kind of job we do, we should have knowledge.

However, quite a few people think that studying is of no use. In their eyes, nothing is more important than money. With money they can buy everything ranging from house to happiness. But studying does not necessarily bring them what they want. Thus they hold such a view that an ignorant old lady who sells boiled eggs can earn much more than a learned scientist who makes atomic bombs.

Is studying really of no use? As a college student, I don't think so. We all know that it is with advanced science and technology that Americans have turned their country into a highly developed one. And if we don't study, if we have no knowledge, we will be left further behind the developed countries like America. So in order to keep up with the developed countries and build ours into a powerful country, we should study much harder and master much more knowledge. Knowledge is wealth. Knowledge is power.

## Unit 13

### Movies of Violence

Nowadays more and more movies of violence are produced as some people enjoy them. According to their opinion, movies of violence are exciting and therefore people easily kill their time by watching them. Some others even go further to argue that movies of violence also encourage us to take the law into our hands and to arm ourselves in case we are attacked.

However, constant exposure to violence in movies does not provide a healthy atmosphere for our society. Movies of violence, sometimes, even serve to teach the people how to commit crimes. Moreover, the unhealthy effects on our younger generation, especially teenagers, can hardly be estimated.

Why can't movies be a better source of information? Movies of violence, I'm afraid, will remain as they are because so many people enjoy them. However, we must voice our protests. Without our voices of protest, there is little hope for change.

## Unit 14

### On Exams

A student has to take various exams in order to eventually graduate. Exams are thought to be the most important and effective means of measuring the qualification of a student. It is true that a well-designed exam can reflect, in some degree, students' ability. So people take it for granted that those who score high in exams are good students and those who do not are poor ones.

However, exams may be unfair or fail to justify the work of a student. As we know, a lucky student may guess the right answer without really knowing the material on objective tests. As for essay tests, students' scores depend too much on examiners' personal feelings and opinions.

Therefore, we must have a right attitude towards exams. We should come to understand that an exam is not everything to us students. There are other means by which we can demonstrate our abilities or capabilities. On the other hand, we should not abolish exams because of their limitations, for we have to have something as a criterion to test students.

## Unit 15

### On Water

Human beings can't exist without water. It is necessary to both our life and production. If there were no water, we should have nothing to drink. And the plants and vegetables disappear on earth. For the same reason, factories would cease production.

Yet water resources are not limitless. It is reported that China is a country short of water. Water is overused by paper mills, textile mills and other factories. Lots of rivers and lakes have been polluted and the waters unfit to drink. Moreover, inhabitants in some mountain areas even haven't enough drinking water.

Therefore under no circumstances should we waste water. On the contrary, human beings should take emergency measures to treasure water. First, we should save with all our efforts water resources and make good use of them. Second, we must try our best to keep water clean and protect it from being polluted. Third, planting more trees is beneficial to water and soil conservation and help maintain ecological balance. Treasure water!